THE GEO. S. LONG PUBLICATION SERIES

WORLD TRADE IN FOREST PRODUCTS

Edited by James S. Bethel

UNIVERSITY OF WASHINGTON PRESS
Seattle and London

Copyright © 1983 by the University of Washington Press
Printed in the United States of America

All rights reserved. No part of this publication may be reproduced or transmitted in any form or by any means, electronic or mechanical, including photocopy, recording, or any information storage or retrieval system, without permission in writing from the publisher.

The following articles were written and prepared by U.S. government employees on official time, and are therefore in the public domain: "Wood in the World's Materials Mix," by R. S. Whaley and Susan J. Branham; "Developing Foreign Markets for U.S. Solid-wood Products," by Vernon L. Harness; "Export Credit Initiatives: Explaining Credit Programs," by Melvin E. Sims; "A Market for U.S. Wood," by John B. Crowell; and "Grades and Specifications in World Trade of Wood Products," by Robert L. Ethington.

"Forest Products Exports--The 1980s in Perspective," by C. W. Bingham of the Weyerhaeuser Company, is also in the public domain.

Library of Congress Cataloging in Publication Data

International Symposium on World Trade in Forest Products
   (1983: University of Washington)
   World trade in forest products.

   (Institute of Forest Resources contribution; no. 48) (The Geo. S. Long publication series).
   "Proceedings of the International Symposium on World Trade in Forest Products, held at the University of Washington, March 22-25, 1983...and sponsored by the University of Washington College of Forest Resources and the Geo. S. Long Fund."--Copyright p.
   Includes index.
   1. Wood-using industries--Congresses. 2. Lumber trade--Congresses. 3. Forest products--Economic aspects--Congresses.
I. Bethel, James Samuel.
II. University of Washington. College of Forest Resources.
III. Title. IV. Series: Contribution (University of Washington. Institute of Forest Resources; no. 48.) V. Series: Geo. S. Long publication series.
HD9750.5.I55 1983   382'.41498   83-17012
ISBN 0-295-96078-7

Proceedings of the International Symposium on World Trade in Forest Products held at the University of Washington, March 22-25, 1983, conducted by the Office of Continuing Education, College of Forest Resources, and sponsored by the University of Washington College of Forest Resources and the Geo. S. Long Fund.

Institute of Forest Resources Contribution No. 48

THE GEO. S. LONG PUBLICATION SERIES

The Geo. S. Long Fund was established in 1975 to promote a better understanding of forestry, natural resources, and conservation. The endowment was made by Miss Helen Long of Tacoma to the University of Washington in memory of her father, a distinguished timber industry executive and natural resources conservationist, who opened the first Weyerhaeuser office in Tacoma in 1900 and served as the firm's manager, vice president, and chairman of the executive committee.

The generous financial support provides an income to the University of Washington to carry out the intent of the founder with provisions that allow the dean of the College of Forest Resources to develop scholarly programs that enlarge and enhance the contribution which the College makes to society.

Since Geo. S. Long was a pioneer exporter of Northwest woods, it is appropriate that the Geo. S. Long Fund support the International Symposium on World Trade in Forest Products convened March 22-25, 1983, which marks the initiation of the College program in international trade. The program, consisting of undergraduate and graduate instruction, research, and public service, will be the first comprehensive effort of this kind in the field of forest resources.

Contents

Foreword: DAVID B. THORUD ... xi

Preface ... xiii

I: THE WORLD'S FOREST RESOURCES

Forest Products Exports—The 1980s in Perspective ... 1
   C. W. BINGHAM

World Supply of Wood—Physical Resource ... 9
   M. A. FLORES-RODAS

Forest Products Trade in the United States and the Pacific Northwest ... 19
   JOHN SPELLMAN

Wood in the World's Materials Mix ... 23
   R. S. WHALEY AND SUSAN J. BRANHAM

World Supply of Wood: Economic Resources ... 44
   ROGER A. SEDJO

The Japanese Timber Market Is Open and Competitive: A View from a Consumer Nation, Japan ... 62
   YOSHIO UTSUKI

II: U.S. OPPORTUNITIES IN WORLD TRADE IN WOOD

U.S. Opportunities in World Trade in Wood ... 78
   BRONSON J. LEWIS

U.S. Opportunities in World Trade in Wood for Roundwood and Chips ... 81
   BRUCE R. LIPPKE

U.S. Lumber Export Opportunities ... 94
   H. A. ROBERTS

U.S. Opportunities in World Trade in Panel Products ... 101
   M. T. FAST

Pulp and Paper and International Trade   111
    IRENE W. MEISTER

Developing Foreign Markets for U.S. Solid-wood Products 121
    VERNON L. HARNESS

           III: THE POLITICAL CLIMATE FOR WORLD TRADE IN WOOD

Environmental Issues and Their Influence on World
Trade in Wood   129
    JOHN LARSEN

U.S.-Japanese Trade Relations and the Pulp and
Paper Industry in Japan   137
    TAKASHI AKUTSU

Export Credit Initiatives: Explaining Credit Programs 143
    MELVIN E. SIMS

Export Promotion, Import Restriction, and Antitrust
Considerations in U.S. Trade with Japan   149
    JOHN O. HALEY

Federal Trade Laws and Their Impact on International
Trade in Wood   157
    JAMES G. YOHO

The Case for Eliminating Japanese Duties on
Wood Products   175
    JOHN WARD

Toward World Trade Expansion: A Canadian Perspective 183
    EUGENE W. SMITH

Timber Market and Timber-processing Industries
in Japan   194
    SHIZUO SHIGESAWA

              IV: GOVERNMENTS AS WOOD PRODUCERS:
                   THEIR INTEREST IN WORLD TRADE

A Market for U.S. Wood   231
    JOHN B. CROWELL

Government as a Wood Producer—Its Interest in
World Trade: British Columbia 240
    DAVID HALEY

The Role of Joint Ventures in World Trade in Wood 256
    RICHARD L. ATKINS

The Market for Wood-based Products as Seen By a
Southeast Asian Producer 262
    PEDRO M. PICORNELL

A Market for Washington Wood: The Challenge of
the 1980s 269
    BRIAN J. BOYLE

We Wooded the World 276
    W. D. HAGENSTEIN

V: THE CHALLENGE OF INTERNATIONAL TRADE IN WOOD

The Product Mix and Flow of Wood Materials in
International Trade 287
    JAMES S. BETHEL AND GERARD SCHREUDER

Production Problems in Lumber and Panel Manufacturing 316
    W. RAMSAY SMITH AND DAVID G. BRIGGS

Factors Influencing the Increased Worldwide Use of Wood 328
    STEPHEN B. PRESTON

Factors Influencing International Trade in Forest
Products, as Seen from a European Viewpoint 340
    T. J. PECK

Transportation Factors in the Movement of Various
Forest Products in International Trade 365
    CHARLES E. DOAN

Structuring an Export Trading Company 375
    RICHARD V.L. COOPER

International Wood Markets: A European View 402
    JOHN WADSWORTH

New Markets: People's Republic of China 416
    VIVIEN F. LEE

Financing Export Sales for the Small- and Medium-size
Forest Products Firm   425
    ROBERT M. INGRAM III

Market Prospects for Tropical Hardwoods from
Southeast Asia   432
    KENJI TAKEUCHI

VI: ORGANIZATION FOR WORLD TRADE

Expanding Corporate Horizons from Domestic to
World Trade   447
    CHARLES L. SPENCE

International Joint Ventures and the Pacific Rim
Wood Industry   451
    J. FREDERICK TRUITT

Bank Finance and Support for Exporters   483
    ROBERT E. BIEHL

Grades and Specifications in World Trade of
Wood Products   494
    ROBERT L. ETHINGTON

Information Needs for Forest Products Trade Strategies   502
    JUAN E. SÈVE

International Trade Strategies   523
    GEORGE E. TAYLOR

Contributors   527

Index   531

Foreword

World trade in wood is in a transitional stage. The trading of wood between countries is certainly not new. It has been going on for centuries. The basis for much of the early trade in wood was the reservoirs of large natural forests that could be rather easily exploited to increase national wealth. As one country's natural timber supplies became exhausted, it gradually faded from the international timber trade only to be replaced by another country that was just beginning to develop its natural forests.

Wood is a renewable material, however, and in the long run the basis for international trade in wood must be the conscious and deliberate effort to manage forests for the continuous production of material. Those countries that are endowed with highly productive forest land and are willing to dedicate large areas of that land to production of timber, and to encourage the development of facilities to convert it to useful products, will be the permanent participants in international timber trade. The countries that emerge as major factors in a permanent international timber trade are becoming increasingly important as the world's reservoirs of unmanaged or lightly managed forests become smaller and smaller, more and more remote, and costlier to exploit.

We have heard much about the loss of tropical forests in recent years. I had occasion a few years ago to serve as co-chairman of a working group that was asked to report to the President of the United States on the status of "the world's tropical forests." The long-term role of the tropical forests of the world as major sources of structural and fiber material was an enigma then and it still is. On the other hand, we see countries like New Zealand and Chile emerging as increasingly important elements in the world materials supply, based upon intensively managed forests.

All of these developments are important to the United States. We are the largest consumer of wood based structural and fiber products among the countries of the world. We are one of the largest producers of these products. We produce more wood than we use but we are a net importer of wood. One of the world's largest importers of wood products, we are also one of the world's largest exporters of wood products. In short, we are deeply involved in trade in wood.

All of the major forest regions of the United States participate in international trade. Ports like New York, Philadelphia, Savannah, New Orleans, and Duluth move large volumes of wood in and out.

The Pacific Northwest is one of the major centers of timber trade in the United States. Here, the Southeast Asian plywood coming into the United States passes the softwood timber products going out. The timber companies, the banks, and the ports are all involved. Because international trade in timber is so important to our state and region, and because it is likely to become more important in the future, we at the College of Forest Resources of the University of Washington have placed it high on our agenda of teaching, research, and public service for the decades of the 1980s and the 1990s. We view the present symposium as a beginning contribution to this program initiative.

As we celebrate the 75th anniversary of our College, we realize that the cargo mills that were such prominent features of the Puget Sound land- and sea-scape when the College was in its infancy were the precursors of today's multifaceted forest products industries.

David B. Thorud
Dean, College of Forest Resources

Preface

International trade in wood is expanding rapidly and there is every indication that it will continue to expand. Countries that are deficient in wood supplies either temporarily or permanently look to their trading partners to meet their needs. Less developed countries with substantial forest resources see trade in wood as a reasonable road to development. And when these countries are deficient in other natural resources, a large supply of merchantable wood or the potential for producing such a supply is an attractive option for earning essential foreign exchange.

This worldwide trade in wood will be fostered by a greater mutual understanding among trading partners with respect to availability of woods, utilization preferences and traditions, marketing habits, transport restrictions and opportunities, and a host of other issues. The number of individuals who are knowledgeable in the domain of international trade in wood is limited worldwide. It is important to foster collegiality among the members of this group to encourage exchange of information and to identify areas of knowledge deficiency. The International Symposium on World Trade in Forest Products represented an effort on the part of the College of Forest Resources to contribute to these objectives. This publication includes most of the papers presented to that symposium. It is expected that it will be the first in an annual publication of contributions to the state of the art in international trade in wood.

The faculty group responsible for organizing this symposium and assembling the papers for this publication consisted of Dean David B. Thorud, Professors James S. Bethel, Vivien F. Lee, Gerard F. Schreuder, and Thomas R. Waggener, and Director Reid M. Kenady.

WORLD TRADE IN FOREST PRODUCTS

FOREST PRODUCTS EXPORTS--THE 1980s IN PERSPECTIVE

C. W. Bingham

This is a truly historic symposium, appropriately convened in this state and under the sponsorship of the College of Forest Resources in this great university.

<u>The state</u>. Washington is the most export oriented of the western states. In fact, in per capita terms, it is the most trade dependent of all the continental states, with at least one job in six dependent on the international movement of goods. Unlike the United States as a whole, Washington State has a very favorable trade balance.

Approximately half of this region's exports fall into three categories. Last year, food products--agricultural and fisheries goods--totaled $3.3 billion; transportation equipment exports--mostly Boeing airplanes, but also trucks and railway equipment--followed closely, totaling $3 billion; forest products exports have on trend been increasing rapidly, and even in the depressed world economy of 1982 and its distorted currency relationships, forest products exports totaled $2.1 billion.

<u>This university</u>. The University of Washington College of Forest Resources was one of the pioneers in training land managers and forest technologists. Later, pulp and paper sciences were added to the curriculum, as integration took place in the Pacific Northwest forest products industry. Still later, the other very important nontimber resources of the forest were recognized in the curriculum. It is perhaps inevitable that the College is giving recognition to the international trade implications of our forest base.

<u>Why Trade?</u>

It is important to remind ourselves of the basic reasons why international trade is so important to the world's population. Trade often precedes, and certainly contributes to, sound political relationships between nation states. Moreover, international trade is an important element in the desire of every national entity to improve its gross national product. The international division of labor, raw materials and capital, plus rising educational levels, have permitted gradual increases in the world's standard of living. Increasingly, however, each national entity has had to become aware of the relative economic advantage which it has, and to concentrate its trade efforts in these areas.

For the United States, these advantages are exemplified by the state of Washington, which has: (1) a favorable land and people balance, good soils, adequate moisture, with temperature and climate providing a large relative advantage to our forest products and agricultural sectors; (2) the commitment of the United States to the defense of freedom and our effort to understand and explore the mysteries of space, providing an equally dramatic relative advantage in the space and high-technology industries; (3) the unique financial and management systems of the United States, as well as our open culture, providing a similar advantage in the service sectors.

The products which originate in the world's forests, like those which come from farmlands, are absolutely essential to improving the standards of living of the world's population. Solid-wood products are used primarily to provide shelter and other structures. The major end products from pulped wood fiber are papers for packaging and for communications. Of course, forests are also important to the retention of moisture. Thus food, shelter, clothing, and literacy are all closely related to farms and forests. Another important factor, often overlooked, is that most of the forest products, as they move in world commerce, are converted in the receiving country. Lumber and plywood are remanufactured or used directly to build houses, office buildings, public works, or furniture. Paperboards are used to make containers and cartons. Pulp is further refined to make a broad range of paper and other products. Newsprint and other papers are used in the communications industries. In fact, a higher percentage of the value added in forest products occurs in the consuming country than is the case with nearly any other industrially traded good.

Forest Products Trade

International trade in forest products has a long history, and has finally reached a level of significant importance. The first recorded reference, documenting the shipment of cedar logs from what is now Lebanon to Egypt, dates back to 2600 B.C. Timbers and naval stores were among the first exports from the English colonies in North America.

Yet, perhaps because of the bulk and weight and the relatively low value of forest products in relation to shipping costs, remarkably little attention has been paid this trade which, in world terms, accounted for transactions worth about $60 billion in 1980. In the European Common Market countries, in fact, forest products imports now rank second, behind oil; and in Japan, wood and forest products imports have averaged $10 billion annually in the past five years. And, while the U.S. exported only 4 million tons in 1960, that volume had grown to 24 million tons in 1980!

Only two major world areas have in recent years had major exportable surpluses of softwood forest resources. Those areas are, of course, North America and the Soviet Union, with North America by far predominant. Within North America, however, Canada is a huge net exporter of forest products, while the United States has continued as a slight net importer, although on trend, that gap has been narrowing.

I believe that North America, in terms of forest products trade, should continue to be viewed as an economic unit. Because of its geographic location, the upper Midwest can reasonably be expected to continue to import products from the Canadian forests to provide the lowest cost to the consumer. Because of national maritime policy, northeastern United States consumers will probably continue to find coastal British Columbia and the Maritime provinces their most competitive sources for many forest products.

Yet the United States commercial forest base is potentially much more productive than the Canadian base. The forest products industry is, in fact, one of the few basic United States industries which can compete in a free market with any other producer in the world.

Our commercial forests are also uniquely accessible. We have, increasingly, a state-of-the-art manufacturing base. We have a skilled, well-educated work force, and additional port capacity. Also, as mentioned earlier, our net export volume has been growing.

In other words, the United States has the capability to be not only a net forest products exporter but a major force in world forest products trade--a capability that is largely untapped. Even though our volume has grown, this country today is the source of only about 20 percent of the exports of forest products in world trade. Why?

1. One reason is because the United States market itself has been so large and, in the past, so relatively easy to serve. American timberland managers and American producers have found it simply much more comfortable to serve this domestic market and to ignore the complexities of world marketing. Exports, in modern history, have most often been a market of last resort to be approached through third parties during downturns in the domestic market cycle. In the past decade, as we in the Pacific Northwest have begun to recognize the rapid competitive deterioration of our position in North American markets, this has begun to change. The region's forest products industry, and particularly our building-products segment, however, has spent far too much time and effort trying to force-fit United States product preferences and standards upon overseas customers, rather than in trying to supply the real needs and preferences of these offshore markets.

2. At the same time, we find both tariff and nontariff barriers to our products in many countries. As we struggle to protect, for national political and social reasons, certain segments of our economy that do not enjoy a relative economic advantage, so have our trading partners tried--with great success--to protect their farm and forest sectors.

    The perceived need to "protect" parts of the U.S. economy is something new arising only late in post-World War II history. We should remember, however, that the fantastic past success of the United States economy has come from two major sources. One, of course, was the tremendous hoard of unexploited resources that led to industrial development in the 19th century. The other was the pioneering of high-volume, standardized production processes early in the 20th century. Although some of the mineral resources were depleted rather quickly, or were surpassed by higher-grade resources discovered elsewhere, the lead in low-cost, high-volume processes continued to give United States industry preeminence even through the decade of the 1960s. Then, countries with lower labor costs demonstrated that high-volume, standardized production facilities requiring minimum skills can be placed _anywhere_ in the world where raw materials are available. In addition, the development of low-cost ocean shipping and unloading meant that most parts of the world could compete for raw materials. World trade became increasingly competitive. Change in our industrial mix, and in the way our industries are organized, has been a slow and wrenching process, but it is reversible, pervasive, and well under way.

The forest products industry is one of those already able to provide needed products and we continue to benefit from that other major historic source of American economic success, a unique supply of natural resources. In our case, it is the soils that are the basic resource; the raw material that grows upon them--wood--is not only renewable but vastly expandable.

Thus, our industry's transition to full capability to compete in today's world markets is perhaps potentially less painful than that for many major United States industries.

3. Unfortunately, there are still some in our industry who feel that competing in world markets will be automatic and easy. It will not. The view from 1970 was much more optimistic than the view today. At that time, we were extrapolating from a base of continued rapid economic growth in the developed world, and accelerating growth in the less developed nations. The projections were for increasing wood shortages by the end of the century, particularly for softwoods. Scandinavia appeared to be reaching its maximum sustainable wood harvest, growth rates were escalating in Europe, and Japan in some years was experiencing real growth of nearly 15 percent.

The view from the early 1980s has looked quite different. World wood demand probably will remain a greater problem than world wood supply through the rest of this century. Floating exchange rates, which benefited United States exports through most of the 1970s, have turned against us with a vengeance, at least temporarily shaking our confidence in many industries. Economic growth has slowed worldwide, following the successive oil shocks of the last decade, and many developing nations have narrowly escaped absolute economic disaster. Household formation rates will begin to slow in the developed world from 1990 on, and wood use per unit of housing has begun to fall. In terms of volume increments, there still are major opportunities for participation in world forest products markets: these opportunities will go to the ones who understand those markets best and who are geographically best located to serve them. The situation will be __highly__ competitive through the rest of this century.

As but one example of the change: recently the Scandinavian harvest level has been well below the level of sustainable harvest. The extreme European wood shortage projected just a few years ago has yet to arrive. In the next few years, we believe, demand growth in Western Europe will be slow, matching low economic growth rates. Currency weakness and an improved Scandinavian industry cost structure may favor local production.

Because of forest products' major impact on trade balances, protectionism and development of local sources or substitutes for forest products imports have been encouraged, particularly in France.

Opportunities

This does not mean that there will be no opportunities for United States exports to Europe. It does mean that growth rates in product use there will be lower than in the past, although growth in absolute tonnage will be significant.

There __should__ be substantially greater opportunities for exports to the Pacific Rim nations. Economic growth rates for countries in the region exceed the world average, and growth of forest products demand also should be significant. In addition, there are likely to be reductions in the amount of raw material available from regions that have been traditional suppliers. I believe the opportunities for North American producers--specifically United States producers--look brightest during the rest of this decade in Japan, Korea, Taiwan, Hong Kong, and China.

The future size and growth rates of Pacific Rim markets, however, depend to a large extent on a growing export trade from these regions and on their unhindered access to world markets. These countries simply must export more if they are to import more. Anti-Japan protectionism in the United States and Europe, and such actions as the recent imposition of quotas against Chinese textiles in the United States, are very ominous signs. Since this symposium took place in the state with the largest per capita stake in exports--one with an economy dependent for future growth upon increasing exports--we are today situated in a locality every bit as threatened by the growth in __U.S.__ protectionist sentiment, as well as that of Europe, as __if__ we were meeting in Tokyo or Seoul.

We also look for moderate export opportunities elsewhere in the world including the Middle East, Persian Gulf, Oceania, and Latin America. Many of the export opportunities will be specific. They will require time spent in developing market knowledge and trade contacts.

Weyerhaeuser Example

Weyerhaeuser Company began exporting modestly in the early 1960s in a search for new markets to help offset the high cost of our Pacific Northwest products in reaching domestic markets in the eastern United States. We opened our first overseas sales office, for pulp, in Tokyo in 1962, and followed quickly with offices in Europe and Australia. Although log exports have had wide public recognition, they are in the minority in our total sales mix. The majority of our exports--to some 62 nations last year--are of finished products. Japan is our largest single export market, but we also export about $300 million to Western Europe and the Middle East, another $100 million to Latin America and Canada, and substantial amounts to other Pacific Rim nations, including China, where we made our first product sale in 1972.

In addition to overseas sales offices, we have made substantial investments to support our exports. Docks and storage facilities have been added at several Oregon and Washington ports. We schedule and operate, on long-term charters, specially designed forest products cargo ships both to Europe and Japan. We have built primary newsprint mill capacity dedicated to exports. Several of our lumber mills have been modified to allow metric-size production. We have overseas wood products distribution centers, and have recently established two warehouses and distribution centers for linerboard in Japan.

I think it is apparent, then, that we do not view exports as a bonus of some sort, or as an outlet for products only in times of domestic downturn. We view them as a long-term economic necessity, particularly for our Pacific Northwest operations west of the Cascade Mountains. We not only see export markets as the key to the _future_ for forest products in the Pacific Northwest; they _today_ take, in one product form or another, nearly half of the annual cubic harvest from our Douglas-fir region forest ownership.

Conclusion

Geography and overland transportation costs dictate that through the future, maintenance and growth of the western industry depend on its ability to compete in offshore markets. We do have major relative trade advantages, if we choose to exercise them. Growth in world populations and standards of living, while slowing, will require major incremental volumes of forest products. Growth in the Pacific Rim economies provides a major opportunity, particularly for the Pacific Northwest. As a nation and as an industry, however, we must make a stronger commitment to serving offshore markets on a <u>consistent</u> basis.

Again, demand--not supply--now appears to be the long-term problem for our products in world markets. We have to <u>earn</u> our way in a highly competitive world market by providing better service and by delivering higher quality products at competitive costs. We have the capability to do just that, if we dedicate ourselves to doing so.

National trade policy must recognize the importance of this industry. Protectionism in our trade policies will only delay the revitalization of the overall United States economy. In trade negotiations, there must be less emphasis on protecting the weaker, historically organized industries from the rigors of the new world economy. Emphasis must shift, instead, toward removal of tariff and nontariff barriers designed to protect the weaker industries of other nations from the competitive capabilities of the United States industries with relative trade advantages.

Forest products is most assuredly one of the industries that provide this nation with a major opportunity to improve its trade balance and to set growth in living standards back on track. It can do so while contributing to basic world needs and living standards. I believe that if we address ourselves to the issues discussed here, we can play a positive role in leading the industry and the nation toward realizing that opportunity.

WORLD SUPPLY OF WOOD--PHYSICAL RESOURCE

M. A. Flores-Rodas

Introduction

The world's supply of wood cannot be dealt with realistically without also considering the demand for forest products. These are two sides of the same coin. Therefore, all relevant economic factors and interrelated political, social, cultural, and historical aspects have to be considered in projecting the wood yield possibility. Merely basing it on the physical potential of world wood supplies would be too simplistic a treatment of the topic.

Background

We are now in an era when most of the world's forests are already in use, and hence where maintenance of the physical stock rests increasingly on investment in management, regeneration, and reforestation, as well as reduction of conversion losses. Replenishment of the physical stock is therefore critically affected by the availability of funds for investment in forestry. Investment for the modernization of the forest industry in order to improve conversion efficiency and productivity is equally important.

First and foremost, in this context, we have to face the fact that the world, as a whole, is at present undergoing a state of economic recession. Even the richest countries are directly affected. There is virtual stagnation--a contraction in economic activity in many spheres and massive unemployment. Money supplies are tight and costly. Inevitably, long-gestation, low-yielding industries such as forestry have growing difficulty in attracting investors at such a time.

Aid and Trade

As always, it is the Third World that is suffering most from the recession. In 1981 a drop in per capita GNP took place for the first time since the 1950s. In that year alone the recession in the developing countries caused production and income losses of more than $50 billion. That is twice the level of public aid to development, which is in turn hardly likely to increase in real terms in the near future.

This means that the chances are bleak for stepped-up resource transfers to the Third World. Moreover, the conditions are among the worst in recent history for using trade to generate the capital for investment. Revenue from trade in tropical timber products has dropped dramatically in the 1980s. Not only has demand dropped, but there also seems to be strong political temptation almost everywhere to become more protectionist, an attitude which promotes such actions as cutting off access to markets in certain cases.

Finance and Investment

If that is not enough for the fragile economies of the world, mounting indebtedness is a burden of unprecedented magnitude. This situation is sending shock waves through money markets, thereby throttling the channels that fuel economies and provide the lubricant for mobilizing supplies. The present level of the debt is now around $700 billion, the servicing of which requires about $150 billion. Prima facie, little, if anything, is then left for investment, especially in long-maturing projects. Given this conspectus, it can be anybody's guess to project how much will go for investment in forestry.

However, forest production is eminently suited to long-range adjustments. For instance, many industrialized countries depleted their forest capital in the past for the purpose of mobilizing investment funds to help their economies develop. They are now advanced enough to invest in the re-creation of forests, and their net area under forests is indeed increasing.

In this context, industrialists and economists should appreciate the philosophy of forest conservation as it relates to questions of the common good. These arise because of differences between private and communal costs and benefits associated with the use of forests and the need for ensuring their renewal. Similarly, the conservationist must concede that natural resources make an important contribution to the economy, and forest resources are a form of capital that must be released from the shackles of puritanic conservation.

To sum up so far: one of the principal features of the present situation is the jeopardy under which the future of the physical resource is, at least temporarily, placed because of difficulties in securing the investment necessary to ensure its renewal.

Let me now turn for a few minutes to considering <u>what</u> this forest resource actually comprises. Most of us probably tend to think about it essentially as a wood and fiber resource--as is implied in the title of this paper. But before I move on to discuss the wood resource in more detail, let us not lose sight of the fact that this wood resource is an integral part of a much larger resource, the forest biomass. Tannins, bark, gums, resins, latex, medicinal plants, dyestuffs, pharmaceuticals, bones, skins and hides, and numerous other animal and plant products, are on many occasions of greater value than the wood from the forest.

Let us for a moment also look a step further, at the underlying resource--the forest land resource. If this basic resource is to be put to its fullest use, it must produce more than trees.

Food production in the forest is seldom recorded. About 300 million landless people practice shifting cultivation in order to obtain food from the forest. The livelihood of another 160 million pastoral nomads and countless rural communities is dependent upon forest grazing and browsing. This, together with fodder tree crops, is a major source of sustenance for livestock, including dairy cattle, in the Third World. In any sustainable forest production scheme, need exists to give attention to wildlife utilization and the management of forests as a perennial source of food. Forest and food crops, including animals, can be raised together. Combined cropping patterns may be a better form of land use in some situations, thereby adding to food production and often increasing income.

Fuelwood

In returning to the narrower focus of the world supply of wood, I should like to make one further basic point. Although practically none of it enters world trade, roughly one half of all wood used in the world is used for fuelwood. In the Third World, where more than 2 billion people depend on wood for the energy to cook their food, fuelwood accounts for about 85 percent of total wood use.

Fuelwood is not only the largest single use of wood, but the nature of its use and the pattern of its demand have very important implications for the wood resource. Most fuelwood is gathered locally by the people who use it. Demand is greatest where population density is heaviest, which means in areas outside the closed forest. Wood supply for fuelwood is therefore often most destructive in those areas which can least afford a heavy drain on a limited resource--the fragile savannah woodlands of arid and semiarid areas, the protective tree cover on hillsides adjacent to agricultural areas, and even valuable fruit and other trees carefully preserved and nurtured in the agricultural lands themselves.

At the same time, fuel and energy needs technically could be met very well by the small-size and inferior quality wood which has no value for timber use, obtained from the more distant closed forests. Energy from the forest could be a logical and efficient co-product with timber and fiber. Making this idea economically viable is one of the great challenges we face.

A Closer Look at the Timber Resource

The forested area of the world covers about 4 billion hectares. Over 1 billion hectares are classified as "other woodland." The bulk of this latter is open woodland and shrub areas, largely located in the Third World. There they are under severe biotic pressure and are being depleted rapidly through, for example, excessive fuel-gathering and grazing in many parts of Africa, South America, and the Near East. Sizable areas of "other woodland" occur also in Canada, the Soviet Union, and Australia, but are likely to be little used at least until the turn of the century.

Given this situation, let us focus our attention on the remaining forest area classified as "closed forest," covering around 2.8 billion hectares. Less than half is in the predominantly tropical Third World, which has mostly broadleaved forests composed of a diverse variety of species, many hitherto commercially unused.

Around 1.6 billion hectares of the closed forest are in the temperate region. These are largely homogeneous and coniferous, occurring mainly in North America, the Soviet Union, and Europe. By the next century there should be a gradual increase in net forest area in developed countries, largely due to continued gains in agricultural productivity, stabilized population, reforestation, and afforestation.

On the other hand, about 7.5 million hectares of closed forest area are disappearing annually in the tropics. In addition, several million hectares of the other woodlands are being destroyed. One could be cautiously optimistic, nevertheless. By the turn of the century, there may be a slow deceleration in losses of tropical forest. Most of the currently accessible areas will have already been depleted by then and agriculture may not be feasible on most of the remaining forest land. The closed tropical forest at the onset of the next century may be about 1.2 billion hectares, half of which will be in Latin America and around 30 percent in Brazil.

Many developing countries have come to recognize the need to create man-made forests. They are estimated to have 13.5 million hectares of plantations, mainly of pines and eucalypts, as well as a few fast-growing hardwoods. That number excludes China, with perhaps an equal if not larger plantation area. By the beginning of the next century, according to current trends, Asia and Latin America may each have around 20 million hectares of plantations, and Africa will have probably 6 to 7 million.

Growing Stock and Productivity

The total growing stock of the closed forest is estimated to be approximately 330 billion cubic meters, of which 145 billion are in the temperate region and the rest in the tropics. The average standing timber volume per hectare ranges from 80 to 100 cubic meters in temperate regions, and 100 to 200 cubic meters or more in the tropics. This difference derives from the biomass growth attainable in the two bioclimatically distinct regions.

This does not, however, indicate the productivity potential of the two main types of closed forest, for an analysis in terms of wood yield possibility calls for close attention to potentially usable wood volumes. These are close to standing volumes in the case of the generally homogeneous temperate forests which are vastly superior to the usable wood volumes obtainable in the tropics.

Also, it is necessary to exclude from our analyses quite extensive forested areas where growing stocks will not be available in the foreseeable future, owing to physical or economic inaccessibility. The same may hold true for other areas because of legal restrictions against wood production, or in view of overriding considerations of protection and preservation of the forest cover. Only the remaining forest area is operable for roundwood production and immediately relevant to our topic.

The operable closed forest area in the developed countries totals about 940 million hectares. Its growing stock, measured as the tree bole volumes, is estimated to be almost 96 billion cubic meters. Of this, North America, Europe, and the U.S.S.R. together account for 90 billion. These volumes represent the productive forest capital of the developed world. Conifers represent about three-quarters of the volume.

The operable area of closed forest in the Third World is estimated at around 1 billion hectares. These carry a growing stock volume of about 161 billion cubic meters, nearly 150 billion of which are concentrated in the tropics. Latin America has more than half of the Third World's operable forest area and a corresponding growing stock. Although Asia's operable area is about twice that of Africa, the growing stock is almost equal thanks to the volumes still available from West and Central Africa.

Supply Potential

In spite of sizable tropical deforestation through the turn of the century, there should remain considerable potential for the expansion of production in the wood-rich countries of Latin America and Africa, especially from presently less accessible areas. Technological progress will make it possible to use a wider range of tropical forest species. Furthermore, the potential supply of industrial wood from plantations in the tropics is expected to increase tenfold during the last quarter of this century, with production reaching about 200 million cubic meters by the year 2000. Latin America should be the leading region, with wood from plantations as an important raw material resource there.

In short, the wood supplies from the world's forests may just be sufficient to meet a demand of around 2.6 billion cubic meters of roundwood for processing into sawnwood, panels, and paper by the year 2000. These supplies, however, will continue to be unevenly distributed with respect to the distribution of demand.

Western Europe and Japan will have to continue to rely increasingly on imports to bridge their growing gap. North American wood removals may be expected to expand by approximately one half by the year 2000, but this increase will contribute little to net exports. The Soviet Union will have sufficient wood supply for its needs, with a substantial quantity available for export, although much is remotely located.

In the tropics, additional industrial wood supplies could come from less accessible areas but particularly from presently underutilized species, sizes and grades, and from plantations. Africa and Asia, where current industrial wood removals are 33 and 86 million cubic meters respectively, may double their production by 2000. That would meet their domestic demand, but their exports would not satisfy the import requirements of developed regions for mechanical forest products or logs for processing. Latin America, on the other hand, may well reach a level of industrial wood removals of around 125 million cubic meters, enough to satisfy its own consumption and to have 15 million cubic meters of roundwood equivalent potentially available for export, if not substantially more.

The expected increase in removals by the year 2000 implies an 80 percent increase in pulpwood removals over that of 1975, but a less than 45 percent increase in sawlog removals. Simultaneously, the use of residues is expected to double and reach over 300 million cubic meters in 2000. The availability and use of nonwood materials is also likely to increase appreciably.

The predictions are, therefore, that the world's forest resources will be able to supply the projected global demand for industrial wood until the turn of this century, although supplies will continue to be unevenly distributed in relation to the distribution of demand. At the same time, however, some parts of the Third World will experience increasingly acute shortages of fuelwood supplies.

Management and Policy Implications

According to the above predictions, industrial roundwood supplies by the end of the century would have increased by about 1 billion cubic meters. But transformation of "projections" into "reality" is not an easy matter. This increase in yields will have to be achieved from a diminishing forest resource base and against a background of rising costs and keen competition for shrinking investment resources.

Undoubtedly, technological progress such as the use of a wider range of species from tropical forests, access to new areas, more intensive forest management, creation of new plantations, reduction of logging and conversion losses, and greater utilization of waste will facilitate the forestry sector in meeting this additional demand for industrial roundwood.

Apart from technological advancements and beyond economic and feasibility studies, efforts in a number of directions need to be intensified if the forestry sector is to meet the challenge in the years ahead. First and foremost, there must be better understanding of the forest resource and of the potential of forest lands. This will entail not just more effort devoted to forest resource appraisal and land capability studies but information on the full range of goods and services that the resource could supply.

The management of the resource needs to be organized in such a way that production of industrial raw materials, output of materials and goals for the work of artisans, and supply of local needs can coexist in the most effective manner. Management which is responsive to local needs must be based on people's involvement and participation.

More efficient management systems will require more effective understanding of human needs. Moreover, forest development policies, strategies, and programs must be closely articulated with overall national development aspirations and goals.

The isolation of the forestry sector in the past from the overall sociopolitical system has been responsible in large measure for the destruction of forest and fauna resources in many parts of the world. Traditionally, forestry programs have tended in the past to concentrate on contributing to the growth of national economies while preserving the ecological balance by assigning priority to the protection of forest ecosystems and making the production of raw materials for industry a secondary consideration.

As a result of the inability of the sector to demonstrate a more active interest in social development, forestry administrations in many countries lack the political relevance necessary to attract the financial resources and authority which are required for them to plan and execute programs which make a real impact on national development.

There is now a growing recognition of the fact that most of the problems of deforestation and of forest resource degradation have a socioeconomic origin. Forest development strategies can therefore no longer disregard social considerations. The forest system is now expected to make a distinct contribution to rural development in addition to its traditional role as a source of raw material for industry and fuelwood for domestic energy.

The Food and Agriculture Organization is advocating the incorporation of a strategy for rural development in overall development policies for the forestry sector. Basic to this strategy is the recognition of three interrelated missions for the forest system addressed to the production, protection, and social objectives. These three missions must be considered concurrently and their interactions harmonized if forest goods and services are to be provided beyond the year 2000 and the perpetuation of the forest resource base and of its productivity is to be ensured.

In order to achieve these objectives, the management of the forest system must ensure the free interplay of its three main institutional components: (1) forest administrations and other public institutions, (2) conventional enterprise institutions including industrial and commercial bodies, and (3) local and rural institutions concerned with community needs and interests.

This approach to forest management does not downgrade the importance of economic growth or the overall role of forestry in the national economy, but aims to orientate the production system and the distribution of forest goods and services to ensure a more equitable share for rural people, particularly those residing in and near forest areas. It can provide both opportunities for these people to become self-reliant and facilities to permit them to perform positively their role as a component in the management system, instead of branding them as "enemies of the forest."

I should like to conclude by qualifying a statement I made earlier on, namely, that the world's forest resources will be able to supply the projected demand for industrial wood _provided_ that forest managers fully recognize that (1) traditional forest policies, even those which continue to be relevant, may need to be recast, to be given different dimensions and new perspectives, in order to update and adjust them to present-day development targets and social values; and (2) development is a process of social change through which an increasing number of human needs, preexisting or created by that change, are satisfied. This implies a differentiation in the productive system that is obtained by the introduction of innovations in the administration and conservation of natural resources, in the institutional structures, and in the industrial conceptions. The forestry sector cannot escape from such considerations.

An imperative prerequisite, therefore, for continued and sustained production of forest goods and services is the existence of forest policies which recognize the multidimensional values of the forests, and of development strategies which ensure the contribution of forest resources to economic growth but also permit and encourage the equitable participation of rural people in forestry and forest-based activities.

Such policies and strategies must be backed by strong political will and must be implemented by able, professional management.

FOREST PRODUCTS TRADE IN THE UNITED STATES AND THE PACIFIC NORTHWEST

John Spellman

Introduction

The economic history of this nation reflects regional diversity. We have been blessed with an abundance of resources capable of sustaining--and indeed enhancing--the economic and social benefits that have made this nation one of the best and most desired places to live. In achieving this prosperity, we have traded among ourselves--first across colony boundaries, then among the states and territories. Goods and services soon found their way across the oceans. Trade became a basic ingredient of our economic growth. This pattern was very evident in the development of Washington State and the Pacific Northwest. Settlers first came here largely because of the natural resources they had heard about from others--the large stands of prime timber, the fertile soils for crops, the salmon migrating to the clear coastal streams and rivers, and the minerals in our mountains. They also came to recognize the abundance of the scenic wonders of the region--a resource that has become the basis for another important export industry, tourism.

Just as the natural resources of the region came to support trade with the rest of the nation and thereby promote our early regional development, we soon engaged in international trade as well. Forest products were an early and primary cornerstone of Pacific Northwest international trade.

Forest Products in State and National Economy

The forest products sector of the Washington economy has been a significant component of both the state economic base and the national forest products sector. The forest industries have directly employed an average of 62,000 workers in the last decade in this state. The forest industries have directly provided about 8 percent of annual payrolls for all employment covered by the Washington Employment Security Act.

Although Washington provides about 1.5 percent of all national manufacturing employment and payrolls, it accounts for approximately 9 percent of national lumber and wood products payrolls and 3.5 percent of paper and allied products payrolls. In total, Washington's forest-based economy provides directly approximately 5 percent of national employment and 7 percent of payrolls within this important sector.

International trade in forest products is a significant and vital part of the total activity in this industry. In the first nine months of 1982, the U.S. exported some $2.1 billion in wood products, led by $898 million in exports of softwood logs and $435 million in softwood lumber. Pulpwood and chips, softwood plywood, hardwood lumber, and other wood products exports were also important. In terms of value, however, the U.S. has been a net importer of wood products. In 1981, the U.S. exported $2.9 billion in wood products but imported just over $3 billion, for a trade deficit of about $122 million. During the first 9 months of 1982, we imported just over $1.1 billion in softwood lumber and some $800 million in other wood products.

Washington State is even more trade dependent than the U.S. as a whole when we look at the forest products sector. First, we look to markets in the rest of the U.S. as a primary source for sales from our forest products industries. Export or international markets, however, have increased in importance for all forest products sectors.

Washington State and the Pacific Northwest have a natural comparative advantage in the area of forest products, including pulp and paper products. Our productive forest land base is second to none. The productive entrepreneurship of our industry attracts attention worldwide. Nevertheless, we must strive to remain competitive in the world's markets if this comparative advantage is going to be maintained as the basis for our economic well-being. We must develop policies which will assure the viability of our present industrial economic activities based on our abundant natural resources, even as we look to new technology and new opportunities for economic development. Our economic roots are largely in our forests. These roots are strong and will provide a firm foundation for a growing regional economic base. International trade will be a critical element in this base.

Programs to Promote Trade

This symposium includes a discussion of the recent export trading company legislation passed at the national level. Much of what we can hope to achieve by state policies and in the private sector depends upon our national trade policies. The export trading company legislation offers a renewed potential for the smaller independent companies in our forest products sector to remain competitive and to engage in increased international trade with greater market specialization. The forest products industry has been traditionally characterized by the family or community-based independent mill.

Another state-supported effort in this regard involves a recent and innovative program undertaken in cooperation with the International Trade Administration of the U.S. Department of Commerce. Through a grant to the Department of Commerce and Economic Development of the State of Washington, this program will assist the state's small plywood producers in the development of export markets in Latin America. Private industry, in cooperation with the American Plywood Association, is also making a significant contribution to this project. The project will involve sponsorship of a wood housing symposium in Colombia, Ecuador, Peru, and Chile this spring. The programs will introduce the advantages of wood-panelized construction systems and other wood products to government housing officials, architects, and developers.

A second phase of this program will bring housing ministry officials from South American countries to this state in June 1983 for visits to forestry operations and wood products manufacturing facilities.

A third phase involves participation in November 1983 in a major Latin American home-building exhibit in Santiago, Chile.

This program represents only one example of the type of public/private sector cooperation which will be necessary to further our mutual international trade efforts in forest products.

Managing the Forest Products Trade for the Future

This symposium properly provides a comprehensive focus on many aspects of world trade in forest products. First, it examines the land and natural resource base for forestry and forest products production. This productive base is, ultimately, the source of our forest products international trade. A global perspective is necessary if our total forest resource is to be used wisely for the benefit of us all.

Developments in the area of production technology are also discussed. Closely related are questions and issues of product specifications and standards. When taken together with market development, these issues will be critical in determining the success of our international trade efforts.

The symposium also focuses attention on the role of public policies which directly affect international trade in forest products. There is a need for those of us in the public sector at both the state and national levels to maintain a business environment conducive to free trade and competition. We are vitally aware of the many diverse and conflicting interests which arise in trade discussions. The temptations are often strong to promote short-run protection as a way of mitigating adverse impacts of changing economic conditions. This has been particularly appealing to some during the recent economic difficulties confronting our region and the forest products industries. Debate over the issue of softwood lumber imports from our good Canadian neighbors to the north, possible adverse impacts on forest products trade with the People's Republic of China because of textile restrictions, and recent proposals for domestic content legislation are all examples of these sensitive concerns.

The forest products industry participated recently in the 4th Annual U.S./Japan Forest Products Committee meetings held here in Seattle in late January. These trade discussions brought together direct representation of those involved in issues of tariffs and product standards which are vital to the important forest products trade between these two countries. Although total agreement was not reached on all these issues, this represents the type of cooperation and involvement which will be necessary as we work together to develop sound and viable trade policies.

We are all producers and consumers, both of wood products and other goods and services. Our mutual interests require that we continue to work together to find solutions to those problems which limit our ability to move forward with strengthened trade ties. Our state and our people will be well served because of these efforts.

WOOD IN THE WORLD'S MATERIALS MIX

R. S. Whaley and Susan J. Branham

Introduction

In this paper we want to go beyond simply looking at past trends in wood products markets and trade, and speculate about some shifting economic and political trends which could have a major influence on the role of wood in the world's materials mix. In summary, we will try to build the case that within the next couple of decades we may well live in a world where world trade flows far exceed estimates derived from the examination of historic trends. But first let us review those trends.

Current Wood Markets: Big and Growing

Over the past three decades demand for forest products has been on the rise nearly everywhere. Following are some examples. Since 1950, world industrial wood consumption has gone up about 90 percent, reaching almost 50 billion cubic feet in 1977 (figure 1).
- Japan experienced the world's fastest increase in wood consumption since World War II. Postwar industrial expansion led to a more than twofold growth in wood products use, and by 1977, over 3.5 billion cubic feet were consumed, led by lumber used in the booming housing industry. Since 1960, Japanese wood demands have grown at a rate of 4 percent a year, and steady growth in demand is continuing for all major products.
- In Europe, lumber use increased one and one-half times, panel products doubled, and paper and board tripled. The roundwood equivalent of European wood products consumption amounted to 13 billion cubic feet in 1977, about a quarter of world production (USDA Forest Service 1982).
- The Soviet Union ranks third in the world in timber products consumption, accounting for over 12 billion cubic feet in 1977. In two decades, wood use there

rose more than 33 percent. Their per capita lumber consumption is relatively high and lumber production takes about a 40 percent share of their wood harvests.
- The U.S. is the world's largest consumer of industrial timber products with consumption in 1977 of 13.6 billion cubic feet in roundwood equivalent, led by softwood lumber, paper, and paper products. This, together with Canada's 1.8 billion cubic feet, made North American consumption nearly a third of the world's total. North American consumption has increased nearly 60 percent in the past two decades.
- Wood consumption patterns in the industrialized nations differ markedly from those in developing countries. Although the developing nations contain nearly half the world's closed forest and half the growing stock of industrial-size wood, industrial wood products are much less important in their economies than is fuelwood for heating and posts and poles for construction. While industrialized nations consume over 90 percent of the world's processed forest products, developing countries consume almost 90 percent of the wood used as fuel. The combination of growth in population and accelerated development has resulted in increased consumption of both industrial wood and fuelwood (Council on Env. Qual. 1980; FAO 1981).

Future Increase in Wood Consumption

Though slowed by the recent recession, demands for timber products have been growing rapidly in the major consuming countries, and dependency on other regions to satisfy needs has also been increasing (figure 2). European wood consumption early in the next century will be double that of 10 years ago. The most dramatic expansion is foreseen for panel products, spurred by the rapid acceptance of particleboard there. In terms of volume, pulpwood will also show large increases while miscellaneous roundwood and fuelwood use continues to decline, particularly in Nordic countries.

Neither expansion of forest land nor afforestation programs appear likely to significantly affect timber supplies in Europe before the end of the century. The problems are confounded by the Nordic situation where removals may be as much as 25 percent above growth in 20 years. Present efforts to bring removals in line with growth could result in shortages well before 2000. Improved management and utilization will have to be achieved to alter

WOOD IN WORLD'S MATERIALS MIX   25

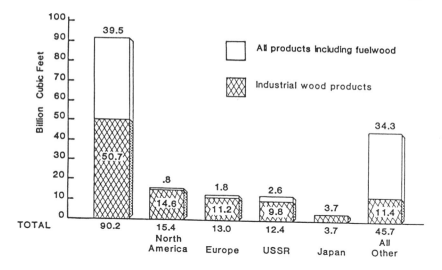

FIGURE 1.   World consumption of wood products, 1977.

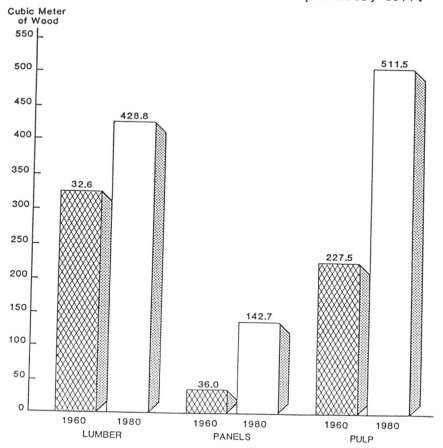

FIGURE 2.   World consumption of wood products.

this picture, and timber deficits in Europe could easily reach a third of demand very soon. The search for wood and fiber will intensify European competition with the U.S. for available supplies from other nations (Council on Env. Qual. 1980).

Japan is faced with similar circumstances. As a result of planned harvest reductions to build up inventories and also land use changes, domestic production of timber has declined. Thus, Japan's expanded consumption in recent times has been made possible through imports. Their Ministry of Agriculture and Forestry foresees continuing rises in demand at a rate of over 2.5 percent per year; however, even with expanded domestic production, Japan will likely remain one of the world's primary deficit areas in timber.

Demands for timber products are expected to rise in the U.S. and Canada, as well as Central and South America, drawing on their substantial forest resources and the considerable investment being made in intensive management efforts to meet prospective demands. The United States' demands for lumber, plywood, and panel products will be influenced by housing starts. In spite of recently depressed housing starts, they are expected to substantially increase in the last half of the eighties in response to the need to house the last of the baby boom generation combined with pent-up demand resulting from the recent low levels of housing construction (Phelps 1982).

Demand for forest products in the developing countries has been rising and this will continue, resulting in a 50 percent increase in consumption in the next two decades. Expectations are that tropical forest output will increase despite concern about the capacity of these countries to overcome the growing drain on domestic forests. The market for paper and board products is expected to expand appreciably as living standards improve. Fuelwood demands, especially in light of increased oil prices, will accelerate with population growth.

The Soviet Union has the largest forest resource base in the world. Current harvests there represent less than one percent of the growing stock, but gaps between production goals and actual output persist. If current trends continue, the U.S.S.R.'s forest industry will be unable to satisfy either domestic or foreign demand.

Wood Supply in International Trade

As wood products consumption has increased worldwide, a growing share has been supplied through international trade (figure 3). Together, the U.S., Europe, and Japan consume over half the industrial wood produced in the world and depend on others—Canada, the Soviet Union, and the tropical hardwood regions of Asia, Africa, and Latin America—for nearly one-fifth of their supply. U.S. reliance on Canada as a source for softwood lumber, hardwood veneer, and newsprint is balanced against our position as an important source for Japan for softwood logs and lumber, pulp chips, woodpulp, and paper and board products, and an important supplier to Western Europe and others for woodpulp, paper and board, and lumber and plywood.

Japan is second only to the U.S. in volume of wood imported. The principal impetus for their softwood log and lumber imports has been housing construction—recently about twice that of the U.S. on a per capita basis. Though heavily forested, Japan is limited in timber resources relative to population, and this, combined with severe forest depletion in World War II, has forced them to rely on imports to meet the bulk of consumption. Rapid growth in trade there has been based on unique circumstances, with Japan transforming from a postwar recovery period to one of a fully mature industrialized economy (USDA Forest Service 1982). It is anticipated that Japan will continue to be a major importer, consuming a substantial portion of the softwood sawlogs and pulpwood production in western North America, and much of the Philippine mahogany and other high-value hardwoods from Southeast Asia.

An important percentage of timber products consumed in Europe is attributable to imports. The Nordic countries, a major softwood source for the rest of Europe, have a potential critical supply situation developing, so while the bulk of European wood products has traditionally come from within Europe, as demands rise, they must continue to rely on outsiders. More and more emphasis is on the Asia-Pacific region for hardwoods and expanding U.S. sources, particularly in fiber-based products, which have increased especially rapidly. Europe's major import, softwood lumber, is supplied chiefly by the Soviet Union, with Canada and the U.S. also important sources. Other major imported products are woodpulp and paper mostly from North America and, since the mid-seventies, wood chips going from the southern U.S. to Scandinavian pulp producers.

The U.S. is currently the leading importer of timber products (USDA Forest Service 1982). More than a fifth of

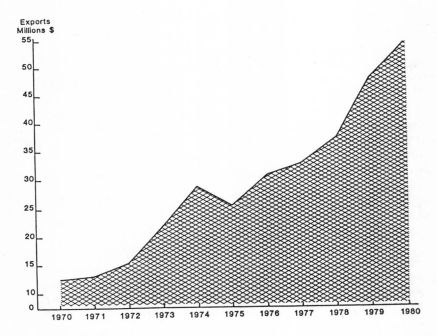

FIGURE 3. World trade in forest products.

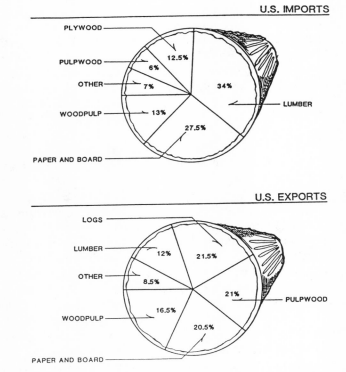

FIGURE 4. U.S. wood products imports and exports.

our consumption of wood products--chiefly lumber, woodpulp, paper and board, and veneer and plywood--has been imported in recent years, much of it originating in Canada and Asia (figure 4). Hardwood plywood imports are mainly from Korea, Taiwan, Japan, and the Philippines, and most of the timber used in its manufacture originates in the tropical forests of the Philippines, Malaysia, and Indonesia. The U.S. is now dependent on Southeast Asian producers for about two-thirds of its hardwood plywood. End products are regularly imported in cross-border trade with Canada.

The Soviet Union is the world's largest producer of softwoods and a major net exporter, ranking second to Canada. Roundwood and softwood lumber account for the bulk of Soviet wood exports. Except for roundwood, which goes mainly to Japan, a rapidly increasing share of Russian timber sales has been taken by East European countries.

If trends in increasing international trade in wood products continue, we can look forward to a growing interdependence in wood products marketing. This will have significant economic and political consequences.

The observations made so far in this paper are those traditionally cited to make the case for increased international trade in wood products. That is, wood products consumption has been growing, will continue to grow substantially, and geographic differences between supply areas and population and industrial consumption areas will result in more international trade. We are convinced that, taken alone, these factors will underestimate the future role of international trade in wood markets.

Trade in Agriculture as a Leading Indicator

Some observations on changing trade patterns in agriculture can be offered here as an example of what appears to be growing global interdependence and the diplomatic repercussions of that interdependence. Twenty-five years ago, trade accounted for only about 2 percent of world food consumption. By the end of the seventies, it accounted for almost 12 percent. As domestic programs were altered to adapt to an expanding world market, U.S. agriculture was transformed. The trade boom caused farm income and prices to climb on a roller coaster sensitive to sharp fluctuations in the international market.

Expansion of demand for some products in the U.S. was almost entirely in exports, which then became the dominant force in total demand. Developed market economies were not the major contributors to this growth in trade. By far the largest increases in imports were by centrally planned economies (the U.S.S.R. and Eastern Europe) and the developing countries, where economic and political policies dictated the actions of government-controlled import agencies (Hathaway 1982).

A number of factors contributed to this changing trade pattern: (1) détente with the Soviet Union and its associated Eastern bloc countries; (2) rapid economic growth of developing and Eastern bloc countries fueled by large-scale international borrowing; (3) exacerbation of the oil crisis by events in Iran, spurring worldwide inflation; (4) normalization of relations with China; (5) marked decline in exchange rates for the dollar.

The export boom in foodstuffs that began with the seventies appeared to end with the eighties (figure 5). Policy actions taken by various governments to combat rising inflation, the end of the credit expansion that had carried economic growth and led to an unprecedented international debt, and various restrictions on exports resulting from Russian intervention in Afghanistan and martial law in Poland dampened the demand for imported foodstuffs. Declines were predominantly where the largest expansion had been a few years before--in centrally planned economies and developing countries. This was the main factor in the significant decline in U.S. farm prices and incomes. While this "internationalization" of U.S. agriculture is not specifically linked to forestry, it does help to demonstrate how one sector found that much of its well-being, if not its survival, both in terms of prices and production, was being determined by global demands and international politics. In a larger sense, it shows how resource use is increasingly interdependent and influenced by world events--a point to which we will return.

As we move into the eighties, events will force trade patterns into even further alteration: in particular, we may expect a ballooning foreign debt, changes in monetary exchange rates, difficulties in trade relations among countries, as well as the growing role of state trading agencies in centrally planned economies and developing countries, and difficulties too in foreign policy. International trade is no longer simply a circumstance of demand/supply. One of the consequences of the "internationalization" just referred to is that resource issues are now irrevocably linked with international issues and thus will have to be sensitive to unstable and uncertain global influences. It appears that in the short run there will be several factors interfering with expanded trade. In the future, this may reverse.

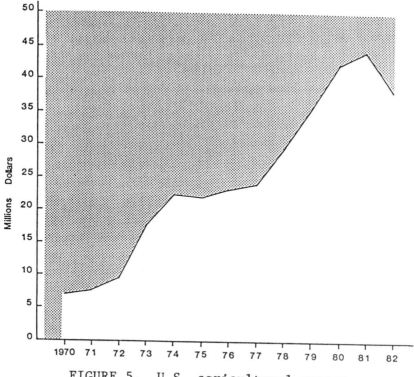

FIGURE 5. U.S. agricultural exports.

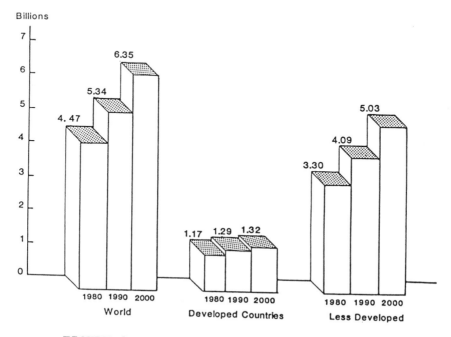

FIGURE 6. Population growth, 1980-2000.

Let us now look at some factors which seem to indicate that over the next few decades we may see substantial changes in the global economic order which will include unprecedented growth in international trade. Three factors will cause this change in the global economic order: (1) increasing impact of geographic imbalances in resources, (2) increasing economic interdependence, and (3) shifting political power.

Resource Inequities--An International Issue

Although there are different viewpoints on the problem of population growth vis-à-vis resource scarcity, there has to be unanimity on the fact that geographical imbalances exist and populations are expanding most in regions where food, fiber, and mineral resources are least abundant. Accidents of geography exert a powerful influence on resource use, and inequality of access has come to the fore as a major international issues. The world oil situation only serves as the most obvious illustration of this point (Castle 1982).

There can be little argument about the pressures of increasing population, already in excess of carrying capacity in some places, with the result that what resources are available are being depleted, forcing many to live at the margins of existence. All estimates show population growing at a rate of nearly 100 million annually, which translates into a 50 percent increase early in the next century (figure 6). As one source puts it graphically--a new India is being created every seven years. Ninety percent of this growth is occurring in Third World countries, where per capita GNP remains below $200 a year, and many areas are resource-poor (Brown and Shaw 1982).

Some specific impacts of such population/resource imbalances can readily be seen. With a 67 percent share of world population, developing countries account for only 38 percent of food production, a situation that will lead to an actual per capita decline in their consumption by 2000 at a time when food production has increased nearly 90 percent worldwide (figure 7). As production barely keeps ahead of population growth in the Third World and prices double, pressures for increased output will be placed on lands already overcultivated, further deteriorating the land base (Council on Env. Qual. 1980; FAO 1981).

In the case of forest resources, the situation is equally disturbing (figure 8). The one-quarter of the world's population that depends for energy on fuelwood--"poor man's oil"--is faced with circumstances where demands will exceed supplies by 25 percent by the end of the century. It is currently estimated that a forested area half the size of California disappears every year in these nations. As populations mount and fuel needs increase, the drain on the forests will continue so that by 2020 virtually all of the physically accessible Third World forests are expected to have been cut (Sedjo and Clawson 1982).

While these resource inequities are tilted against the less developed countries, the one commodity they have in relative abundance, nonfuel minerals, is not in demand locally but it is in industrialized nations which use three-fourths of the world's output. This is partially because the less developed countries lack either the money or the technology to be heavily involved in the world industrial processes that utilize the bulk of nonfuel minerals. But these tradeable commodities are the life blood of the Third World economies and the source of foreign funds available for trade purposes.

There are other geographical imbalances that might be touched on in passing: (1) the most productive fishing waters border industrialized nations with strong central fishing fleets; (2) more than half the world's solid fuel resources are in less populous Western nations; (3) although industrialized countries with the highest consumption rates have the smallest oil reserves, their favorable financial position enables them to command enough oil despite price increases to meet most energy needs. Conversely, poorer Third World nations will increasingly have difficulty meeting demands, thus putting added drain on shrinking forests for fuel as well as lessening the prospects for increased food production because most agricultural processes required for higher yields rely on oil and gas (Council on Env. Qual. 1980).

So what we have is a sort of "Catch-22" of resource imbalance problems. The upshot of all of it is, of course, that much of the world's population will be forced either to pay steep prices for resource goods and services or to do without, with the inevitability of deterioration in the quality of life. This makes it obvious that improving access to resources is probably the most important priority for social change, but at the same time, the most sensitive economic and political issue. On a broad scale, unequal distribution of resources can be expected substantially to affect international relations and make economic interaction almost a certainty, a point that will take on added emphasis as this discussion continues.

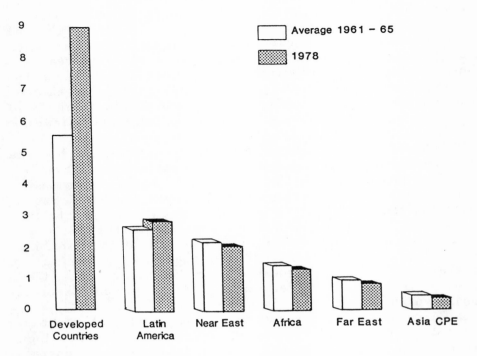

FIGURE 7. Arable area per agricultural worker per hectare.

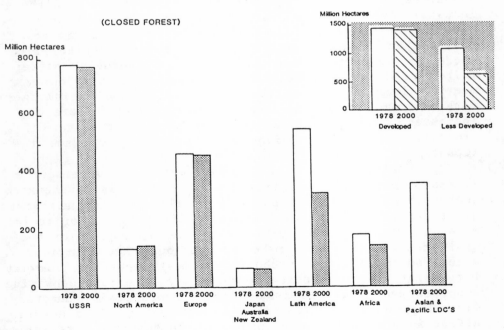

FIGURE 8. World forest resources, 1978 and 2000.

New Pressures Affect the Global Economic Picture

The seventies ushered in the largest international lending boom the world has ever known. Never before in history have so many nations been so deeply in debt and the banking system so haunted by the specter of default. At the core of it is a $700 billion lien held by the world's financial community against some hard-pressed developing and Eastern bloc countries (figure 9). Compounding the problem is the fact that the nations that have borrowed so heavily are consumers of the industrialized world's goods. Thus, we have the proverbial "house of cards," whereby a default anywhere along the line could have far-reaching economic and political consequences. The global stake in this crisis vividly illustrates the interdependence of the world's financial institutions (Palmer et al. 1983; Gall 1982).

It all dates back to the first major oil price increase a decade ago which saw a number of oil-rich nations suddenly earning billions of dollars and depositing their excess wealth in the world's banks. Consequently, oil-poor developing countries hard pressed to pay higher energy costs and oil-rich nations with ambitious development plans found these flush financial markets a fertile field for loans. At first, it looked like an ideal venture. From Wall Street to Western Europe, banks began to take a share in a bonanza of international lending induced by the premise that governments, unlike corporations, do not go bankrupt. Networks of leading U.S., European, Japanese, Arab, and Latin American banks put together huge syndicates that negotiated billion dollar deals almost overnight without considering the potential risks. Lured by easy credit, some countries jumped on the merry-go-round even for the most questionable projects, and corruption was rife.

The party was great while it lasted. Then came the reckoning. A second oil price shock in 1978 and tighter credit enacted by some countries to dampen inflation forced interest and debt service rates to skyrocket. This, together with a recession-sparked decline in imports from Third World countries, left these countries in a financial bind prompting several to ask for rescheduled payments on their debts. In short order, Mexico, Brazil, and Argentina announced that they could not meet payments due on borrowings. So opened the "Pandora's Box" of global credit overextension. Within a year, the International Monetary Fund listed 32 countries in arrears on their debts. Threatened defaults and the fear of the problem avalanching prompted even further bank investment to rescue seriously troubled nations, often through the cooperation among banks of several nations. Many of these banks had gotten too deeply involved to refuse such cooperation.

Only now have the vast dimensions of this worldwide financial involvement become apparent. Indeed, about 1,400 U.S. and foreign creditor banks are said to be tied up in loans to Mexico alone. The largest financial rescue ever constructed, described by some as a Marshall Plan for Mexico, involved the U.S. Treasury, several central banks, the Bank for International Settlements in Switzerland, and some private banks under an agreement with the International Monetary Fund.

How the world's financial institutions handle this crisis will be crucial to the economy, though it is not the subject of this discussion; but the pervasiveness of the problem does portray the interdependence of the world's banking systems, and how the economic and political stability of both borrowing and lending countries hangs in the balance.

Another indication of the world's changing economic picture has been the rise of multinational corporations. Multinational corporations, according to Richard J. Foster, "pose both the gravest danger to a more just and equitable world and potentially the most powerful force for realizing a more just and equitable world" (Foster 1980). Aureho Peccei, a director of Fiat and organizer of the Club of Rome, sums it up by saying that the multinational corporation is "the most powerful agent for the internationalization of human society" (Barnet and Miller 1974). One recent estimate indicates that by 1986, two or three hundred global corporations will control 80 percent of all the productive assets of the noncommunist world (Green and Massie 1980). Perhaps the most recent actions by the industrial sector to be receiving considerable attention from the popular press have been the joint ventures between U.S. and foreign automobile interests. _Time_ magazine recently described with tongue in cheek the new California plant jointly owned by General Motors Corporation and Toyota, whose product will be a new automobile that industry members are already dubbing the "Toyolet."

The growth of multinational corporations means that within the global economic scene there are firms which have the means to finesse restrictions from individual governments or pressures from labor or owners of particular resources. The era of global corporations has the potential for a new form of imperial domination or a golden age of peace and prosperity. That multinational corporations will have a growing impact on the economies of the world, there is little doubt. It remains to be seen whether they will be

WOOD IN WORLD'S MATERIALS MIX   37

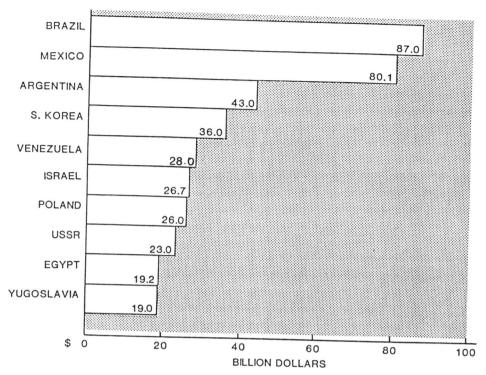

FIGURE 9. International debt at the end of 1982.

FIGURE 10. Changing pattern of U.S. immigration, 1920-1980.

the genesis of global economic order or parasites on global economic chaos.

A Changing Political Order

The interdependence of nations is visible not only in a changing economic climate but in a changing political climate. To discuss intelligently the changing dimensions of the world political scene would tax the scope of this paper, the experience of its authors, and the attention of this audience. But the central argument of this discussion would be incomplete without a reminder about the obvious tie between economic stability and political stability; or, expressed negatively: differences in the economic welfare of the nations of the world will manifest themselves in political differences between these nations. Briefly to make this case, let us refer only to three prominent influences on the political scene: changes in the actors on the world political stage, transformations due to migration, and the intermingling of political and trade issues.

As the United Nations grew from the 29 initial signatories to 152 members, the debate shifted from east-west to north-south, and the Third World nations became a stronger collective voice for the world's poor. In a less fortunate way, international terrorism has given smaller nations a prominent place in international issues. Rather than our dwelling on the broadening of world political concerns, let it suffice to point out that political interdependence has grown along with the growth in economic interdependence.

The migration of people from one place to another to improve their living conditions is as old as time. But greater global economic development as well as political upheavals have generated increases and changes in immigration patterns that are marked departures from tradition and have international ramifications (Castle 1982).

Today, the flow of migration is from less developed to more developed nations, with the result that, in the case of the U.S., for example, Asia and Latin America have replaced Europe as the origin of most immigrants (figure 10). Part of this has been the huge influx of displaced groups such as the refugees from Vietnam and Cuba (Imm. and Nat. Serv. 1980).

Social, political, and economic pressures can readily be imagined when one considers the large numbers of Vietnamese and Cambodians inhabiting Southeast Asia; Indians and Middle Easterners in Great Britain; Haitians, Vietnamese, and Puerto Ricans in the U.S. More and more, low-income countries are emerging as potential sources of immigrants to the industrialized nations. Movements of people to improve access to the world's resource goods and services can thus be seen as further links in the chain of international interdependencies that is the subject of this discussion.

At its furthest extreme, the intermingling of political and trade policies previously referred to has been discussed in the context of a "resource war" between superpowers. Though no attempt is made here to address that topic, it does stand out as an exaggerated example of the link that is now perceived between the two. Trade dependence has been for so long a way of life for so many nations that political sanctions are doubtless the likely outcomes of ever-changing worldwide economic problems (Hathaway 1982).

Just two manifestations of the impact of foreign policy actions on materials trade illustrate the point:

1. The 1980 U.S. grain embargo enacted as a protest against the Soviet Union's invasion of Afghanistan, which had far-reaching political and economic consequences for both nations; and

2. In the wake of Soviet involvement in Polish affairs, the ban on sales of U.S. equipment and U.S.-licensed European products for use in constructing the Soviet gas pipeline to Europe, an action which has had not only major implications for U.S. businesses but also created foreign economic policy differences with Western Europe.

Political arguments for self-sufficiency in food, energy, or wood are spurious at best, especially when one considers American reliance on foreign oil, Soviet reliance on American grain, and Japanese reliance on both for wood. Obviously, sustained political turmoil anywhere in the world could have serious repercussions in the form not only of supply interruptions but also of financial chaos for exporting countries. Thus, the "Gordian knot" that now ties trade to political policies can be expected to increasingly restrict resource flow between countries, with obvious economic consequences for all.

So What?

Two points have been brought forth in this paper. First, a traditional look at current and future wood products demands would lead one to conclude that it is reasonable to anticipate increased international trade. The second point is that you have not seen anything yet. It is the latter which is important in looking at the future of international trade in wood products.

What one makes of the brief observations about the changing global scene probably depends upon whether one takes the pessimistic view of a Cassandra and concludes that the world is sitting on the precipice of economic and political collapse or is of the optimistic mind of a Pollyanna warranting that the physical and human resources are available to substantially improve the welfare of us all, and what is lacking is the institutional framework to benefit from the vast riches of the world.

Though the path will be a bumpy one with economic peaks and troughs greater and more frequent than in the past, the overall direction will be toward a greater economic and political interdependence which will lead to greater justice in the allocation of the world's resources (figure 11). For us here at this conference, it is significant that most of our official projections of changes in international trade over the next two decades are probably substantial underestimates, making this conference of timely importance and the beginning of a major challenge to those of us interested in tracking and forecasting changes in the U.S. wood resource picture and its implications for the U.S. economy. The major force influencing the products and direction of trade will be the comparative advantage of each supply region to offer those goods and services for which it is favorably endowed. In the case of the U.S., wood fiber will undoubtedly be one of the major contributions to the world materials mix.

We will leave you with an exhortation from Tennyson, something of a futurist himself:

"Come, my friends, 'tis not too late to seek a newer world..."

WOOD IN WORLD'S MATERIALS MIX   41

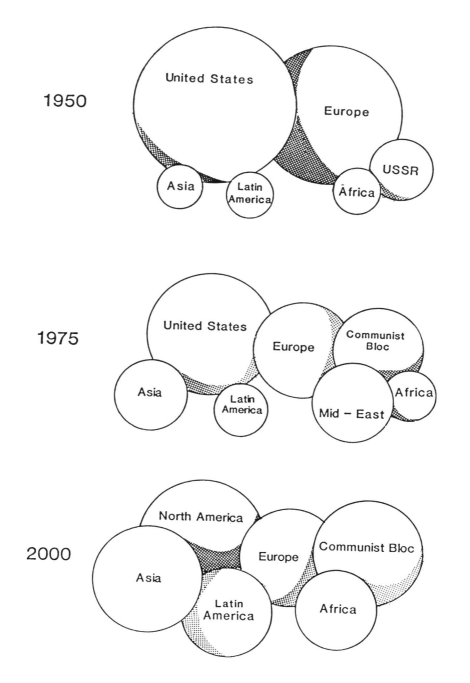

FIGURE 11. U.S. perspective of the commercial world.

Graphics for this paper by Sherman Gillespie, USFS.

References

Barnet, R., and R. Miller. 1974. Global reach: The power of the multinational corporations. Simon and Schuster, New York.

Brown, L. R., and P. Shaw. 1982. Six steps to a sustainable society. Worldwatch Paper 48. Worldwatch Institute, Washington, D.C.

Castle, E. N. 1982. Resource adequacy, global development and international relations. An address to the Annual Meeting of the American Association for the Advancement of Science, January 8.

Council on Environmental Quality and the Department of State. 1980. The Global 2000 Report to the President. Vols. 1 and 2. U.S. Government Printing Office, Washington, D.C.

Food and Agriculture Organization of the United Nations. 1981. Agriculture: Toward 2000. Rome, 1981.

Food and Agriculture Organization of the United Nations. 1982. World forest products demand and supply 1990 and 2000. Rome.

Foster, R. J. 1980. Freedom of simplicity. Harper and Row, San Francisco.

Gall, N. 1982. The world gasps for liquidity. Forbes, October 11, pp. 150-154.

Green, M., and R. Massie, eds. 1980. The big business reader. The Pilgrim Press, New York.

Hathaway, D. E. 1982. U.S. food policy and international affairs. Draft Report, RFF Forum, October.

Immigration and Naturalization Service, U.S. Department of Justice. 1980. 1979 Statistical Yearbook. U.S. Government Printing Office, Washington, D.C.

Palmer J., et al. 1983. The debt-bomb threat. Time, January 10, pp. 42-51.

Phelps, R. B. 1982. Outlook for timber products. An address to the 1983 Outlook Conference, Session 23, Washington, D.C., November 30.

Sedjo, R. A., and M. Clawson. 1982. The world's forests. Resources, No. 71. Resources for the Future, Washington, D.C. October.

USDA Forest Service. 1982. An analysis of the timber situation in the United States, 1952-2030. FRR-23. U.S. Government Printing Office, Washington, D.C.

WORLD SUPPLY OF WOOD: ECONOMIC RESOURCES

Roger A. Sedjo

The world's supply of wood as a <u>physical</u> resource has been examined by Dr. Flores-Rodas (this symposium). In this paper I will examine the world's supply of wood as an <u>economic</u> resource and attempt to cover the following issues: the distinction between the physical resource and the economic resource; the nature of the forest resource with particular focus on how the resource is experiencing a fundamental transition through time; methods for expanding the supply of the resource; and the potential availability of the current and future economic supplies of the timber resource.

The Economic vs. the Physical Resource

Resource economists have found it useful to distinguish between resources and reserves of a natural resource (Landsberg and Tilton 1982). Very simply, the term <u>resources</u> refers to the physical concept of what exists, both what is known to exist and what appears likely to exist. There is uncertainty here, of course, particularly with respect to "hidden" resources such as petroleum and underground minerals. Hence, the estimate of the amount of the resource is likely to change through time. The concept of reserves is much more limited and has an economic component related to that portion of the resource that is exploitable with current technology and at acceptable costs. In forestry then, the forest reserve would refer to timber stands that are accessible at acceptable costs and that have timber that is presently merchantable.

Forests as a Renewable Resource

The forest resource has a dimension that is often lacking for other resources, that is, renewability. Also, most stands are expanding in volume through time. Of course, other resources are also renewable; e.g., fisheries, water, and even soil have natural regenerative capacities. Thus, while at any time the potential supply of timber or other renewable resource is limited by the existing stock, in the longer term the supply is dependent upon the initial inventory, the rate of draw-down of that inventory, and the rate of regrowth and regeneration. Therefore, while the short-term supply of forest resources and other renewable resources has characteristics like those of nonrenewable resources, the long-term supply is critically dependent upon the renewability feature of the resource. The renewability feature is generally subject to human management and manipulation. This characteristic generates some especially interesting economic behavior patterns with respect to the utilization of renewable resources through time (Lyon 1981).

Resource Availability

The availability or scarcity of a resource in a physical sense is simply a question of accurately inventorying that resource. In the case of certain minerals this may be difficult. In a simplistic sense, if society is aware of an increase in a resource due to growth or discovery, its physical availability has increased. Conversely, if less is physically available due to net depletion for whatever reason, the resource is physically more scarce.

The issue of economic resource availability, i.e., the economic supply, is more complex. It is often assumed that the world faces increasing economic scarcity for most natural resources due largely to what is viewed as their inevitably decreasing physical availability. Contrary to this "common sense" view, however, the historical evidence suggests a decrease of economic scarcity for most natural resources. An RFF study by Potter and Christy (1962) showed that the real price, which is a measure of economic scarcity, of almost all natural resources had been on a steady and continuing downward, albeit cyclical, trend for periods of up to and in excess of 100 years. This implies an increasing economic abundance of resources. A follow-up study by Robert Manthey (1978) confirmed that this trend continued into the 1970s. Although at first many find such a result perplexing, this result is clearly consistent with a world in which real per capita incomes and living standards have been experiencing long-term improvements. The essential reason for such a result appears to be technology. Of course, the fact that the historical trend has generally been toward lower real resource prices (greater economic abundance) need not imply that this trend will continue indefinitely. It does, however, tell us rather definitely that the economical availability of natural resources has been increasing for at least the past 100 years.

Many observers were quick to note that the 1970s brought an abrupt turnaround in this trend for certain energy resources, most notably petroleum. When viewed from our perspective in 1983, however, with the OPEC cartel in disarray and with real petroleum prices experiencing significant declines and under pressure for further erosion, it may well be that history will view the energy-scarce period of the 1970s as an anomaly in the long-run trend to greater resource availability.

Technology has provided humankind with the ability to expand its economic resource base both by allowing for the exploitation of previously inaccessible resources and also by devising ways whereby "plentiful" poor-quality resources are substituted for "scarce" high-quality resources. To use the energy example again, the demise of the "energy crisis" is due in large measure to the role of technology which, together with appropriate economic incentives, resulted in the development of alternative economic supplies.

Nevertheless, while the preponderance of natural resources have been experiencing increased economic availability, forest resources have been the one major exception. The studies cited above reveal that the real prices of forest resources have been rising (e.g., sawlogs) or constant (e.g., pulpwood) over the long time periods examined. This implies that the important forestry resource has _not_ become economically more abundant. Thus, despite the development of new technology (e.g., allowing for the use of previously unusable species, developing wood panels to substitute for lumber, and developing resource-saving mill operations) the resource has nevertheless become economically more scarce.

The Forest Resource Through Time

Forestry is currently experiencing a transition similar to that experienced by agriculture two or three millennia ago (Sedjo 1983a). Human food needs that once were provided by hunting and foraging are now being provided by cropping and livestock raising, thanks to modern agriculture. Similarly, wood needs that traditionally have been provided by utilizing the natural forest resource are now increasingly being provided as the produce of forest management and ultimately by forest plantations where trees are planted, grown, and harvested in a manner akin to agricultural cropping. While in agriculture the transition essentially has been completed for centuries, in forestry the transition is just beginning in earnest. Of course, in some regions of the world the forestry transition has been under way for some time and in parts of Europe and East Asia the transition is largely completed.

Still, the major part of the world's current wood supplies originate in natural, essentially unmanaged forests. The future promises to be different from the past, or even from the present. Artificially regenerated forests constitute about 90 million hectares worldwide (Lanly and Clement 1979) and have considerable economic potential (Sedjo 1983b). For example, in South America, which contains some of the world's largest physical volumes of the physical resource, about one-third of the current industrial wood production originates from only 1 percent of the forest area, the industrial plantations. Furthermore, projections for the year 2000 anticipate that about 50 percent of the much larger production of South America will come from man-made plantations (IDB 1982).

Old-Growth Forests and the Transition to Plantations

In order better to appreciate the short- and long-term supply potential of a forest in transition and its price implications, it may be useful to examine figure 1, which presents the time profile of harvest volumes for a hypothetic old-growth forest that is experiencing a transition to plantation forest management under an economically optimal harvest management (Lyon and Sedjo 1983). You will note that the old-growth forest is harvested rather quickly--in this case over a period of perhaps fifty years. In this example, where two land classes are forested, the higher land class is harvested much more quickly than the lower site. This reflects the higher productivity of the land site and, hence, the economic desirability of quickly establishing new vigorous stands as well as the superior economics of harvesting the higher quality forest resource. As can be seen, the economically optimal time path for the old-growth forest consists of a fairly rapid draw-down of the inventory and then movement to an even-aged, sustained yield forest. You will notice that concepts such as nondeclining even flow play no role in the optimal management regime. Figure 2

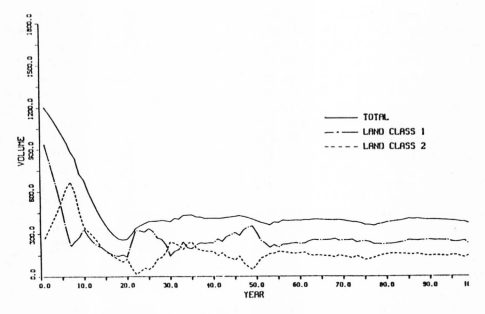

FIGURE 1. Forest volume over time.

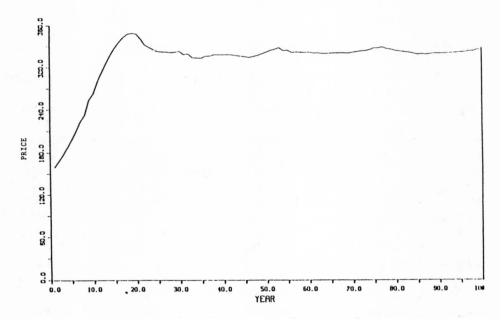

FIGURE 2. Price trend over time.

presents the timber price trends consistent with the harvest time profile just presented for a closed system with a given level of long-run demand. The price trend of timber is in some sense simply the mirror image of harvest levels. Initially, the high harvest levels imply low prices. As the old growth is drawn down, prices rise. This is consistent with the actual historical experience described above. Eventually, in this system, prices will level out to reflect the achievement of a system-wide sustained yield forest.

Figure 3 presents the time profile of timber supplies (harvests) for alternative levels of demand. Two striking features emerge from this analysis. First, a higher level of demand implies a more rapid optimal rate of old-growth draw-down. Second, a higher level of demand implies a higher level of sustained yield harvest. This second feature is due, of course, to the higher management inputs associated with higher demand and higher prices.

It may be instructive to compare the idealized old-growth forest that has been harvested and managed regularly for generations. Figure 4 presents a time profile for such a forest. In this case the basic underlying inventory and growth data are those of Sweden. The time profile in the figures assumes that Sweden would move to a harvesting rotation that would be economically optimal, i.e., a substantially shorter rotation than currently practiced. Movement to such a regime, given the current level of Swedish inventories, would result in an initial draw-down of inventories that would then become gradually built up over time as this different type of forest completed its transition to a long-term sustained yield mode.

The construct just presented can be a useful tool in looking at both the regional and worldwide transition from old-growth to plantation-type forests and in the process provide insights into both short- and long-term economic supply. Initially, with low prices the old-growth resource is drawn down at a rapid rate. In this context there is little incentive for forest management or planting. The price of the resource is low and the availability great. As this process continues, however, the availability diminishes and the price rises. At some point the higher prices generate incentives for management and artificial regeneration. Industrial forest plantation begins to become a sensible economic proposition.

Over the long term, the principal source of the forest resource supply will shift from old-growth to second-growth and plantation forests. Within the context of managed forests, inputs can be applied and the productivity of the

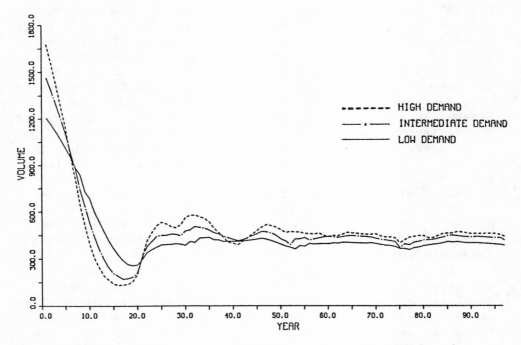

FIGURE 3. Volume over time at various demand levels.

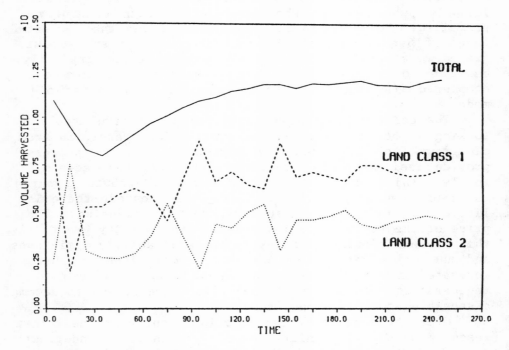

FIGURE 4. Volume over time, intermediate demand.

forest can be greatly increased. Forests can also be relocated to regions of high productivity. In addition, as technology develops through time, higher yields can be expected in forestry, perhaps akin to the yield increases that have been achieved in agriculture. Thus we might expect to see a long-term upward trend in harvests (not presented in our profile) as higher yield technology is put in place. Finally, even with a given technology we would expect higher levels of demand to elicit higher levels of management and thereby higher growth and harvest.

Dimensions for Increasing Supply: Technology

Within the context of the gradual draw-down of the old-growth forest and the transition to plantation forestry, there are a number of means for expanding the quantities of wood available to world markets. As mentioned above, for a resource to be economic it must be accessible and merchantable. Accessibility can be increased by changing to a technology which lowers costs of harvesting and transport. Similarly, anything that increases the real price of the resource will allow additional costs to be profitably incurred in the process of harvest. Again, this truth has been dramatically demonstrated recently in the petroleum industry where resources inaccessible at the lower product prices of a decade ago--e.g., the Overthrust Belt--are now being developed at a rapid rate. Finally, it should also be noted that higher product prices encourage research and innovative technology which lowers accessibility costs.

It is well recognized that timber is not a homogeneous commodity but rather has a variety of quality characteristics. In the solid-wood products we have both hardwoods and softwoods. In the fiber products these translate to long-fiber and short-fiber characteristics. Of course, there are numerous other features that are commercially important. For a given technology there is some--albeit limited--ability to substitute wood with these different qualities. New technologies can substantially enhance our ability to substitute, however. For example, the development of new woodpulping processes in the early post-World War II period allowed southern pine fiber to be used for the production of newsprint. Similarly, the recent development of a new product, waferboard, has created an entirely new market for Lake States hardwoods such as aspen. One might also consider the ramifications of an economically viable process which would allow short fibers to be used in the production of papers with strength and tear characteristics comparable to those of long-fiber papers. I understand that substantial progress is being made toward the development of such a process. Of course, technological improvements are also occurring in forest management. While still in the primitive stages (Staebler 1978), numerous innovations ranging from genetic improvements to mechanical planting are enhancing the economics of forest plantations.

Dimensions for Increasing Supply: Economic Margins

In addition to improved technology, there are at least four economic margins on which supply can be increased. First, intensive management can be introduced and intensive forestry can replace natural regeneration. Management techniques such as artificial regeneration, fertilization, weed control, and thinning are economically rational for much of the world's better forest lands and new land can be brought into the production of timber. Second, the forest land base can be expanded. Although forest lands have traditionally been viewed as those lands that are submarginal for all other uses, this is becoming less true today. For example, in regions such as New Zealand and Chile, plantation forests are replacing pasture lands and some croplands as the best and highest value use of the land. And we are all aware that forest plantations in the U.S. South occupy lands that were once used for agriculture. Third, in the short and intermediate term supply also can be expanded by lowering the rotation and thereby drawing down existing stocks more rapidly. Finally, total supply can be increased by a more rapid draw-down of old-growth timber. It is now well recognized that when dealing with mature old-growth forests, timely liquidation of inventories increases total output by allowing for the utilization of the old growth while simultaneously allowing rapidly growing young stands to replace the slower growing old-growth stands.

These vehicles--intensive forest management, increased total land area in timber production, improved management of rotation lengths, and old-growth inventories, as well as technological change--are all capable of expanding the economic supply.

Current Economic Supplies

Let us now move out of the conceptual plane and on to a discussion of current supplies and some judgments and speculations about future supply. I will focus upon the regions that are major wood producers, with particular attention being paid to regions that have large surpluses, either currently or potentially in the future, available for export to wood-deficit regions. The number of current major wood-surplus regions (exporting regions) is relatively small. The most important of these are the Pacific Northwest (PNW) region of the U.S. and the British Columbia (B.C.) region of Canada. These two adjacent regions provide forest resources in some form to virtually all of the major markets of the world. A third major supply region of the world is the East Indies region of Southeast Asia, encompassing Indonesia, Malaysia, the Philippines, and Papua New Guinea. Like the PNW and B.C. regions of North America, the East Indies provide large volumes of wood, in this case largely unprocessed, to most of the world's major markets (Takeuchi 1974). A fourth major supplying region is the Soviet Union. In fact, it might be better to characterize the Soviet Union as two separate supplying regions: one in the European portion of the U.S.S.R., providing local markets and exports which go largely to Europe, and the second the Far East region of the U.S.S.R., providing supplies to the Pacific Basin--almost exclusively Japan (Solecki 1982). Another important supply region is the Nordic region of Europe with its production directed almost exclusively to the markets of continental Europe and the U.K. Also, we have the important North American supplying regions in eastern Canada and the U.S. South. Both of these regions provide very large volumes of forest product outputs, most of which are consumed within the industrial centers of the East and North Central regions of North America. Finally, we have the emerging forest resource producers of the Southern Hemisphere, most notably South America and Oceania.

Let us now look briefly at the nature of the resource which predominates in each of these regions. The timber supplies of the PNW, B.C., the East Indies, the Soviet Union (both European and Far East), and eastern Canada come predominantly from old-growth forests. In these regions, we are still primarily in the process of drawing down old-growth inventories that have been provided by nature. In two other regions discussed, the Nordic countries and the U.S. South, the resource is drawn largely from second-growth forest--some of the regrowth which has occurred naturally, and other regrowth being the result of human management, including man-made plantations. Of course, in both of these regions the degree to which the regrowth represents conscious decision making has increased through time as the economics of timber-growing investment have improved. Thus, the incidence of artificial planting of forests has been greater in recent years in these regions than in the more distant past. Finally, we have the potential production of the emerging Southern Hemisphere regions.

Future Supplies

Any speculations as to future availabilities of economic supplies of forest resources are fraught with difficulties. Nevertheless, some expectations of future availabilities, either explicit or implicit, are necessary to both private and public decision makers regarding investments in forestry. Since I am currently undertaking a major study examining the long-run timber supply potential of major world supply regions, I will venture some very preliminary and tentative judgments as to the possible supply potentials of some of the major regions.

Pacific Northwest: The present Forest Service view seems to be that the private forests in the PNW will be forced to reduce harvests in the 1990s due to the draw-down of the old growth (although I understand that this view may be currently under reassessment). An alternative view, held by some in the forest industry, is that the drop-off in harvest from private lands will not be as precipitous as the Forest Service has projected, since high-yield forestry combined with a shifting to the harvest of smaller logs will mitigate against some of the more extreme harvest-reduction forecasts. The preliminary findings of my current work in process tend to support this view. Harvests from federal lands are, of course, constrained by the "nondeclining even flow" rule which limits harvests from old-growth forests to levels that are indefinitely sustainable. This does not reduce the fluctuations, however, but merely shifts them to the private harvests. In the long term, the introduction of intensive management should increase the future levels of sustainable harvests above those levels currently projected.

One final point in this regard ought to be made. The Forest Service projections of demand over the 1980s for forest products in general and lumber in particular appear in retrospect to be too bullish. This is not a defect found solely in Forest Service projections, but is also found in the projections of a variety of forecasters. The high forecasts appear to be more than simply an anomaly resulting from the current severe recession. While a discussion of this topic is beyond the scope of this paper, suffice it to say that to the extent that the demand projections made in the late 1970s were systematically too high, the draw-down of the forest inventories envisioned in forest projections will be high, and the actual abruptness and timing of the fall off in old-growth harvests on private lands--and hence in the aggregate--will be reduced.

British Columbia: Due to lack of comprehensive data, we know substantially less about B.C. than we do about the PNW. Recently, concerns have been expressed about the long-run availability of the Canadian resource in general and the B.C. resource in particular (Reed 1978). It is undisputed that the more accessible forest resource of the coastal region is gradually giving way to the less accessible resource of interior B.C. Also, growing conditions in B.C. are generally less favorable than those of the Pacific Northwest of the U.S. and access is also generally inferior and more costly. Therefore, we would expect the long-term potential of harvests from second growth and plantations in B.C. to be substantially below that of the PNW.

East Indies: The third major region of economic old-growth forest resources is the East Indies. While this region has provided only about 5 percent of the world's industrial wood, it is the dominant source of tropical hardwoods, with most of the production being for world markets. The major sources of industrial wood have been the Philippines, Malaysia, Indonesia, and more recently, Papua New Guinea. Over time, the dominant supplies have shifted from the Philippines and Malaysia to Indonesia and, more recently, Papua New Guinea has taken on increased importance. In recent years also the major log exporting countries increasingly have examined ways in which they could participate to a greater extent in the processing. Various types of bans and restrictions on exports of unprocessed wood have been discussed and in some cases introduced.

I must admit that when I first investigated this region I expected that it would fit the pattern of the old-growth draw-down region discussed above. This would be consistent with the conventional wisdom that the old-growth forests of the East Indies are being depleted at a rate far in excess of a sustained yield rate and that beyond some point in the not-too-distant future, the world will have to rely upon the far less plentiful resource provided by second-growth forests. Such a view does appear to be an accurate description of the experience of the Philippines and also that of Peninsular Malaysia.

While these rich timber resources were being depleted, however, the vast resources of Kalimantan and Irian Jaya in Indonesia and the resources of Papua New Guinea are only beginning to be tapped. For the East Indies as a whole, the current harvest is still well below what appears to be the sustained yield potential of the forests under very passive management. If we assume growth of only one cubic meter per hectare per annum of merchantable timber from the secondary growth forests of Malaysia and Indonesia and growth of one-half cubic meter per hectare per annum from Irian Jaya and Papua New Guinea (a growth rate well below that normally expected in this region), the sustained yield production is conservatively about 100 million cubic meters per annum. This is well above the peak harvest volume of the region of about 70 million cubic meters in 1979. Of course, we are all aware that the sustained yield potential is more a biological than an economic concept that in many cases is not attainable in an economic environment. Nevertheless, it appears feasible that the economic sustained yield of the region, at real prices similar to those experienced in the late 1970s, may be 70 million or more cubic meters per annum.

I should stress that the above supposition for potential supply from the East Indies is not intended as a forecast. Certainly, there are a variety of impediments that the individual governments have considered and could construct that would preclude timber flows of the level that I suggest might be economically feasible. These include: (1) improper logging procedures; (2) governmental restrictions which choke off the flow of wood that is economically attractive to grow and harvest;[1] and (3) a reduction of the forest land base.

<u>Soviet Union</u>: The fourth major area which has been an international source of industrial wood utilizing old-growth inventories has been the Soviet Union. Information about these forest resources is not readily available in

---

1. Recent restrictions of log exports by some countries in this group need not necessarily adversely effect the long-term wood flow if processed products exports replace the log export flows.

the West; however, it is well known that the forest resource is vast and includes heavily forested regions in European sections of the Soviet Union as well as vast areas of forests in Siberia and the Far Eastern region. Logistical considerations suggest that Soviet exports to Europe will probably be drawn largely from the European forests. Substantial commercial forests exist in western Siberia, but it is likely that these resources will be used largely for domestic markets within European Soviet Union. This process may free forest resources further to the west for export markets. Eastern Siberia and the Far East regions both have vast inventories of old-growth forest. Processing facilities are being developed in eastern Siberia to draw upon the pine and spruce resources found there (Solecki 1982). These forests appear to have the greatest potential to move into world markets through the Pacific Basin and continue or perhaps supplement the already substantial log flows currently moving from the U.S.S.R. to Japan.

It is difficult to estimate the extent to which timber from this source will affect future world supplies. As a centrally planned economy, the Soviet Union has numerous considerations other than economic. Furthermore, even with an expansion of rail facilities in the region, the opportunity costs of utilizing those limited facilities for the export of timber may be large. On the other hand, the Soviets have experienced a chronic shortage of foreign credits and forest products exports may alleviate that shortage.

A final consideration with respect to the long-run economic potential of the Soviet forest resource relates to considerations of species. The dominant species in eastern Siberia and the Far East region is larch, but under current technology larch is viewed as of inferior merchantability. Thus the short-term supply prospects from this resource appear limited for this reason alone. Over the long term, however, technology or the lack of a readily available alternative species could enhance the merchantability of larch.

Supply from Second-Growth Forests

In addition to a continued flow of wood from largely unmanaged timber stands, over the next twenty to fifty years we can expect most of the major inventories of old growth to continue to have a major influence upon the total supply of industrial wood worldwide. In addition we would expect second growth and plantations to provide a growing source of the world timber supply. I would expect the forests of the Nordic countries to continue indefinitely as important suppliers, with their focus being primarily upon the European market. The U.S. South can be expected to experience substantial expansion as the naturally regenerated second-growth forests of the post-World War II period mature and become increasingly available. In addition, the man-made forests of the U.S. South, which were begun initially as a conservation program and subsequently continued as industrial forest plantation, will provide an increasing flow of wood fiber resources. It is also possible that the secondary forests of the Great Lakes region may play an expanded role in future timber supply due to the influence of technology. The development of waferboard technology may result in this Lake States product's displacing southern plywood in some important North Central and North Eastern markets. The result could be a greater availability of southern forest products for foreign markets. Finally, the second-growth forests of eastern Canada may begin to supplement old-growth flows from this region and play a role in global timber supplies over the next twenty to fifty years.

Supply from Plantation Forests

In addition to wood supplies from what I have called the traditional timber supply regions, over the next twenty to fifty years I expect to see an increased role being played by supplies from plantation forests in regions that have not traditionally been important industrial wood suppliers. The region that is likely to have the largest role on world markets is Latin America (Ryti 1983). Also, parts of Oceania--especially New Zealand--and perhaps South Africa are likely to participate. The beginnings of this phenomenon are seen in the increasing flow of market wood pulp out of Brazil, and the exports of logs and processed wood products from New Zealand and Chile. The amount of wood in New Zealand available for the export market will increase by a factor of ten from current levels after the year 2000 solely on the basis of the age and volume of existing forest plantations. Given the existing level of forest plantations in Brazil, that country will likely become an important factor in world woodpulp markets before the turn of the century (Sedjo 1981). As noted above, an IDB-sponsored study has forecast that by 2000 about one-half of a much larger volume for production will originate on industrial plantations.

While much of the increased roundwood production of South America will be absorbed internally, both Chile and Brazil have the potential to become important exporters and Venezuela also has such potential.

Conclusions

Through the end of this century the major sources of the world's supply of wood are unlikely to change dramatically. The principal sources of old growth—the PNW and B.C., the East Indies, and the Soviet Union—should continue to dominate, with their old-growth inventories. In addition, wood from second-growth and plantations will continue to flow from Nordic forests at about the current rates, and increasingly from forests of the U.S. South. Also, industrial wood from nontraditional suppliers, particularly plantations of the Southern Hemisphere, will play a growing role in the supply of industrial wood and provide a major new supply source by the beginning of the 21st century. Major new supplies are also likely to result from improvements which allow for increased utilization of forests both through greater utilization of merchantable timber and by allowing for formerly unused species to be merchantable. Beyond the year 2000 this trend will continue with a gradual transition toward much larger portions of the world's supply of industrial wood being supplied from intensively managed industrial plantations.

References

Inter-American Development Bank. 1982. Forest industries development strategy and investment requirements in Latin America. Technical Report No. 1. Prepared for IDB Conference on Financing Forest-Based Development in Latin America, Washington, D.C.

Landsberg, H. H., and J. E. Tilton. 1982. Nonfuel minerals. In: P. R. Portney, ed., Current issues in natural resource policy, pp. 79-84. Resources for the Future, Johns Hopkins Press, Baltimore.

Lanly, J. P., and Clement, J. 1979. Present and future forest and plantation areas in the tropics. FO:MISC/79/1. Food and Agriculture Organization, Rome.

Lyon, K. S. 1981. Mining of the forest and the time path of the price of timber. Journal of Environmental Economics and Management 8:330-334.

Lyon, K. S., and R. A. Sedjo. 1983. An optimal control theory model to estimate the regional long-term supply of timber. Forest Science, forthcoming.

Manthey, R. 1978. Natural resource commodities--A century of statistics. Johns Hopkins Press, Baltimore.

Potter, N., and F. T. Christy. 1962. Trends in natural resource commodities. Johns Hopkins Press, Baltimore.

Reed(F.L.C.) and Associates Ltd. 1978. Forest Management in Canada, Vol. 1. Information Reprint FMR-X-103. Canadian Forest Service and Environment Canada, Vancouver.

Ryti, N. 1983. Trends and likely structural changes in forest industry worldwide. In: A. Anderssen, M. Kallio, A. Morgan, and R. Sappala, eds., Systems analysis in forestry and forest industries. TIMS Studies in Management Sciences, forthcoming.

Sedjo, R. A. 1981. Forest plantations in Brazil and their possible effects on world pulp markets. Journal of Forestry 78(1):702-705.

Sedjo, R. A. 1983a. The potential of U.S. forest lands in the world context. In: R. A. Sedjo, ed., Governmental interventions, social needs and the management of U.S. forests, pp. 53-77. Resources for the Future, Washington, D.C.

Sedjo, R. A. 1983b. The comparative economics of forest plantations: A global perspective. Resources for the Future, Washington, D.C., forthcoming.

Solecki, J. A. 1982. Resources of the Eastern USSR. In: H. E. English and A. Scott, eds., Renewable resources in the Pacific, pp. 91-96. International Development Research Center, Ottawa, Canada.

Staebler, G. R. 1978. Assessment of technology in fiber production. In: J. S. Bethel and M. A. Massengale, eds., Renewable resource management for forestry and agriculture, pp. 57-66. University of Washington Press, Seattle.

Takeuchi, K. 1974. Tropical hardwood trade in the Asian-Pacific region. World Bank Staff Occasional Paper No. 17. Johns Hopkins Press, Baltimore.

THE JAPANESE TIMBER MARKET IS OPEN AND COMPETITIVE:
A VIEW FROM A CONSUMER NATION, JAPAN

Yoshio Utsuki

Japan's Forests

Sixty-eight percent of Japan's total land area, amounting to about 25 million hectares, is covered with forests which contain about 2.5 billion cubic meters of growing stock. Nevertheless, forest area per capita is only 0.22 hectares--which is quite small compared to 1.32 hectares in the United States, 13.77 hectares in Canada, 1.70 hectares in Malaysia, and the world average, which is 0.95 hectares.

Japan is doing its utmost to utilize its domestic forest resources effectively, but Japan's forests cannot possibly satisfy the domestic timber demand. The present ratio of timber self-sufficiency is only about 36 percent and, according to FAO statistics, Japan imported 64 million cubic meters of timber out of the total world export of 274 million cubic meters in 1980.

Japan is the biggest timber importer in the world, and has no hope of becoming self-sufficient.

As Japan is a mountainous country with steep, rugged topography, the rivers are short and the inflow is rapid. The population--115 million people--live mostly on the less than 15 percent of the land surface that is level. Population, economy, and industrial activities are heavily concentrated in this small, level area.

Therefore, the forests of Japan, covering 68 percent of the land area, have the very important role of protecting the land by preventing erosion as well as conserving water.

About 10 million hectares, or 40 percent of the forests, are plantations established after World War II, and need careful tending and thinning. Because of the depressed timber market, however, forestry activities are sluggish and, as a result, the number of unhealthy plantations is increasing. We are afraid that plantations may be destroyed by such dangers as disease or typhoons, if proper tending and thinning are not practiced.

Sound forest management with appropriate harvesting is very important not only from the economic viewpoint but also from the view of conserving land and water. We cannot import land conservation and water preservation; therefore, we have to look after our forests.

Timber Demand in Japan

Japanese timber demand sharply increased during high economic growth and reached a peak of 118 million cubic meters in 1973. After the oil crisis in 1973, it decreased significantly. Although it recovered to reach about 100 million cubic meters, it again dropped to about 90 million cubic meters in 1981, and has stayed there since.

The "long range demand/supply projection for important forest products" approved by the Cabinet Council in May 1980 indicates the future prospects for timber demand.

The projection assumes certain conditions, such as an average annual GNP growth rate of 5.5 percent until 1985, 5 percent until 1990, and 4.5 percent until 1996, with annual new housing starts of 1.4 to 1.5 million units.

Based on these assumptions, the demand/supply projection concluded that timber demand will increase at an average annual growth rate of 1.2 percent--from 104 million cubic meters in 1976 to 118 million cubic meters in 1986 and 133 million cubic meters in 1996.

Lumber (log equivalent) demand is also predicted to expand 0.65 percent a year--from 57 million cubic meters in 1976 to 63 million cubic meters in 1986 and 65 million cubic meters in 1996.

When assessing the recent trend in new housing starts, however, which is the major end use of lumber, it does not seem that the new housing starts will expand as they did during the 1970s. The actual new housing starts were 0.42 million units in 1960, 1.91 million units in 1973, 1.50 million units in 1979, and 1.15 million units in 1982. Reasons for this change include the facts that the number of marriages has decreased from the peak years of 1973 and 1974 and urbanization has slowed down.

Other factors to be considered include: (1) the number of existing houses exceeds the number of established households; (2) recently those who want to build houses are mostly young people or those with low income; (3) nonwooden houses are expanding their share of the market.

Considering these points, the future lumber demand by the housing sector does not appear bright for the time being. Some even say that if new housing starts were about

1.2 million units a year, the annual lumber (log equivalent) demand would be only about 55 million cubic meters. On the other hand, the demand for better housing is becoming stronger, and there are still a lot of people who want to own a separate wooden house. Also it is expected that rebuilding and remodeling will increase the demand for lumber. In order to meet such demands, it is very important to develop high-quality wooden houses and to supply high-quality lumber. As a result, the timber industry is strengthening its activities to expand the timber demand in the housing sector as well as the nonhousing sector.

Pulp/chip demand, which shares about 30 percent of the total timber demand, steadily increased until it reached 36 million cubic meters in 1980, but then declined to 29 million cubic meters in 1981. The recent drop in demand is considerable and can be attritubed to: (1) the depressed market--due to the slowdown in economic growth rate; (2) the rapid change in demand after the two oil crises (e.g., wider use of low-grade paper, a change in the practice of excessive wrapping, and replacement by petrochemical products in the industry sector; and (3) in the cultural sector, fewer purchases of books by young people, thus decreasing the demand for paper used in publishing.

It has been forecast that the demand for paper and paperboard will increase only 2.6 percent annually, growing from 17 million tons in 1981 to 21.5 million tons in 1990.

Even if we accept the most pessimistic projections for lumber, paper, and paperboard demand, however, it is still necessary to emphasize that the total Japanese demand for about 100 million cubic meters will continue, and that Japan will remain one of the biggest timber importers in the world, although domestic supply will gradually increase.

The Open Japanese Timber Market

The Japanese timber trade was liberalized during the 1950s, and there are no import restrictions. Tariffs on forest products were reduced or eliminated between 1955 and 1970. The average tariff rate on forest products has decreased from 0.26 percent before multilateral trade negotiations to 0.24 percent after MTN. This rate is substantially lower than the average Japanese tariff rate for all products--4.9 percent before MTN and 4.1 percent after MTN (actual tariff base is the aggregated mean tariff rate for 1976 import values). This shows how open the Japanese timber market is in relation to tariffs. We hear that the average U.S. tariff rate for all products decreased from 5.7 percent before MTN to 4.0 percent after. The average Japanese tariff rate on forest products is lower than that.

There are many opinions on timber trade among forestry- and timber-related industries in Japan. Some forest owners insist that the main cause of the depressed domestic market is excessive imports of foreign timber and, therefore, timber import restrictions should be imposed in order to boost the price and stimulate the demand for domestic timber. On the other hand, some of the port workers urge companies to import as many foreign logs as possible so that employment opportunities will be secured. As for the national government, it has been standing firm against pressure from protectionists in order to maintain free timber trade.

After the complete liberalization of post-World War II timber trade, timber import volumes skyrocketed. Imported timber (industrial timber, log equivalent) increased from 7.5 million cubic meters in 1960 to 49 million cubic meters in 1969, when it became the major portion of the total timber supply. It jumped to 75 million cubic meters in 1973, and reached 76 million cubic meters in 1979--the highest import volume in history. It later dropped to 61 million cubic meters in 1981, however.

The value of timber imports ranks third, following oil (41 percent of the total) and coal (4 percent), sharing about 4 percent of the total import values, and amounting to 8.1 billion U.S.$, because of the depressed market (logs, lumber, chips, veneer, plywood, etc.).

Japan imports from many sources in the world. The value of 1981 imports by country is as follows: U.S., 1.8 billion U.S.$ (33 percent of the total imports); Malaysia, 1 billion U.S.$ (19 percent); Indonesia, 0.7 billion U.S.$ (12 percent); U.S.S.R., 0.5 billion U.S.$ (9 percent); and Canada, 0.5 billion U.S.$ (8 percent).

One of the main reasons why imported timber has quickly occupied the major part of the market is that, in addition to the competitiveness of foreign timber versus domestic timber (which could not satisfy the demand), the big trading companies established a stable supply system of imported timbers in both quantity and quality and gave favorable payment terms to the domestic users.

There has recently been a significant change in the timber importing business, however. The exchange rate of currency is unstable and the big trading companies have lost their leading role. Sawmills and wholesalers have also been joining the timber importing business.

Therefore, it is said that foreign timber has been losing its advantage in terms of stable supply, and short-term market fluctuation has become very significant. This situation is not favorable to timber exporters either.

Japanese-U.S. timber trade represents a huge credit for the U.S. and helps to reduce the overall trade imbalance between the two countries. As I mentioned before, the U.S. is the biggest timber supplier to Japan, sharing more than 30 percent of Japanese imports. On the other hand, Japan is the biggest U.S. customer and buys about 70 percent of U.S. timber exports. Therefore, the sound development of the Japanese-U.S. timber trade is really important for both countries, and in particular for the U.S.

The values of U.S. timber exports and imports to and from Japan during 1979-81 were as follows:
- In 1979, U.S. exports were 1.79 billion U.S.$, imports were 70 million U.S.$, resulting in a surplus of 2.54 billion U.S.$.
- In 1980, U.S. exports were 2.70 billion U.S.$, imports were 50 million U.S.$, and the surplus was 2.65 billion U.S.$.
- In 1981, U.S. exports were 1.79 billion U.S.$, imports were 50 million U.S.$, with a surplus of 1.74 billion U.S.$.

As you can see the U.S. surplus is very significant.

The major portion of Japanese timber imports consists of logs. The value of the log imports, however, has decreased from 76 percent of the total timber imports in 1979 to 67 percent in 1981. On the other hand, the value of the lumber imports has increased slightly from 12 percent in 1979 to 13 percent in 1981.

What is called "American" timber includes both U.S. and Canadian timber, which are the same species and compete with each other. Let's look at the volume of Japanese imports of North American timber as a whole.
- Log imports decreased from 12.7 million cubic meters in 1979 to 10.8 million cubic meters in 1980 and 7.7 million cubic meters in 1981.
- The volume of lumber imports increased from 3.6 million cubic meters (4.7 million cubic meters in log equivalent with conversion rate of 0.76) in 1979 to 4.1 million cubic meters (5.4 million cubic meters in log equivalent) in 1980 and then dropped to 3.0 million cubic meters (4.0 million cubic meters in log equivalent) in 1981. In terms of total imports, lumber imports (in log equivalent) have risen from 27 percent in 1979 to 33 percent in 1980 and 34 percent in 1981.

From these figures, it is apparent that Japan buys not only logs but also a large quantity of sawn lumber.

Canada bans, in principle, log exports and is promoting sawn lumber exports. Canadian exports to Japan in 1981 were 0.32 million cubic meters of logs and 1.82 million cubic meters of lumber. During the same year the U.S. exported 7.4 million cubic meters of logs and 1.13 million cubic meters of lumber. I would say that the U.S. is about to catch up to Canada in lumber exports.

Marketable Timber in Japan

Recently, nonwooden houses have been rapidly penetrating the housing market. The percentage of wooden houses dropped sharply from 70 percent in 1970 to 57 percent in 1981. The first objective of the lumber industry is to promote construction of wooden houses in order to increase timber demand. It is also important to expand timber demand in the nonhousing sector. There will be room for both Japan and the U.S. to promote greater utilization of wood. As an example, almost all of the desks in our offices are made of steel. We recently installed a wooden desk in our office to show our visitors that wood has a warm feeling and is also very practical for manufacturing such items. The recent decline in demand for timber is a critical problem for wood-related industries and will require special efforts to ease the current situation.

We have two kinds of wooden houses in Japan. One is the traditional Japanese-style house using post-and-beam construction, which has a history of more than 1,200 years and accounts for the major portion of the market. The other style is the platform-frame construction house, which was recently introduced in Japan and is not yet widely used.

The Japanese-style house requires many kinds of parts of various sizes and shapes. Since these parts are used without being painted and the surfaces are exposed in the rooms, the wood surface appearance is very important. People very much appreciate beautiful hues, deep luster, fine grain, and also fragrant smells. Therefore, knot-free and straight-grain wood is very precious and costly. Lumber with many knots, crooked grain, a bit of rot, discoloration, or the like, cannot be considered for use in the Japanese-style house market, and thus cannot be sold for a good price.

Just to illustrate my point, the shape of the sausage in a hot dog does not matter, but when we eat sashimi (raw fish), the shape and color give us an appetite. Love sees no faults; however, knots are only knots. We feel that Japanese cypress (hinoki) is the finest timber in Japan, because it has a wonderful luster and a particular fragrance. People say that taking a bath in a cypress tub is one of the most luxurious pleasures, because the fragrance is so delightful.

Let's take an example: the price of Japanese cedar (sugi). It will normally cost about 260 U.S.$ per cubic meter, but if it is knot-free, it will cost about 2,400 U.S.$ per cubic meter--almost ten times more, even though the strength is the same.

The standard modular measurement of a Japanese-style house is 90 cm by 180 cm (3 feet by 6 feet), which is the size of a Tatami mat. The standard size of plywood in Japan is 3 feet by 6 feet, whereas the most common dimension in the U.S. is 4 feet by 8 feet.

In 1981, 650,000 wooden houses were constructed and 98 percent of them were Japanese-style houses. This naturally leads to the conclusion that most of the timber demand in the housing sector is for Japanese-style houses. The number of platform-frame construction houses has recently increased and reached about 14,000 units in 1981, but is still a minor market compared with traditional style houses.

A number of small and medium-scale sawmills in Japan are manufacturing various housing parts for Japanese-style houses. It is the Japanese sawmillers who best know the tastes of the Japanese people. They carefully consider what kind of lumber can be cut from each log. They may spend half a day discussing how the first cut should be made in order to produce the most valuable lumber with the most beautiful surface appearance. They do this because it will pay in the long run. It is because of this know-how that Japanese sawmillers can cope with expensive logs imported from the U.S. They are paying much more for certain logs than Americans will pay.

The Japanese coniferous market--primarily the housing parts market--is very open, and competition among American, Canadian, Russian, New Zealand, Chilean, tropical, and domestic coniferous timber is very keen.

- In regards to Canadian lumber, we notice that COFI has been actively pushing frame construction methods in Japan in order to sell Canadian dimension lumber. I recently saw a beautiful pamphlet, entitled "Beautiful Cedar World" in Japanese, and heard there was a good response from house designers.
- It seems that the U.S.S.R. is very keen on selling Siberian logs to Japan. They have developed a firm log demand of about 6 million cubic meters a year. It is

said that they have recently introduced a compensatory system for foreign exchange fluctuations. They want to maintain a stable business by reducing the foreign exchange risk on the part of the Japanese importers.

- Considering rich forest resources of radiata pine that are coming up in the future, New Zealand is carefully studying the Japanese market as one of its main outlets. Chile is also promoting radiata pine sales in Japan.
- The Japanese domestic plantations are not yet mature, but the supplying power is gradually increasing. In fact, along with the promotion of thinning operations, thinned timber is coming into the market and the technologies for utilizing small-diameter logs have been developing. A stable supply in terms of quantity and quality is being established for the domestic timber.

The Need for Stable, Competitively Priced American Timber

At present, American logs occupy a big share of the Japanese coniferous market. They account for more than half of the imported conifers and about 35 percent of the total coniferous logs. As was described before, the Japanese market will become more and more competitive in the future. Therefore, it would be advisable that the U.S. exert the utmost effort to maintain the current demand for American logs.

There are several restrictive measures being proposed for log exports at the federal and state levels, as we have from time to time drawn to the attention of the U.S. These measures are not appropriate in light of maintaining free trade and improving the present overall trade imbalance between the two countries. In addition, if such restrictive measures were further strengthened, the American timber-related industry in Japan, which is in the forefront for selling the American timber in Japan, may be seriously affected.

There seems to be a contradiction. While the U.S. government is advocating free trade, it is calling for log export restrictions at federal and state levels in order to protect the lumber industry. The Statue of Liberty might weep at such a dichotomy.

About three years ago the Bonker Amendment passed, and unprocessed redcedar logs were banned from export. When it became clear that a stable supply of western redcedar was not assured, the demand was met by substituting other species in the Japanese market. Japanese imports of American redcedar logs almost ceased, and at the same time the volume of imported processed redcedar did not change.

It may be that the U.S. has completely lost a market in Japan and, since this demand is being filled by a replacement product, has gained nothing. If there is not a stable supply, a stable market cannot be maintained.

The U.S. wishes to expand the export of processed products such as lumber and plywood, while reducing the export of logs. To be successful, the most important consideration is the establishment of a production and sales system that will meet Japanese requirements.

As was shown before, Japanese imports of processed products have been increasing. It is presumed that such a trend will continue, and in order to penetrate the Japanese market for processed forest products, it is essential that products that meet Japanese tastes have a stable supply.

Those products will have to satisfy the consumer's demand for quality, with no knots, a straight grain, good color, and fine luster, and meet the prerequisites of exact size for Japanese-style houses.

Great difference in tastes can occur between Americans and Japanese, and in order to sell in the Japanese market it is important for suppliers to understand Japanese preferences through active marketing programs in Japan, using the Japanese language. In an effort to promote cooperation between the U.S. and Japan in this sector, the U.S.-Japan Lumber Trade Promotion Committee was established jointly by the U.S. and Japanese lumber industries. The committee has met twice, once in Palm Springs in November 1980, and once in Tokyo in July 1981. I am sure the American industry has learned a lot about the real situation of the Japanese processed-wood market through these meetings.

The Japanese plywood market is dominated by 3 by 6 plywood made of lauan, a wood with a very smooth surface. In some areas, planed plywood is very popular. Plywood with a rough surface might not be marketable. Such requirements seem excessive and planed plywood may be a luxury, but it is a fact of the market. Since the Southeast Asian log resources are decreasing, there is some movement toward manufacturing softwood plywood in Japan. It is in a trial stage and the development of a softwood plywood market is a future, difficult task.

It often appears that the American timber industry looks at only the local U.S. market when the market is strong and is not interested in the overseas market. When the domestic market is depressed, the industry looks to the export market to take up the slack. It appears that in the 1980 to 1982 period, the U.S. timber industry put more emphasis on export to Japan to compensate for the depressed market in the U.S. Unfortunately during this same period, the Japanese market was also depressed and some resentment toward American friends was built up during that time. Good customers can be kept only by maintaining long-term stable supplies; therefore, even during a period of rather strong demand in the local U.S. market, Japanese customers should not be neglected.

The U.S. says that the only substantial barrier that hinders U.S. forest products from gaining access to the Japanese market is tariffs, and that Japan should unilaterally reduce them. I would like to reiterate here that the most important thing for the U.S. to do in order to expand processed products trade with Japan is to produce products that meet the Japanese market requirements and commit a long-term stable supply.

In regard to tariffs, they are a subject for government-to-government negotiations, with mutual concessions. Tariffs on certain forest products in Japan are higher than those in the U.S., but they are the results of many tariff negotiations between our two countries. I would point out that tariff reductions will have a serious adverse impact on Japanese forest industries, which are presently undergoing reorganization due to long depression and to surplus capacities. Since they are facing difficulties in securing raw materials in terms of volume and price because of log export restrictions, and undergoing reorganization, they are not at the same level as the Americans are in the race.

United States industries indicate that tariff reductions would be beneficial to Japan. They may mean beneficial to consumers who buy houses because tariff reductions would reduce the cost of construction, but since land costs are a major part of the total cost and the average tariff rate of forest products is very low, i.e., 0.24 percent in Japan, the argument is not very persuasive.

Summary

The U.S.-Japanese forest products trade is important to both countries, and in particular to the U.S. Trade has developed basically within the free trade system, on a commercial basis, and this principle should be maintained in the future.

It is desirable that problems, if any, should be solved through continuing dialogues between the two countries. It is fortunate that such channels can freely be used at the government-to-government level and the private industry level as well. These forums should be utilized to the maximum extent possible.

Some may say that "resources are power," but one of our famous singers in Japan often says that "guests are the master."

I hope you will take these comments in light of an old Japanese saying, "A good medicine tastes bitter."

APPENDIX

Table 1. U.S.-Japanese Trade in 1981 (million U.S.$).

	U.S. Exports (A)	Japanese Exports (B)	(A-B)
Total	25,297 (100%)	38,609 (100%)	-13,312
Forest products	1,787 (7.1%)	50 (0.1%)	+ 1,737

Table 2. Japanese Imports of Major Forest Products (logs, lumber: million cubic meters; plywood, veneer: million square meters).

		1978		1979		1980		1981		1982	
Logs	Total	42.7	100%	44.8	100%	37.5	100%	29.2	100%	30.4	100%
	U.S.	10.3	24.2	12.4	27.7	10.3	27.4	7.4	25.3	8.1	26.6
Lumber	Total	3.9	100%	5.1	100%	5.6	100%	3.9	100%	5.0	100%
	U.S.	1.0	25.4	1.5	30.1	1.6	27.8	1.1	29.0	1.4	28.7
Plywood	Total	11.8	100%	13.2	100%	14.9	100%	5.0	100%	4.9	100%
	U.S.	0.3	2.5	0.5	3.6	0.7	4.5	0.5	10.1	0.7	14.2
Veneer	Total	8.1	100%	14.2	100%	19.3	100%	23.4	100%	42.7	100%
	U.S.	0.1	1.0	0.3	2.3	0.2	0.8	0.5	2.1	5.8	13.6

Source: Customs statistics, Ministry of Finance.

Table 3. Forest Products Trade with the U.S. (1981 and 1982).

	1981 (Jan.-Dec.)			1982 (Jan.-Dec.)		
Item	Quantity*	Against previous year	Value (million yen)	Quantity*	Against previous year	Value (million yen)
Imports						
Logs	7,412	72%	263,597	8,088	109%	273,599 67
Lumber	1,130	73	50,647	1,420	126	63,832 16
Wood chips	5,929	72	72,952	4,930	83	66,463 16
Veneer	707	79	425	5,830	825	1,035 ⎫ 1
Plywood	504	76	543	691	137	709 ⎬
Others			176			71 ⎭
Total			388,340			405,709 100
Exports						
Veneer	100	204	34	125	125	56 0
Plywood	9,459	99	6,238	8,767	93	6,083 60
Others			3,861			4,032 40
Total			10,133			10,171 100

*Quantity units:
Logs ⎫
Lumber ⎬ 1,000 cubic meters
Wood chips ⎭
Plywood ⎫ 1,000 square meters
Veneer ⎭

Source: Customs statistics, Ministry of Finance.

Table 4. Demand and Supply of Industrial Timber in Japan (1,000 cubic meters).

Year	Demand						Supply										(B)/(A)
	Total (A)	Sawn Lumber	Plywood	Pulp	Others	Total	Domestic Production			Mill residues	Imports						
							Sub-total	Logs	Logging Residues		Subtotal (B)	Logs	Sawn Lumber	Plywood Veneer	Chips	Pulp	Others	
1955	45,278	30,295	2,297	8,285	4,401	45,278	42,794	42,794	--	--	2,484	1,969	112	--	--	403	--	5.5
	(1,543)			(1,469)	(74)	(1,543)				(1,543)								
1960	56,547	37,789	3,178	10,189	5,391	56,547	49,006	48,515	491		7,541	6,674	211	--	--	656	--	13.3
	(4,307)			(4,087)	(220)	(4,307)				(4,307)								
1965	70,530	47,084	5,187	14,335	3,924	70,530	50,375	49,534	841		20,155	16,721	1,115	2	270	2,036	11	28.6
	(6,737)			(6,737)		(6,737)				(6,737)								
1970	102,679	62,009	13,059	24,887	2,724	102,679	46,241	45,351	890		56,438	43,281	3,957	548	5,031	3,509	112	55.0
	(7,299)			(7,299)		(7,299)				(7,299)								
1971	101,405	59,801	13,362	25,715	2,527	101,405	45,966	45,253	713		55,439	43,909	2,792	200	5,946	2,472	120	54.7
	(7,371)			(7,371)		(7,371)				(7,371)								
1972	106,504	63,613	14,309	26,202	2,380	106,504	43,941	43,114	827		62,563	47,697	3,222	380	8,076	2,962	226	58.7
	(7,797)			(7,797)		(7,797)				(7,797)								
1973	117,581	67,470	17,151	30,415	2,545	117,581	42,209	41,584	625		75,372	52,485	4,666	1,600	12,094	4,061	466	64.1
	(7,653)			(7,653)		(7,653)				(7,653)								
1974	113,040	60,734	14,481	34,957	2,868	113,040	39,474	38,874	600		73,566	48,453	4,287	882	13,580	5,440	924	65.1
	(6,448)			(6,448)		(6,448)				(6,448)								
1975	96,369	55,341	11,173	27,298	2,557	96,369	34,577	34,155	422		61,792	42,681	2,964	335	11,340	3,688	784	64.1
	(7,281)			(7,281)		(7,281)				(7,281)								
1976	102,609	57,394	12,939	29,639	2,637	102,609	35,760	35,271	489		66,849	45,118	3,821	207	13,025	3,798	880	65.1
	(7,251)			(7,251)		(7,251)				(7,251)								
1977	101,854	56,564	12,717	29,841	2,732	101,854	34,231	33,793	438		67,623	44,561	4,125	118	13,820	4,002	997	66.4
	(7,282)			(7,282)		(7,282)				(7,282)								
1978	103,417	57,560	13,585	29,597	2,675	103,417	32,558	32,145	413		70,859	46,158	4,467	138	13,116	5,954	1,026	68.5
	(7,721)			(7,721)		(7,721)				(7,721)								
1979	109,786	60,314	13,915	32,137	3,420	109,786	33,784	33,270	514		76,002	46,950	5,656	172	15,003	6,413	1,808	69.2
										(7,275)								
1980	108,964	56,713	12,840	35,868	3,543	108,964	34,557	34,051	506		74,407	42,395	6,136	199	15,936	7,670	2,071	68.3
										(7,275)								
1981	91,829	48,718	11,086	29,056	2,969	91,829	31,632	31,370	262		60,197	35,932	4,162	122	12,508	5,857	1,616	65.6
										(6,449)								
1982	90,900	48,000	10,300	29,250	3,350	90,900	33,150	32,850	300		57,750	33,250	5,200	150	11,150	6,000	2,000	63.5
										(6,400)								

Note: Figures in parentheses are for chips made of mill residues.

Table 5. Timber Imports of Japan (Logs and Lumber) (1,000 cubic meters).

Year	Total	From Southeast Asia						From North America					From U.S.S.R.	Others
		Sub-total	Philip-pines	Malaysia	Indo-nesia	Others		Sub-total	U.S. Logs	U.S. Lumber	Canada Logs	Canada Lumber		
1960	6,378	4,691	3,471	1,158	12	50		553	378	149	22	4	921	213
1965	16,798	9,333	5,618	3,474	148	93		4,237	3,237	269	165	477	2,636	593
1966	21,949	11,935	6,743	4,851	202	139		5,498	4,183	110	275	630	3,607	909
1967	28,279	13,674	7,229	5,737	528	180		8,435	6,233	585	526	1,091	5,073	1,097
1968	33,567	14,878	7,437	6,079	1,088	274		11,183	8,582	719	535	1,347	5,861	1,645
1969	35,807	17,814	8,320	6,419	2,723	352		9,782	7,956	646	233	947	6,151	2,060
1970	42,366	20,678	7,907	6,300	5,942	529		12,511	9,526	736	538	1,711	7,095	2,082
1971	40,325	21,689	6,332	6,166	8,531	660		9,332	7,110	661	648	913	7,071	2,243
1972	44,762	21,898	5,265	7,082	8,918	633		12,523	10,430	873	252	968	7,922	2,419
1973	52,280	26,969	6,251	8,657	11,336	725		13,313	10,525	1,353	93	1,341	9,155	2,843
1974	47,633	25,512	4,188	8,363	12,179	782		11,469	8,746	1,345	171	1,207	8,306	2,346
1975	38,262	17,628	3,109	6,633	7,433	453		11,625	9,359	1,094	182	991	7,872	1,137
1976	44,890	22,388	1,952	10,204	9,695	536		12,848	10,051	1,085	257	1,454	8,168	1,486
1977	45,465	21,678	1,738	9,713	9,632	595		13,264	10,201	975	430	1,657	8,833	1,690
1978	46,511	22,364	1,908	10,679	9,231	546		13,434	10,325	981	312	1,816	8,961	1,752
1979	49,902	23,078	1,539	10,884	10,043	612		16,365	12,400	1,536	340	2,089	8,013	2,446
1980	43,083	19,656	1,359	8,571	9,030	696		14,865	10,279	1,550	472	2,564	6,297	2,265
1981	33,118	15,493	1,595	8,482	4,629	787		10,673	7,402	1,130	323	1,823	5,770	1,177
1982	35,359	15,865	1,622	10,430	2,950	863		12,225	8,088	1,420	361	2,357	6,120	1,148

Note: Customs clearance basis.

Table 6. Housing Starts.

Year	Housing Starts Total	Housing Funds Private	Housing Funds Public	Structure Wooden (Thousand Units)	Structure Nonwooden (Thousand Units)
1965	843	626	216	647	196
1970	1,485	1,123	362	1,036	449
1971	1,464	1,056	408	967	497
1972	1,808	1,357	451	1,112	696
1973	1,905	1,500	405	1,120	785
1974	1,316	919	397	870	446
1975	1,356	949	407	907	449
1976	1,524	1,128	396	993	531
1977	1,508	1,079	429	946	562
1978	1,549	949	600	958	591
1979	1,493	885	608	910	583
1980	1,269	724	544	751	518
1981	1,152	614	538	654	498
1982	1,146	569	578	667	479

Source: Construction Ministry, Annual Report on Housing Starts.

Table 7. Forecast of Wood Demand/Supply (million cubic meters).

Supply/Demand	Uses and Sources	Result of 1976	1986	1996	2026
Demand	Sawtimber	57.4	62.6	65.4	
	Pulpwood	(5.4) 29.6	(7.0) 36.1	(7.5) 44.9	
	Plywood, fiberboard, particleboard	(2.2) 12.8	(3.0) 14.9	(3.3) 17.6	
	Shiitake cultivation logs, fuelwood	2.9	3.6	4.1	
	Others	1.7	1.2	1.2	
	Total	(7.6) 104.4	(10.0) 118.4	(10.8) 133.2	
Supply	Domestic supply	(7.6) 38.2	(10.0) 46.2	(10.8) 57.7	87.9
	Imports	66.2	72.2	75.5	
	Total	(7.6) 104.4	(10.0) 118.4	(10.8) 133.2	
Proportion of import percent		63.4	61.0	56.7	

Note: Figures in parentheses indicate mill residue.

U.S. OPPORTUNITIES IN WORLD TRADE IN WOOD

Bronson J. Lewis

One clear lesson has emerged from the sharp up-and-down cycles of market demand through the past decade in the United States wood products industry. We now realize how dangerously reliant we had become on our domestic markets, particularly on the U.S. housing segment. The rapid expansion of that market following World War II--and on into the seventies--virtually dictated concentration on grades and systems to serve the American home builder and home buyer. But we learned to our cost that there was a finite limit to the "American dream" of home ownership.

The "boom or bust" cycles in U.S. housing demand that have occurred in recent years, and the emerging trend toward smaller homes, have prompted the manufacturers of finished wood products to accelerate their market development efforts in other parts of the domestic economy, such as nonresidential construction and the diverse industrial markets. The results have been quite rewarding, and the prospects down the road look even more attractive; but no matter what we do in our domestic markets, the greatest long-term growth potential undoubtedly will be found in the international arena.

Our pulp and paper industries have long been major contributors to U.S. foreign trade, as is discussed in another part of this symposium. Exports also have been part of the market mix--albeit at varying levels of commitment--since the early years of the century in the case of plywood, and for much more than 100 years in the case of lumber.

We need to acknowledge that one of the complaints we have heard most frequently from our international customers concerns commitment. The allegation is that the U.S. wood industry, with the world's largest consumer market in its own backyard, tends to regard foreign markets as a dumping ground for its products when domestic demand is low. Then when domestic markets recover--or so goes the charge--the foreign markets get low priority or none at all. It needs to be said that accusations of this kind have had some validity in the past. While we still hear the issue raised periodically, I am glad to say that our wood industry has made substantial progress, particularly in the last ten years, in establishing its international image as a stable, permanent supplier.

We still have some distance to go in developing the full market potential--much still to learn and many new fields to conquer--but there is an increasingly favorable climate of acceptance for world trade in all segments of the U.S. wood industry today. The recognition is growing that we cannot just export our U.S. marketing practices, along with our products, and expect foreign customers to fall at our feet. Successful marketing overseas means long-term <u>commitment</u>. It means:

- <u>government-to-government</u> contact on both tariff and nontariff barriers, as well as other broad policy issues
- <u>cooperation</u> with our federal and, wherever appropriate, state governments in export market development
- <u>on-the-ground</u> involvement by shippers and their associations--direct representation overseas
- <u>study</u> and practical working knowledge of the unique market conditions, trade practices, social/political customs, and consumer preferences in each of dozens of countries
- <u>application</u> of the study results in promotional programs, technical initiatives, and market tools tailor-made for individual foreign markets
- <u>examination</u> of opportunities for joint ventures with foreign companies in capital-intensive industries like pulp and paper
- plus <u>dozens</u> of other factors, many of which receive attention in this symposium

Our industry's experience to date provides incentive for further advances. Western lumber exports have increased from about 3 percent of total production in 1970 to 8 percent in 1980, and 9 percent in 1981, according to the Western Wood Products Association. In 1964, American plywood exports totaled only 37.5 million square feet. By 1981, American structural panel exports (predominantly conventional plywood) had risen to 686 million feet. This 18-fold increase in 18 years is directly attributable to generally increasing investment in overseas promotion--through good times and bad, domestically.

One of the factors most benefiting U.S. lumber and plywood is increasing awareness of the suitability of cost- and energy-efficient wood systems in countries where masonry is traditional. Adaptation of our North American systems to fit a wide variety of housing needs in Europe and around the world will be one of the key priorities in the 1980s. The lumber/plywood potential, which is being greatly enhanced by the assistance we are receiving from the Foreign Agricultural Service, U.S. Department of Agriculture, is just one part of the picture.

The Far East, for instance, is the fastest growing paper and board market in the world. Opportunities such as these will have special relevance for the Pacific Northwest wood-based industry, which is well positioned to increase its market share in the Pacific Basin, or Pacific Rim. The Pacific Basin, as you know, is a regional trading area, which, with the United States and Canada, includes the countries of East and Southeast Asia, Australia, and New Zealand.

Richard G. Landis, President of Pacific R. J. Reynolds Industries, and Deputy International President of the Pacific Basin Economic Council, said recently that the nations around the Pacific have enjoyed such rapid economic growth that they now account for approximately one-third of the world's exports and imports. I agree with Mr. Landis that the closer we can come to a complementary system of international trade, the more stable the world system will become, and the more trade opportunities will increase. Such a bedrock principle implies a reciprocal spirit--greater equity and less unilateralism--in international trading relationships. We need the broad perspective that world trade is a two-way street. With improved communication between the U.S. wood products industry and all the nations and diverse global customers it serves, I believe we can move closer to that goal.

U.S. OPPORTUNITIES IN WORLD TRADE
IN WOOD FOR ROUNDWOOD AND CHIPS

Bruce R. Lippke

Domestic and Foreign Supply and Demand, Processing Efficiency, and Politically Motivated Regulations All Impact Raw Material Exports

Roundwood exports grew in a trend sense until about the late 1970s and appear to have fallen off since then. What caused the growth? What caused the decline? We may not know all the answers, but we can identify several important factors that suggest that the explanation is not a simple one.

To begin with, just consider what the international market would be like if logs and chips were freely traded, but no trade occurred in manufactured products. Under such a market regime, we would be able to deliver logs to foreign processing ports whenever our logs are cheaper than they could be obtained from other domestic or foreign sources. In other words, if our costs of managing and harvesting are enough less than our Japanese or other competitors to offset our transportation cost disadvantage, we could directly compete with those alternative sources. The problem with such a simplistic view, of course, is that it assumes some things that are not really true:
1. That Japanese processing has a significant advantage over U.S. processing, so that our final products are less competitive than our logs (actually, this is fairly true for some products, but certainly not for all).
2. That our log quality and species mix are interchangeable or substitutable with foreign raw material in both manufacturing and in end uses (unfortunately there are serious limits to quality and species substitution).
3. That the U.S. domestic market is overloaded with supply and is not willing to outbid offshore markets for the export material (we clearly are

not going to export material that is more profitable in the domestic market).
4. That regulations do not restrict our raw materials relative to those potentially available from competitors (there are restrictions).

My point is that you cannot really assess the roundwood opportunities from either simple trends or simplistic economics. We will have to talk about the economics of processing as well as the economics of raw material flows. We also must talk about the demand, supply, and substitutability of logs and products, as well as regulations.

In the end, we can only come to the general conclusion that the opportunities may be very large or very small depending on your assumptions. My own guess is that there will continue to be an increased opportunity for exports but that slowly, over time, more of the opportunity will be in product form. Raw material exports will decline as we increase our product exports.

Foreign Demand Exceeds Supply

The Pacific Rim and European countries saw their demand for wood exceed their supply capabilities starting in the 1960s and 1970s. The broad balance in wood fiber trade is shown in figure 1. With foreign raw material costs much higher than our own in those days, and the dollar overvalued, we could compete more easily in raw materials than in finished products (figures 2 and 3).

In the last few years, even though Japanese demand has declined, their currency has been strengthening, a fact which began to reduce their own processing competitiveness. Their timber competitiveness has always been marginal, but they do have a wood inventory that could be used to increase their own harvests if necessary. Under these conditions, the market share of North American products increased, especially in Japan, even though the demand was declining (figure 4). The price spread between U.S. logs and Japanese logs does not always respond instantly to changes, and timber inventory management decisions occur even more slowly, so it takes time for these adjustments to take place. Nevertheless there has been an economic response to changes in relative costs.

I rarely hear anyone argue that changes in the value of the Canadian dollar have no effect on Canadian market share, but people frequently assume that currency changes have no impact on Japan. Japanese harvest decisions may be complex

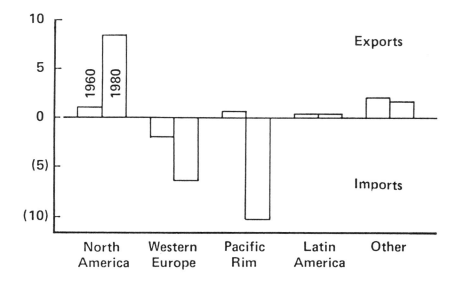

FIGURE 1. Raw materials trade reflects regional imbalances.

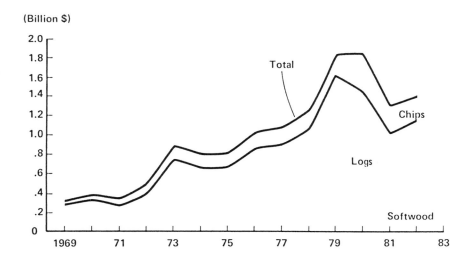

FIGURE 2. Large revenue growth from export logs and chips.

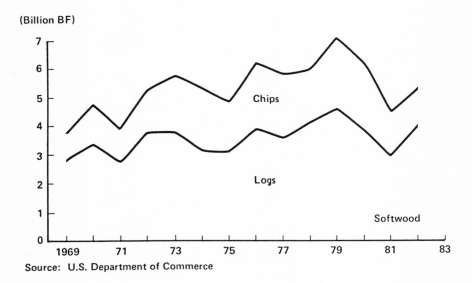

FIGURE 3. Export volumes have started to decline.

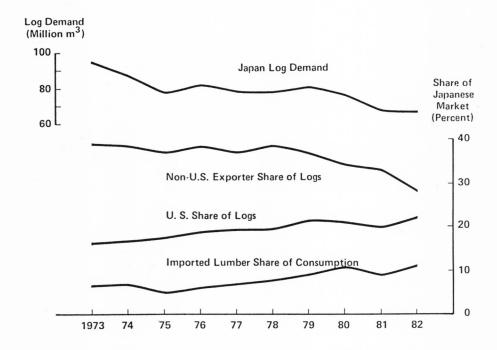

FIGURE 4. Lumber and U.S. logs have increased share in a declining market.

and not as immediately price sensitive as those of some U.S. private suppliers, but it is a mistake to assume there is no economic response. Economic conditions also are continuing to change.

Economics Favoring Solid-wood Product Exports Are Gaining

In spite of some serious ups and downs in the yen/U.S.$ exchange rate, we have seen more product flowing in lumber form. Unfortunately for the U.S., the weakness of the Canadian dollar in recent years has made Canada the preferred lumber supplier. The recent, and--we hope--temporary Canadian currency advantage has slowed the improvement in the U.S. net trade position of solid-wood products (figures 5 and 6). In any event, the triangular trade flows with large lumber imports from Canada into the U.S. and large log and product exports from the U.S. still show a favorable trend in dollars.

Japan's lumber production has been declining steadily for several years, in response both to reduced demand and to increased product imports (figure 7). The number of Japanese sawmills has declined by 10 percent in five years, and sawmill employment has declined 3 percent (figure 8). These declines are far greater than anything experienced in the U.S. Ironically, it has been the small mills in the U.S. which have demanded subsidies and protection in the form of set-asides, export bans, and tariffs on imports.

The structural changes taking place in Japanese processing have been under way for some time and should be expected to continue. Their converted product imports should continue to grow while their domestic production declines, and the share of raw materials in our exports should become smaller as our converted product exports increase. Recently, Canadian producers have enjoyed most of the benefits of this transition due to their exchange rate advantage.

Solid-wood Exports Are More Specialty than Commodity Oriented

Our export share of total chip usage is much smaller than our export share of total roundwood uses and, while there are several explanations, the most important structural differences between solid-wood and fiber exports is often forgotten. Pulp, linerboard, and newsprint are essentially international commodities. Their end-use

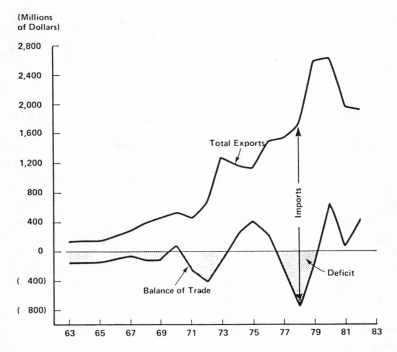

FIGURE 5. U.S. solid-wood products trade balance is improving.

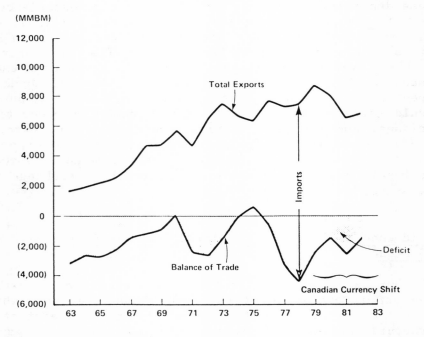

FIGURE 6. U.S. solid-wood products trade volume improvement slowed by Canadian devaluation.

FIGURE 7. Increasing Japanese lumber imports in a declining market.

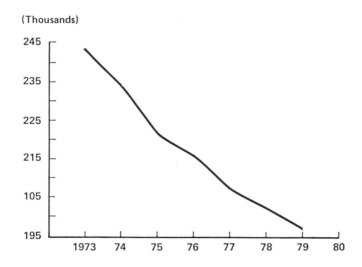

FIGURE 8. Japanese sawmill employees declining rapidly.

converting facilities are quite similar worldwide, so that foreign-produced commodities are nearly interchangeable with U.S.-produced commodities.

This is simply not true for lumber. Construction methods are very different in most countries, and especially in the Far East. The lumber used is, to a great degree, custom-ordered to the construction site. It is difficult to process efficiently a whole log in the U.S. and supply foreign specialty products in competition with foreign mills sitting right on top of their local distribution system. There is little likelihood that construction techniques will become completely standardized rapidly, so the ability to export processed lumber takes a bigger commitment in order to compete in a mostly specialized market, with much less flexibility to distribute products. You cannot efficiently produce side cuts for the U.S. market in U.S. sizes out of a plant designed to service the Japanese market. The marketing channels for more export products are continuing to grow, but it is likely to be an evolutionary change, not a rapid or revolutionary change.

So the economics of foreign markets and conversion are likely to lead to more conversion at the level of pulp, newsprint, and paperboard on the fiber product side, but not to boxes or much value in specialty papers or converted paper products. Those processes are quite specialized in each country and have transportation cost disadvantages as well. There will be continued inroads for solid-wood export products, primarily to Japan, but with declines of raw material exports.

Politically Imposed Restrictions Reduce Exports and Their Benefits

The impact of product specialization on the product form of exports is substantial, but you also have to consider the impact of market restrictions. After we banned the export of cedar logs from state lands in the Northwest, the Japanese turned to the Canadians and to substitute products. As a result, our cedar products exports actually were reduced rather than helped by the restriction.

After Indonesia banned the export of logs, they found they could not sell enough of their finished products in world markets to compensate; their foreign exchange revenues are lower, and jobs have been lost that could have been supported by those revenues. Canada's log export restrictions clearly give the U.S. an advantage, producing more revenue for the U.S. and less for Canada.

If you restrict access to your highest valued markets for export material, you simply have less cash to reinvest in productive facilities. Investment from the cash flow of profits creates jobs. It is a myth that we should be able to add value by processing raw material when, in fact, we lose margin in the process. When export revenues are reduced by restrictions, we clearly have more losers than gainers.

The most detailed studies showing who loses and who gains from a restriction of trade were provided for the case of a ban on Washington State Department of Natural Resources timber exports in a study by Richard Parks at the University of Washington; Jim Youde, a consultant and former professor of Oregon State; and Wes Rickard, a timber consultant. They conducted a thorough evaluation of the gainers and losers from such a ban on exports, and showed very convincingly that there would be a serious net loss to the state and the nation in both jobs and revenue.

The world's current economic crisis is leading to more trade restrictions, not less. In general, these restrictions will increase the cost of goods to consumers by protecting high cost domestic suppliers. In addition, exporters are denied the foreign revenues needed for imports essential to their economy. Since this trade warfare is likely to be more intense at the processed product level, it may have the impact of maintaining some flow of raw material products as processed products run into even greater restrictions.

The Chinese market is also a big unknown. They are importing intermediate commodities on the fiber side and logs and a limited amount of lumber on the solid-wood side. If we are really interested in gaining as large and profitable an export share in that market as possible, we will have to let it develop through its full maturity curve, much as we did in our trade with Japan. First, they have to learn how to use the material most effectively for themselves, and we have to learn how to service those needs in the most economic way that we can. Customer service and supply dependability are the order of the day, just as they are in most markets.

If domestic demand is very high and the supply response low, forget export growth. U.S. consumers will bid it all back to the U.S. market. If, on the other hand, domestic demand falls from the peaks of the 1970s with lower housing needs, <u>and</u> our supply response is high due to past improvements in growth, and more efficient technology continues to be put in place, then we will be pushing hard to capture more export markets--and we should succeed if political restrictions do not get in our way.

It is unfortunate that some of our domestic users forget that without exports there is no production to swing away from other markets to serve domestic needs. If there were no export markets developed, prices would have to go even higher to cause more investment in U.S. production. That is hardly what some U.S. consumers have in mind when they complain about exports and want to restrict them. The most stable market for U.S. consumers would involve the greatest economic development of foreign markets because that would also provide the largest potential product availability for domestic markets. Our current agricultural surplus is a graphic example of how well export expansion has worked to the benefit of the U.S. consumer.

So the only hard conclusion you can make is that you can get almost any answer you want on the size of the export opportunities, depending upon where you come out on the assumptions.

In a trend sense, I think the prospects are excellent for a steady increase in total forest product exports, but with slow declines in both log and chip exports. But unless a lot more time is spent by all of us to better understand the existing data and their limitations, and the political environment, assumptions will probably look quite different from mine.

U.S. Demand Growth Relative to Supply Can Make or Break Exports

Last, but clearly not least, is the impact of the biggest of all world markets, the U.S. We have heard claims of a shortage of wood in the U.S. for 70 years and we certainly are not going to be exporting much for long if there are better economic advantages to markets in our backyard. Any time a domestic market is stronger, it can bid away a portion, if not all, of the export products.

In the long term, if you believe the Forest Service's shortage scenarios as published in the Forest Service Timber Assessments, our prices are supposed to double and triple. It seems unlikely that we will be exporting very much under those conditions; but there are many tenuous, if not unbelievable, assumptions in those scenarios.

We have done our own supply-and-demand modeling of the U.S. at Weyerhaeuser over the years and have studied many of the assumptions used in the timber assessment market model used by the Forest Service. We do not see a scarcity scenario at all. We see a good chance for a surplus for additional exports rather than a shortage. Why do we get such a different result? We start by adjusting the U.S. demand <u>down</u> to levels we think are reasonable, the mill conversion efficiencies <u>up</u> slowly over time to reach <u>current</u> best-practice mills in 25 years, the growth and utilization of timber to match what we believe to be consistent with current management practices, and give the Canadians a reasonable share of the U.S. market.

As an interesting example of the U.S. supply/demand balance, we have estimated what we consider to be the reasonable price responses of private U.S. supply regions and Canada based on the historical period. We add those supply increments to the nonprice-responsive public volumes and get a supply curve. We compare that with our views of demand, which of course moved up sharply over the last decade. So these are, in effect, mid-term or three-to-five-year demand and supply responses. The plots look consistent to us. We see a movement of the equilibrium point, along a fairly stable supply curve based on changes in demand (figure 9). That supply curve is consistent with the elasticities that we estimated.

In contrast, the Forest Service's projections show that the price response on the supply side will become much less in the future. Figure 10 shows the Forest Service's projection from their last assessment for Canada's supply response. In effect, they assume that the Canadian supply curve starts shifting back over time so that their share of the market begins to decline. While there is plenty of uncertainty on Canadian supply, I think there are many good arguments to support higher Canadian production at much higher prices. I have more trouble accepting the argument that there is less wood available at higher prices. Regarding the overall timber supply projections by the Forest Service, figure 11 shows how consumption of U.S. private, plus Canadian, sawlogs (all of which have been price responsive in the past) becomes nonprice responsive in the future.

While the range of uncertainty is still large, these examples should at least demonstrate to you that, in the end, your outlook will be determined by your assumptions.

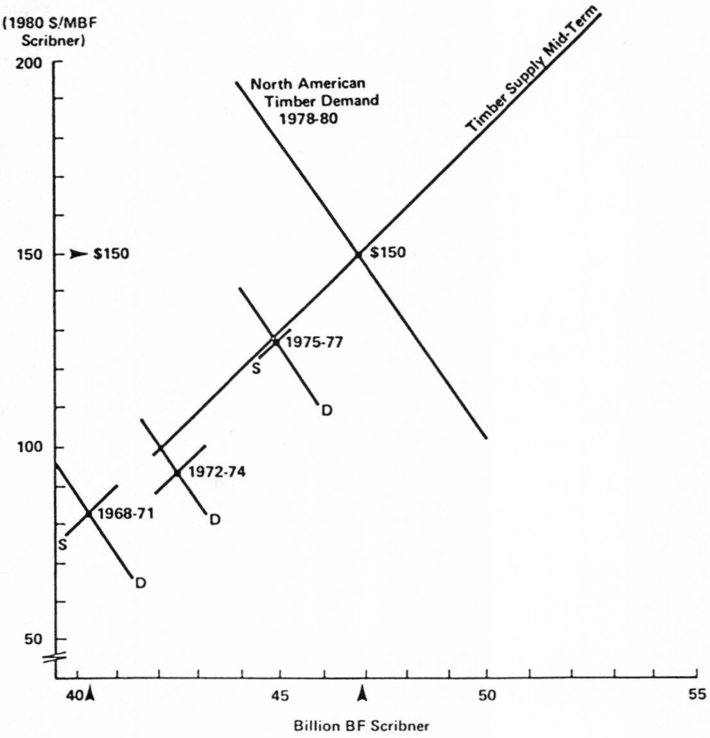

92 LIPPKE

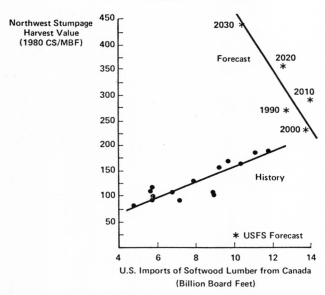

FIGURE 9. Supply has responded to the 1970s demand increase.

FIGURE 10. Forest Service forecasts of Canadian shipments to U.S. are opposite of history. Why?

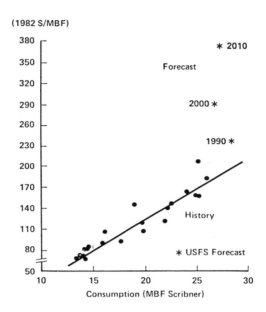

FIGURE 11. Forest Service forecasts an end to sawlog consumption response to price. Why?

U.S. LUMBER EXPORT OPPORTUNITIES

H. A. Roberts

The future bodes well for U.S. products in many of our world markets; however, new overseas market expansion will occur only if the specifics of each market are understood, adequate market development work is done, and producers understand that they must make a legitimate commitment to serve export markets. Volumes exported today could be doubled in future years. Additionally, as softwood lumber replaces some of those exports that today are shipped as logs, further increased volumes can be realized.

Perhaps it is well to mention that my comments here will be rather provincial. The focus of this paper will concern western softwood lumber opportunities only--not the broad range of forest products that are or will be exported. U.S. lumber exports are mainly <u>western</u> lumber, and western lumber likely will continue to be dominant in U.S. softwood lumber trade.

Western producers have enjoyed a tradition of shipping lumber to world markets for more than 100 years. Even though western exports approach 2 billion board feet a year, the United States still does not qualify as a major softwood lumber exporter--especially when measured against other exporting countries such as Canada, Scandinavia, and the U.S.S.R. Through the years U.S. exports typically have been made up of specialized products to a few foreign markets. Most exports have been products that overseas customers cannot find elsewhere--such as Douglas-fir and hemlock clears--from the coastal areas of Oregon and Washington. Times and markets have changed, however; competitive and cost factors not only make it desirable to extend our market base, they make it imperative.

Let me explain. Over the past several years, western producers have seen their traditional share of regional domestic markets erode. Only 12 years ago, 10 percent of all western lumber shipments were destined for the northeastern region of the United States. Now the Northeast accounts for only 4 percent. In 1971 the north central states took 31 percent of western shipments. By 1982, that figure had fallen to 12 percent. Shipments to the South have declined during the same period, although the change has not been quite so dramatic, probably because the so-called sunbelt states have been hot-spot areas for recent home-building and lumber use.

While western producers have been losing eastern markets, fortunately some of those losses have been offset by shipments to the West--our own back yard. Shipments to the West increased by 50 percent over recent years because there has been a robust home-building market here also. Having most of one's eggs in the home-building basket isn't comforting, however, as the past three years have proven. Additionally, there are forecasts that yearly numbers of homes built will be fewer in the future and that the trend will be toward smaller houses that consume less lumber. This, too, suggests that industry needs to look at export opportunities. Industry knows it has the ability simultaneously to supply products to meet domestic needs and meet offshore demand. The combination of these two markets has the potential to make our mills and forests fully productive.

Today, western producers are actively investigating the needs of foreign markets. We are educating potential export customers about our industry's capabilities and promoting the quality of U.S. species. We are cultivating the type of relationship that, if allowed to grow, will provide benefits both to overseas customers and to our domestic producers. Those are the whys of our export rationale. Now let's look at where the opportunities are.

There are five principal consuming markets in the world--six if you include the U.S.S.R. In fact, the U.S.S.R. claims to be the world's largest single user of softwood lumber. Whether or not this is true makes little difference. The Soviet Union is self-sufficient and cannot be considered a potential market. North and South America constitute the world's largest consuming market.

In normal years, lumber consumption amounts to as much as 120 million cubic meters, or 51 billion board feet. About 90 percent of that total consumption, however, is represented by use in Canada and the United States. Other countries in Central and South America also have wood supplies that partially meet their needs and even export minor volumes. Some export opportunities to the Caribbean Islands and South America will continue, albeit at a modest volume level. Western lumber opportunities there may be for timbers used in heavy construction or remanufacture.

The second largest consuming market is the Pacific Rim--including Japan, China, and Korea. In 1980 these countries consumed 50 million cubic meters, or 21.2 billion board feet. Western lumber producers supplied 4 percent of softwood lumber consumed there.

Western Europe, including the United Kingdom, rates as the third largest consumer of softwood lumber, and in 1980 used 38 million cubic meters, or 16 billion board feet. Our present participation in this market reaches only 2.5 percent of consumption, and is composed mostly of clears and timbers going to Italy, West Germany, and the United Kingdom. The remaining softwood markets are relatively small by comparison to these three major market areas, but signs of growth and new demand are stirring.

Fourth in line in present consumption is the Mideast, which includes Egypt, Saudi Arabia, the Emirates, and Eastern Mediterranean countries. 1980 consumption in the region was 5 million cubic meters, or 2 billion board feet. We moved 95 million feet into this market in 1980. This market will grow.

The other marketing area is what we call Oceania, and our interest there is centered in Australia and New Zealand. Consumption was 3 million cubic meters in 1980, or 1.3 billion board feet. The U.S. supplied 12 percent of Australia's consumption.

Excluding the United States and the U.S.S.R., the apparent consumption in these five areas totaled 118 million cubic meters in 1980, or 50 billion board feet, nearly 4 percent of which came from the U.S. Virtually every country in Europe, the Pacific Rim, Oceania, and the Mideast can be considered by our standards to be timber-poor and normally will use as much, if not more, lumber in future years as they have in the past.

In every one of these areas, western lumber company representatives or product associations have been researching market opportunities. In most areas, we understand which products are currently used and in many instances have determined how U.S. grades and species can substitute for competitive species and traditional needs.

More and more we see our mills taking on special cuts for special sizes and grades needed in foreign markets. While most manufacturers would like to see consuming areas change their specifications to fit the products we use here in the United States, we have to accept the fact that traditional foreign grades and sizes will be the norm for most importing countries. I personally believe one of the greatest opportunities for exporting softwood lumber lies in the willingness of mills to produce grades and sizes to importing customers' specifications.

Now let's look at what kinds and sizes of softwood lumber are in demand in these five trading areas, starting with Oceania--the smallest consuming area. Australia is the major user and typically takes large timbers from which they manufacture smaller sizes. In the near term, this trading pattern will most likely continue, inasmuch as Australia has imposed very discouraging duties on smaller dimensions. In the long term, Australia hopes to become more self-sufficient as its forests mature. So, while Oceania normally accounts for 10 percent of U.S. exports and will continue to do so for some years, the long-term growth opportunities are minimal.

The fourth largest consuming market, the Mideast, is another study. It is and likely will continue to be an emerging buyer of North American lumber. Partially this is due to oil riches that have led to unmatched construction activities. The United States has been increasing its share in that area, probably because of the inability of other sources to be able to supply that market. Currently the United States supplies as much as 17 percent of the apparent consumption in some of the Mideast countries.

Turning to Europe, we can predict that market expansion will result from promotional activities and the inability of competing countries to continue to supply all of the volumes that will be needed. Let me explain. The United Kingdom has slowly but steadily adopted a more favorable attitude and trend to wood construction--they call it timber frame construction--in lieu of masonry and other nonwood materials. England's timber frame homes are built in factories and their designs already accommodate sizes and grades that fit nicely with U.S. production. Timber frame construction will more rapidly penetrate the British housing market than in the past, with the need for increasing volumes of softwood lumber to satisfy the demand.

The obvious question is--where will that lumber be purchased? At present, most of the U.K.'s current needs are being met by Scandinavia, Poland, and the U.S.S.R. Scandinavian countries report that they have reached sustainable yield limits. The U.S.S.R. asserts that it will need more and more of its production to meet domestic needs and this may be true. Outside observers postulate that even now Soviet domestic needs are being sacrificed to obtain favorable trade balances that accrue from lumber exports. Should the U.S.S.R.'s domestic needs take precedence, there could be a significant shortfall in the volumes that customarily go into the European market. Thus supply problems, coupled with a growing demand for wood, make it likely that the U.K. and other European countries will be turning to North America to meet lumber needs. Of course, Canada and other competitors also will be wooing this market.

What do we see on the Pacific Rim? The opportunities there are vast and exciting. It was in the late 1960s that the Western Wood Products Association began to court the Japanese market. We saw then a dimension lumber potential, if the Japanese could be convinced to build platform frame homes such as we do in the United States. The Canadian industry also joined that effort, and by 1974 the Japanese had approved platform frame construction methods and lumber grading rules virtually identical to those used in North America.

We expected immediate and significant increases in exports. Quite frankly, our Japanese efforts have not netted us the volumes or results we or the Japanese government originally anticipated. But we have gained important knowledge--that building export markets can be a long-term proposition. It has also taught us that we cannot rely on other countries to adapt rapidly to our sizes, grades, and building methods. We didn't give up, however, and we have continued promotional efforts in Japan. For the past seven years, WWPA has had a full-time Japanese representative. His mission is quite simple: to promote western lumber by working with builders as well as the lumber trade and code and government officials. He provides western lumber manufacturers with market intelligence as to how best to serve the needs of the Japanese. He also spearheads WWPA's promotion of platform frame construction, selling its benefits as a safe, durable, and more economical system than traditional Japanese methods of home building. As you are aware, the Japanese have used a post-and-beam system.

The fruits of our promotion and commitment to this market are beginning to show. During 1981, 11,000 platform frame houses were built in Japan. Last year, with the Japanese economy in the doldrums and overall home building down, platform frame construction still increased to 16,000 units. This year the Japan 2x4 Home Builders Association expects 25-30,000 units will be built. That translates into dimension lumber export opportunities.

Another lumber potential exists in Japan. Japanese sawmills presently process 1.5 to 2 billion feet of United States logs each year. That equals a lumber volume of as much as 3 billion board feet. Japan's converting facilities are slow and inefficient compared with U.S. mills, and their labor costs are approaching those of the United States. Their industry has acknowledged that log costs plus conversion costs in Japan exceed costs of buying finished products here. That, of course, is the reason that 400 to 800 Japanese sawmills are leaving the business each year.

We therefore anticipate a growing market opportunity in Japan. It will require that more of our mills be willing to manufacture products to Japanese sizes and lengths. Such producer attention to customer needs is already taking place at an increasing number of our mills.

China is the one market we know least about. But its potential, based upon population and past wood uses, gives us reason for further investigation. Just recently, our Marketing Director, Bob Hunt, returned from China. Let me share with you some of his observations. Prior to 1940, China was a major western lumber importer. Signs of renewed interest started in 1980, and shipments of lumber in 1981 reached 36 million board feet, along with a substantial volume of logs. Shipments fell off in 1982, reflecting port, cash, and transportation problems. It is also clear that, with a void of some 40 years in product specification and purchasing knowledge, considerable product buying and use education will be mandatory to help the Chinese to be informed buyers and users. The opportunity exists because China is far short of being self-sufficient. Forestry officials in China estimate that, even with their reforestation programs, China will be faced with serious fiber shortfalls for as much as 50 years. And unlike Japan, which is overendowed with processing facilities, there are few such facilities in China. Their lack of primary production facilities and the costs of building new ones bodes well for potential lumber exports.

Housing needs in China are substantial, but presently lumber is used there only for doors, windows, trim, furniture, etc. Their homes--mostly multifamily--are built with brick, steel, and concrete. Away from housing we do see a potential product demand in industrial and commercial construction areas, for highway use, for factories, and for mining uses.

Critical to future buying and selling there is lumber education and development of standard grades and sizes. We are involved with the Chinese government in such development and education. Present activities will result in payback in the long term and are the key to opportunities in that market.

Progress has and will continue to be made in all of our export markets. And we will continue to improve our position in the future if we have the resolve to commit ourselves to a strong relationship with our export customers. Common sense, backed up by market research, has pointed out that one way U.S. manufacturers can build strong relationships with foreign customers is by being consistent suppliers. Sellers here must roll with market prices just as customers overseas must adjust. Producers cannot develop consistent business by being "in-and-outers"--in only when overseas markets are strong, and out when U.S. domestic markets are strong. This doesn't mean, however, that western mills should not take advantage of opportunities in our own domestic markets. Rather, mills will need to seek a compatible mix--sticking with good customers as good customers stick with reliable suppliers. By adopting this type of business relationship, I am confident that U.S. producers can gain very satisfying results in offshore sales.

U.S. producers can learn from Canada's strong points. In a way, they have already charted the course. If we profit by their pioneering and use our U.S. manufacturers' ingenuity, the world's demand for lumber makes our future bright. Our producers have the willingness and aggressiveness to compete and successfully meet the market challenges, and commit to the export responsibilities. We will succeed and prosper in those offshore markets.

U.S. OPPORTUNITIES IN WORLD TRADE IN PANEL PRODUCTS

M. T. Fast

Refer to structural wood panel products in your conversation today and most people think immediately of plywood, made by cross-laminating veneer. Technology, however, has advanced to the point where the term structural wood panel products not only embraces plywood but composites (veneer-bonded to reconstituted wood cores) and nonveneered panels (waferboard, oriented strand board, and structural particleboard) as well.

Until recently, plywood was manufactured following a prescription formula according to the provisions of U.S. Product Standard PS 1-74/ANSI A 199.1 for Construction and Industrial Plywood. This is a detailed manufacturing specification, or commodity standard, developed cooperatively by the plywood industry and the U.S. Department of Commerce.

Now, however, an increasing number of plywood, composite, and nonveneered structural use panels are manufactured under the provisions of American Plywood Association Performance Standards which provide performance criteria for specifically designated construction applications. The Performance-Rated approach to panel manufacture represents an important new trend in the wood products industry that promises to benefit all segments of the trade from producer to consumer. All gain with better and more efficient use of the raw material and with increased manufacturing and product innovation. Baseline criteria for qualifying new products provide a means for their recognition by code bodies while the descriptive APA panel trademark completes the package, identifying the end use of the structural panel product.

APA Performance-Rated panels at present include APA-Rated Sheathing and APA-Rated Sturd-I-Floor, both designed for residential and other light frame construction and industrial applications. Research and development of Performance Standards for exterior siding and concrete form panels are also under way by the American Plywood Association.

There are three basic criteria for qualifying a wood-based panel product under APA Performance Standards: structural adequacy, dimensional stability, and durability. Testing for structural adequacy includes verification of the panel's ability to sustain uniform and concentrated static and impact loads, to hold fasteners, and to resist wall racking. Dimensional stability tests measure the product's resistance to expansion. Excessive linear expansion may lead to warpage and buckling when exposed to moisture during construction or service. Finally, bond durability tests measure the ability of the adhesive system to retain its bonding capability under adverse exposure, thus assuring long-term structural integrity of the product when exposed to moisture during construction or service. A product must satisfy all of these end-use performance requirements to qualify as an APA Performance-Rated Panel.

With new products available, our sights are then set on an area of pure market development opportunity--exports. APA has been active in export market development since 1964, and from that time to the present our international representatives have operated offices in Tokyo, Cologne, Frankfurt, Hamburg, Brussels, Antwerp, Utrecht, and London. Currently, we maintain offices in London, Antwerp, and Frankfurt--the latter in the process of transfer to Hamburg. With the USDA Foreign Agricultural Service (FAS) in 1967, APA pioneered the first contract wherein industry and a government agency teamed together for forest products export market development. We remain as the largest cooperative partner with FAS, although 14 other forest products associations have now joined ranks within the FAS Forest Products Commodity Division.

Since 1976, principal producers have increased their commitment to international market development; some have direct ties with agents or importers and some operate a network of foreign offices to better serve their overseas clients.

Export shipments of American softwood plywood totaled 451.7 million square feet (3/8" basis) in 1982 (399,749 cubic meters). The 20 leading customers for direct shipments are spread all over the map and are ranked as follows:

		square feet (3/8" basis)
1.	United Kingdom	108,136,000
2.	Belgium/Luxembourg	89,593,000
3.	Denmark	49,686,000
4.	Netherlands	41,887,000
5.	West Germany	34,738,000
6.	Canada	30,598,000

7.	Saudi Arabia	16,865,000
8.	France	15,654,000
9.	Bahamas	8,480,000
10.	Japan	7,113,000
11.	Mexico	5,464,000
12.	French Pacific Islands	4,917,000
13.	Italy	4,391,000
14.	Trinidad	4,015,000
15.	Dominican Republic	3,137,000
16.	Trust Pacific Islands	2,347,000
17.	Ireland	2,053,000
18.	Leeward/Windward Islands	1,925,000
19.	Sweden	1,918,000
20.	Panama	1,876,000
	Total:	434,793,000

(square feet 3/8" basis)
96% of total shipments

International markets are in a sense microcosms of the U.S. domestic market, each country representing an opportunity at a stage of development similar to the U.S. 20 to 40 years ago. Each is a unique market with individual practices, preferences, and wood products supply characteristics. Some have their own production. Others must rely completely on outsiders for the bulk of their wood products. All have needs which translate into an array of opportunities for American timber product producers. These opportunities can best be capitalized on by systematic market development--that is, market research to identify targets and needs, followed by planned step-by-step promotion to the specifiers, users, and distribution trade. Along the way, tariff and nontariff barriers must be identified and analyzed and then the most effective course for change can be pursued through appropriate federal agencies and code bodies in the receiving country.

As an example, early in 1976 APA contacted the various timber trade federations in the European Common Market countries addressing the issue of the EEC Coniferous Nil Duty Quota on plywood. Immediately following this, the initial contact was made with the American Mission to the Common Market in Brussels on the same subject. Active liaison with both agencies continues focusing on the Association of Common Market Nations which import panels (Union pour la Commerce d'Importation de Panneau), which addresses the EEC administrative body at the working level. In addition, the U.S. timber industry has petitioned through the U.S. Trade Representative for assistance in gaining expansion and modification of the European Community's quota for softwood plywood. At this time, the European market offers the greatest opportunities for our products and it is here that we are concentrating our effort. Let us look at primary markets individually for a cross section of applications.

United Kingdom

The U.K. consumes over one billion square feet of plywood from all sources annually. Last year they purchased 108 million square feet (95,575 cubic meters) or 24 percent of U.S. exports of softwood plywood. Principal uses are for concrete forming (shuttering) and for crating and pallets. Timber frame construction is the developing market as the system gains in popularity. Approximately 30 percent of the housing now being built is timber frame.

Civil engineering jobs (highways, government, and commercial construction) are the big users of American 3/4" B-C Exterior as concrete forming (shuttering). It competes favorably with Canadian production. Some C-C P&TS is also used for this application, as are limited amounts of APA-Rated Sturd-I-Floor.

The automotive industry utilizes APA-Rated Sheathing, C-C Exterior and C-DX P&TS for pallets, crates, and slip sheets where components and subassemblies travel between final assembly points. Many of these crates become long-term dockside storage units, "warehousing" parts until needed in production--a real test of glueline integrity.

Breweries also are using American plywood pallets for handling kegs, bottles, cans, and cartons in varying degrees of automated storage.

Approximately 90 percent of timber frame house construction is sectionalized, requiring plywood wall sheathing. Plywood in this case becomes the structural diaphragm of the unit, providing the necessary racking strength. Structural values for American plywood grades have been accepted and will appear in the new British Standard BS 5268, Part 2, Structural Use of Timber, when published.

Scotland, though an integral part of the United Kingdom, appears a bit more progressive in its transition to timber frame construction. Plywood is used as decking on gabled roofs, whereas in England it is primarily relegated to flat-roof structures. In both regions, however, there is very limited use of plywood siding (cladding) except for accent panels.

There are nearly 100 U.K. ports through which wood products may enter. Shipments are translocated to trucks for shipment to importer's warehouses, usually within a 100-mile radius. Sales are handled through timber agents.

Belgium and Luxembourg

Belgium and tiny Luxembourg use American plywood in much the same manner. Aggressive trading is the byword and much of the product originally invoiced to the port of Antwerp, for example, is transshipped to other West European terminals.

There is no basic restriction on the use of American plywood in housing; however, only about 15 percent of new construction is timber frame. Each home is individually designed by an architect for approval by community officials. Belgium's housing reflects this uniqueness of design, and good progress is being shown in wood used for its decorative and insulative values. APA 303 Sidings are used for accent panels on some elevations. Quality construction is the norm, and at least one fabricator uses overlaid American plywood as structural sheathing behind brick veneer in the houses he builds. Component construction, folded plates, stress skin panels, and box beams are used in both residential and light commercial designs.

The use of American plywood in automotive crating and pallets parallels that in England; however in Belgium, the fabricator may consign all of his production to a single plant such as Ford tractor. Household goods containers made particularly for American military use also require plywood.

Concrete forming uses plywood from all sources, though 3/4" B-C Exterior is making inroads. Rated Sheathing is used as wall and roof sheathing and for crating. CD X P&TS is used for exposed ceilings and for a range of industrial applications.

Denmark

Denmark is the most sophisticated market for structural plywood in Europe, with a wide spectrum of end uses requiring a large range of grades and thicknesses. The Danes "think" and use plywood in their construction very much like Americans.

The Danish Port of Esbjerg is a major receiving and transshipment point, as Danish agent/importers also supply the remaining Scandinavian countries. It is also the scene of remanufacturing operations where American plywood blanks are textured and milled for a variety of applications. Per capita consumption of structural plywood in Denmark is the highest in Europe. Housing fabricators in Denmark have also shipped sectionalized housing to the Middle East and North Africa.

The range of thickness of APA A-C Exterior is used in construction, furniture, and industrial applications. Rated Sheathing is used as roof sheathing, for stress skin panels and components. Rated Sturd-I-Floor is used for combination subfloor/underlayment, and 303 Sidings are used for exterior siding, paneling, and door skins.

Denmark establishes the pattern in codes and standards development and administration for the Nordic region. Continuing liaison with the Danish Ministry of Housing is very important to code acceptance there. The technology of new product development must be well documented to obtain approvals. Important research is being carried out by the Danes utilizing plywood components with thick insulation for greater thermal efficiency. The "zero energy" house, or one with minimum heat input, is the goal.

The Netherlands

Over 80 percent of the American plywood currently being sold in Holland is underlayment or Sturd-I-Floor. The trade approach followed by some in Holland is to stock the one grade and offer it for a variety of applications. Development of the stressed skin insulated roof deck and the government drive toward more wood in construction are both favorable to American plywood. Design values for American plywood have been approved by the Dutch Timber Research Institute (TNO) and our promotion builds on this recognition.

Sanded B-C Exterior and touch-sanded C-C Plugged Exterior are used for concrete forming and for some industrial applications. Rated Sheathing is also used for crating while C-D Exterior P&TS is used for ceilings and components. APA 303 Sidings are used to a limited degree as paneling and for exterior siding.

Mass merchandisers have evolved in Holland to serve the do-it-yourself market. They feature paneling, insulation, kitchen cabinets, hardware, electrical plumbing, lumber, and most materials required for building or remodeling a home. Some merchandise Sturd-I-Floor and in some cases have offered American waferboard cut to 4 foot by 4 foot size for underlayment applications. Do-it-yourself usage of plywood sidings and sanded will also increase.

West Germany

There has been good acceptance of American plywood in crating, packaging, and household goods containers in Germany. The automotive industry is the demand factor for crating, while the American military community in Germany requires the household goods containers. Concrete forming takes the range of grades, from C-DX P&TS for single-use applications to B-C Exterior—the universal form panel—to overlays. American overlaid plywood in competition with phenolic-faced finish material is being evaluated for automotive bus floors and other industrial applications.

Plywood bearing APA trademarks has been approved for industrialized housing through the Institut für Bautechnik, the federal building agency of Germany. A quality certification agreement will be signed with FMPA Stuttgart and the product can be trademarked to show conformance.

Approximately 50,000 units are prefabricated in Germany each year, or 10 percent of total starts. Having penetrated the market that has been traditionally reserved for the local particleboard industry, we have the opportunity to broaden the applications for our products in a number of ways. Meetings and seminars for architects and designers are under way, and, with a German national soon to be added to the APA staff, we can extend the training in products and applications to all levels of the distribution trade.

France and Italy

A market research study will be conducted in France and Italy this year to identify market applications and opportunities. The indication is that France has increased its imports of American plywood rather dramatically in the last couple of years, which could be a positive sign for future market development.

The market for American plywood in Italy is potentially good. Total plywood needs will double over the next 10 years and American plywood will play an important part in that growth. Although the major portion will continue to be concrete forming and crating, the earthquake of 1980 brought new interest in timber frame housing. Plywood structural diaphragm performance has been well documented in Alaska, California, and other areas of high seismic loading.

Scandinavia

Denmark has been covered as a major market; however Sweden and Norway, and--to a lesser degree--Finland, also offer ready markets for American wood products. To succeed in Sweden, appropriate layups and species combinations must be assigned engineering values to stress level P-30. This evaluation work is now under way. In Norway, the immediate opportunity is in floor construction which meets the stiffness requirements of the Norse code. Certain layups and species combinations are now accepted. Technical seminars are under way in Norway and Sweden now.

Opportunities to work with the Finnish forest industry in cooperative promotion to increase per capita consumption of plywood should be explored. Some Finnish firms marketing throughout Western Europe have already broadened their portfolio of products by merchandising American structural panel grades along with their more specialized Finnish overlaid products. We must increase educational emphasis on American products and how they are used. Broadening the base of literature in the language and vernacular of each of the Scandinavian countries is required.

Japan

Technical work has been under way in Japan for several years, with significant developments occurring recently. Only a single technical issue related to white pocket in old-growth veneer must be resolved to permit unrestricted recognition of U.S. Product Standard PS 1 plywood in Japan's structural plywood standard.

While progress has been made on the standards issue, Japan's high tariffs on American plywood severely restrict our market access. The basic problem is that Japan has been overly protective of its own domestic plywood industry, which is second only in size to the U.S. manufacturing base. As long as raw materials are available for plywood production, the Japanese prefer to utilize their own facilities and supply their own plywood.

Middle East

The oil-rich countries of the Middle East and North Africa reportedly have substantial need for new housing as well as construction of roads, bridges, dams, and industrial complexes. While some shipments have been made into these countries, they have apparently come from indirect promotion--that is, orders from firms that have obtained contracts to build in these countries. Not much is known about the distribution trade for wood products there, and in some cases it is not even known if trade channels exist. Our first priority is to obtain reliable information from selected Middle East and African countries to determine the most productive course for future market development in this region.

Caribbean/Latin America

Caribbean and Latin American countries have substantial housing needs. Oil wealth has been added to the agricultural and mineral assets of some of the countries, giving them the wherewithal to pay for increased imports. Political and economic problems continue in the region, but despite these an opportunity exists to develop new, profitable markets for American products.

The early development pattern in Venezuela and subsequently in Chile has been the establishment of a link to the country's political heads of housing. Prefabricated housing units have been erected to show how they can be used to satisfy the extensive housing requirements for both the poor and the middle-income groups.

In Chile, the APA, working with the State of Washington Department of Commerce and Economic Development and the U.S. Department of Commerce International Trade Administration, will erect an all-wood model home for display during the International Trade Fair in Santiago, Chile, in October.

A Trinidad Housing Symposium held last fall was very successful and a Dominican Republic Wood Housing Symposium is now under way. Additional seminars are scheduled this year in Lima, Peru; Bogota, Colombia; and Quito, Ecuador.

There are substantial tariff barriers to be overcome in this region, and the assistance of our federal agencies is being sought to negotiate these changes.

Conclusion

With this brief review I have attempted to show that real opportunities exist for panel products market development in several regions of the world. A systematic approach is necessary to capitalize on these opportunities. We must identify targets and needs and then develop a step-by-step promotion to specifiers, users, and the distribution trades. Tariff and nontariff barriers must be overcome to insure unrestricted entry of our products into the markets.

A look at any chart of export shipments since 1965 shows the volatility of the international market. The timing of demand varies, as does the availability and cost of products from other sources. Despite these obvious ups and downs in exports, the trend is upward.

In five years, provided that industry retains the continued promotional support of our federal agencies, we can anticipate that the demand for our panel products will reach close to one billion square feet annually.

PULP AND PAPER AND INTERNATIONAL TRADE

Irene W. Meister

How important is international trade to the U.S. paper industry? The direct exports of pulp, paper, and paperboard in 1982 amounted to just under 10 million tons or $4.3 billion. The significance of foreign trade to the paper industry would be highly underestimated, however, unless we took into account the impact of indirect exports on the industry's health. I define indirect exports as domestic sales by the paper industry made only because of the export demand for products of another industry. Boxes, in which various exported American goods are packaged, are a clear example of indirect exports. Companies that sell boxes often do not know that they are a part of export trade.

There are many other examples of such indirect exports which enter foreign trade as components of other products, such as filters and insulating or impregnated papers. Domestic sales of printing and writing papers used for export documentation are an important indirect export and these sales increase proportionately to the growth in our country's total industrial and agricultural exports.

All of these are sales that are made to domestic customers but are dependent on that particular customer's export trade. In 1980, we estimated that such indirect exports for the paper industry account for about $10 billion in domestic sales, and they may be even higher now. This key economic multiplier is very little understood. So if you add indirect exports sales of some $10 billion to our direct sales of over $4 billion, the impact of foreign trade on our industry is substantial.

So far I have referred only to the export trade. We also must take into account the fact that a significant number of U.S. paper makers depend on imported pulp from Canada for the production of their particular grades.

Perhaps an even more important question would be how important foreign trade will be for the U.S. paper industry in the future. Let us look first at the indirect exports for the paper industry. Those would rise dramatically if overall U.S. trade increases. If a company such as Boeing sells more aircraft abroad, its numerous domestic contractors, many of them small companies, would use our industry's packaging materials to ship the various parts to Boeing. All types of packaging would certainly benefit if exports of manufactured goods were to increase, as would sales of component parts made from paper. There would also be greater use of printing/writing papers because more export documentation would be required. Thus, if the United States were to succeed in attaining a larger share of the existing world market or if the total level of world trade were to expand substantially, then of course the paper industry would benefit greatly from the impact of indirect exports.

But what is the future of our direct exports? It might be useful to look at the historic evolution of our direct export trade in the last decade before we talk about the future. Exports of three products, if we relate them to total production--woodpulp, kraft linerboard, and, more recently, bleached board--are of major importance to the paper industry. In the case of woodpulp, exports account for about 60 percent of market pulp production, and the corresponding figures in 1982, when related to total production of linerboard and bleached board, are 14 percent and 14.5 percent, respectively. (See figure 1.)

Where are these exports going and is there a change in the share of different regions? While export tonnage of pulp increased by more than a million tons, the regional distribution has shown little variation over the past decade. Western Europe remains the largest consuming region for U.S. pulp exports, but exports to the Asian countries have increased their share as well. Within Asia, Japan is the largest consumer, although both South Korea and Taiwan are important and growing markets. (See figure 2.)

Kraft linerboard exports reached their peak level in 1980, and 1982 export tonnage was only slightly above 1972. We can see a dramatic shift in distribution between 1972 and 1982, however, as the Asian share of U.S. shipments has increased more than threefold, while Western Europe's share has fallen sharply. Prior to the late 1970s, U.S. exports to Asia were relatively insignificant. In 1980, however, China came strongly into the market with purchases of 319,000 tons. But in 1981 and 1982, these exports declined sharply. (See figure 3.)

It is worth noting the steady growth in exports to "Other Asia"--principally the countries of the Association of Southeast Asian Nations (ASEAN) and Hong Kong. These dynamic economies hold out considerable potential for the future. Finally, even though Japan is the most

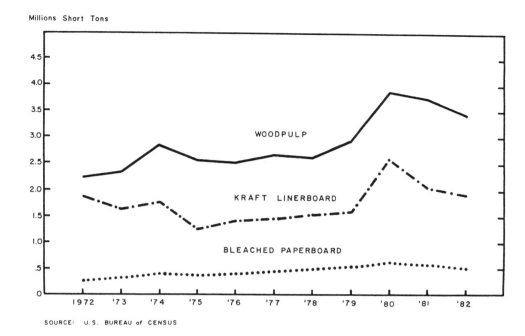

FIGURE 1. U.S. paper industry exports, 1972-1982 (major grades).

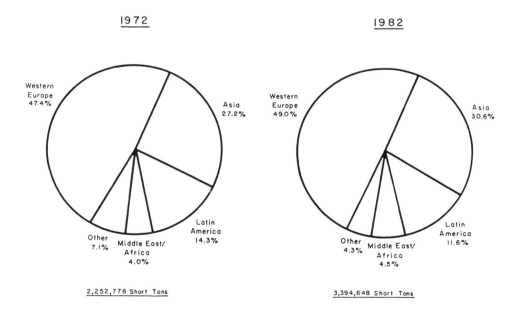

FIGURE 2. Percentage distribution of U.S. wood and pulp exports by major regions.

industrialized economy in the region, and its GNP is far larger than the combined GNP of ASEAN countries, Japan's purchases of linerboard are still rather low, especially when compared to its total consumption. (See figure 4.)

In total paper and paperboard, tonnage has increased between 1972 and 1981 by over 1.5 million tons. In terms of regional distribution of total U.S. industry exports, Asia and Europe are now roughly equal, with Europe's share halved compared to 1972, and Asia's share doubling. (See figure 5.)

A few words should be said about the other major product categories. Exports of printing/writing paper have increased from 210,000 tons and $125 million in 1972 to 383,000 tons and $355 million in 1981--a good year for exports of printing/writing papers--although in 1982 these exports declined to 247,000 tons and $239 million. Exports of specialty grades have risen sharply and for many smaller companies in the specialty field now amount to a very important part of their income.

So much for the historic review of the last ten years. But what about the future prospects for U.S. exports? This question should be examined within the framework of worldwide demand/supply developments and, last but not least, attitudes of U.S. top management.

The recent recession played havoc with all previous forecasts and growth trends. We do know, however, that demand growth in this decade is likely to be considerably less than what the Food and Agriculture Organization (FAO) had forecast in 1977 when they released their study on supply/demand and trade to 1990. Both capacity and worldwide demand are clearly growing at a lesser rate than was forecast. A new supply/demand study by the FAO to 1995--a very much needed one--is now on the way, and, with the help of the worldwide industry Working Party, should be ready in 1985. In the meantime, without the firm data, and in the light of a much longer and deeper recession worldwide, one can make only a few conjectures. (See tables 1 and 2.)

The fact that demand trends will show slower growth than anticipated for the period from 1980 to 1990 does not necessarily mean that U.S. exports will suffer a similar decline because capacity growth in many importing regions will also decline. Here we have to look at such factors as competitive strength in costs, quality, and willingness on the part of the U.S. industry to make a major effort to sell in the world market on a sustained basis. This latter factor is to my mind as important as our cost

PULP AND PAPER   115

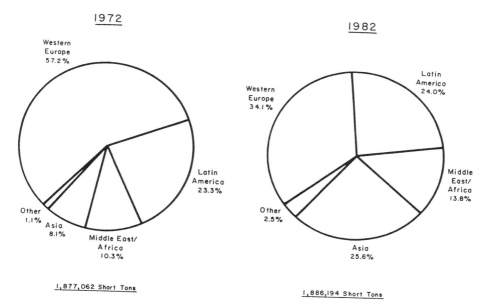

FIGURE 3. Percentage distribution of U.S. exports of kraft linerboard by major regions.

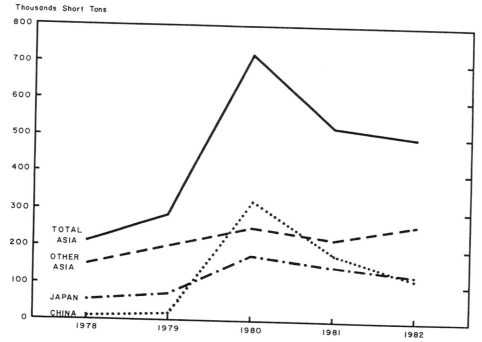

FIGURE 4. U.S. exports of kraft linerboard to Asia, 1978-1982.

competitiveness and, for the industry as a whole, is still the area where much progress needs to be achieved.

In this coming decade our competitors for the traditional U.S. exports--pulp, kraft linerboard, and bleached board--will be the same ones as in the 1972 to 1982 period, with Brazil a factor in pulp exports. At this point I do not see any new, large international competitors in addition to the existing ones. Competition with domestic producers in Europe and Japan can intensify, however, if protectionist trends within individual countries should increase. In Europe, the Scandinavians have a considerable advantage in tariffs over the United States and Canada, and competitive devaluations, such as we have seen recently in Sweden, play havoc with any forecast that is made on a rational basis. A more disciplined world monetary system is a must and should be a key objective of the U.S. government.

On the basis of our cost competitiveness, there appears to be no reason why the U.S. paper industry should not be the leader in exports. We have natural resources that are accessible. We have solved most of our environmental problems. We have developed major savings in energy. That combination of factors ought to allow us a prime position as world competitors in paper and paperboard. In terms of quality of special requirements, I have no doubt that our fine technical people can meet these challenges.

When it comes to the willingness to make a major effort to export, much will depend on the attitude of the top management of individual companies. Those companies which perceive that exports will be a key area for growth--and their numbers, in my view, are growing--will lead this industry into greater world prominence as a supplier. We already have a number of companies that view their future within the total global context of the marketplace. They have learned that to be successful they must be reliable suppliers in good times when U.S. demand is high, because that would allow them to stay in the world marketplace during the bad times when exports become especially important in maintaining a more acceptable production-to-capacity ratio. No doubt these companies will pursue their future development with the view of the world as one market.

We still have a number of companies that are interested in exports predominantly during periods of domestic oversupply and who put relatively little effort into cultivating customers for the long term. Their temporary success in overseas markets is usually a result of their ability to compete in the very short term through price reduction, and they withdraw from foreign markets as soon as

PULP AND PAPER 117

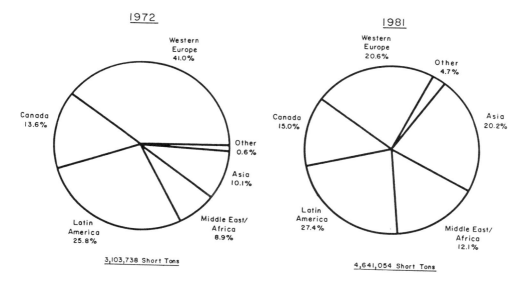

FIGURE 5. Percentage distribution of U.S. exports of paper, paperboard, and converted products by major areas.

Table 1. Capacity Growth for Paper and Paperboard by Major Area (total increase in million metric tons).

Area	1970-75	1975-80	1980-85 (forecast)
World Total	33.8	23.0	27.5
North America	7.5	5.8	8.6
Western Europe	8.2	4.7	6.0
Japan	6.8	1.8	2.8
Less Developed Countries	7.5	6.6	6.2
Other	3.8	4.1	3.9

Source: FAO world pulp and paper capacity surveys.

the U.S. market strengthens. A sustained export effort requires considerable investment of time and money which these companies do not feel is justified for the results obtained. I believe, however, that compared to ten years ago, there are now many more companies considering exports as a part of their long-term strategy. It is because of these developments that I am optimistic about the future of the U.S. paper industry's exports.

Which of our products are going to be moving in worldwide commerce, and to which areas? I believe that the three products which are already heavily exported--that is, pulp, linerboard, and bleached paperboard--will continue to dominate the export picture in the next five to eight years. I also firmly believe that if we look beyond that period, that is, the next eight to twelve years, the U.S. can become a major exporter of commodity grades of printing/writing paper because we can be fully cost competitive, and because this country will need to utilize its domestic potential and economy of scale by going outside of its borders.

Europe will remain an important export market for U.S. producers of kraft linerboard and bleached board, but the Nordic countries will continue to have an important advantage in terms of distance and tariffs. In 1984 there will be tariffs for Nordic exports to the Common Market, while the U.S. and Canada will still face significant tariff barriers. On the other hand, I am much more optimistic about U.S. opportunities in the Asian market, as well as some of the markets of the less developed countries.

Despite its limited fiber resources, Japan is the second largest producer of paper and paperboard in the world. It has high quality products, but their costs are also very high, and the market is protected by tariffs, and by cultural and, some believe, governmental nontariff barriers. In 1982, imports of kraft linerboard from all sources constituted only 7.4 percent of the total consumption of linerboard in Japan. Since Japanese costs are much higher than ours, many U.S. producers believe that if that market were fully open to them, they could substantially expand U.S. exports, meeting Japanese demands for special quality and long-term supply reliability. Recently, Japan's Industry Structural Council of the pulp and paper industry, an advisory body to MITI, scaled down earlier projections of demand to 1990. It remains to be seen whether they are right or wrong. Also, it remains to be seen what impact the newly proposed Depressed Industries Law will have on imports of pulp and paper in Japan.

Even the most liberal trade advocates believe, however, that if the U.S. market is to remain open to a large inflow

Table 2. Capacity Growth for Paper and Paperboard by Major Area (average annual percent increase).

Area	1970-75	1975-80	1980-85 (forecast)
World Total	4.4	2.5	2.6
North America	2.4	1.6	2.2
Western Europe	3.9	1.9	2.2
Japan	8.8	1.8	2.3
Less Developed Countries	10.5	6.1	4.4
Other	5.0	4.3	3.4

Source: FAO world pulp and paper capacity surveys.

of Japanese exports, Japan, in turn, will have to open its market much wider to the cost and quality competitive U.S. industries of which forest products could be in the lead.

I believe that demand for pulp, paper, and paperboard in China will grow and become an important market for many of our products, especially if the U.S. is able to increase its imports of their goods. China's demand for capital will be enormous, and since the paper industry is so capital intensive, rapid growth of China's domestic industry, which has to start from a rather low base, is unlikely to be dramatic. Here again, the U.S. will compete against other countries, but we ought to be able to hold our own, even though lots of effort will be required.

Earlier, when I referred to figure 4, I spoke about the countries of the ASEAN region. At one point these countries had envisaged a rapid growth of their domestic paper capacity. One of the FAO studies has examined the possibility of regional development where one country would produce kraft products, another printing papers, and a third perhaps some specialties. The cost of this regional program was over $6 billion. This development has not taken place, even though some countries will show capacity expansion. On a regional basis, the countries could not agree--perhaps because of politics--to the division of production, and there is little hope that they will do so in the near future. The cost of capital will be the key inhibitor to any dramatic growth in capacity, and if demand continues to expand, prospects for U.S. exports should be considerably enhanced.

Changes in supply and demand plus the vagaries of

economic cycles are normal business risks. But what are other major obstacles to the expansion of U.S. exports? Some obstacles are of U.S. governmental origin. I am thinking of the problems that all exporters face in the application of the U.S. Foreign Corrupt Practices Act, the uncertainties in the enforcement of antiboycott laws, and others. Currently, the U.S. trend toward protectionism can become a self-defeating obstacle to trade. The pulp and paper industry is certainly quite vulnerable when it comes to the retaliation which inevitably follows protectionist binges. With unemployment as high as it is now, the temptation to blame all domestic ills on foreign imports is likely to be high. The exporters' voice in Congress is often much weaker than that of import-impacted industries.

There are, however, other barriers that are not of our making. Here I am referring to existing tariff and nontariff barriers, and protectionist restrictions imposed by other countries. I already mentioned the impact of tariffs on our exports to Europe and Japan. Red tape, licensing, and currency restrictions are serious barriers when it comes to exports to the less developed countries. We, as a country, will have to do much more to bring these countries into a more disciplined world trading system. Currently, their debt problems are sweeping away all rational approaches to foreign trade. But again, generalizations are a dangerous game, because there are developing countries that are relatively stable financially and provide excellent opportunities for exports. More effort will be needed on the part of the U.S. government to protect U.S. exporters against some predatory practices of our competitors who on occasion subsidize their exports in order to maintain production and employment.

The tariff and nontariff barriers are an area in which the American Paper Institute has been especially active and in which the industry's success is in direct proportion to its unity in pursuing the objective of more liberal trade. There is no doubt that individual companies can have an influence on governmental actions, but our strength lies in a united approach to trade obstacles. As an industry, we have been consistently advocating more liberal trade and therefore have been in tune with the basic objectives of our government. We fought protectionism in Congress and provided leadership vis-à-vis other industries in that area. U.S. exports (and here I am speaking of direct as well as indirect exports) cannot prosper in isolation. They can prosper only if the United States government develops a strong and effective trade policy. In that endeavor we need the participation of every one of API's member companies.

DEVELOPING FOREIGN MARKETS FOR U.S. SOLID-WOOD PRODUCTS

Vernon L. Harness

Introduction

To a U.S. mill, successful exporting can mean survival of the firm. But getting into the export market is not for all firms and is never easy. In this paper, I will try to indicate where opportunities for exports exist, outline some of the problems associated with exporting, and describe United States Department of Agriculture assistance to exporters. These comments are made to help you determine your best course of action.

Traditionally, compared with other segments of American agriculture, the solid-wood products industry has not been very export minded. Until recent years this industry was a net importer, with exports taking only about 3 to 5 percent of total production. As much as two-thirds total production of other major crops has traditionally been exported.

In recent years, wood products exports have increased, reaching $3.7 billion in 1980. Since then, interest in exports has remained high, although widespread recession in market countries dropped shipments to about $2.9 billion in each of the past two years. In 1983, we expect total exports of about $3.2 billion. Logs continue to be the largest segment exported, but the greatest growth has been in lumber and plywood. Top markets are Japan, Canada, and West Germany.

World Trade Opportunities

World wood imports increased from 2.4 billion cubic feet in 1960 to 6.4 billion cubic feet in 1980. The United Nations Food and Agriculture Organization estimates that

---

Appreciation is given to the National Forest Products Association for much of the data in this paper.

world lumber consumption, excluding North America, will increase 30 percent between the years 1980 and 2000. Consumption of panel products may rise 70 percent. World trade should increase at an even faster rate. Analysis by the National Forest Products Association suggests that world trade in wood products will grow faster than the U.S. domestic market during the 1980s and 1990s.

This country's major competition in foreign markets is from Canada, Scandinavia, and the U.S.S.R. Both Canada and the U.S.S.R. have larger acreage in timber, but productivity is lower than in the United States, and costs are rising more rapidly. Scandinavia is at its limit of sustainable yield production.

The above factors suggest that export opportunities are favorable in the years ahead and that the United States is in a strong position to increase market share. Total U.S. exports could approach $7 billion by 1990.

Let us look at some of these markets and their potential over this decade.

Asia: Japan will continue as our largest market. Shipments reached $2 billion in 1980 and should exceed $3 billion in 1990. Currently, logs make up two-thirds of the total, but the percentage will decline. Softwood lumber shipments could triple, with attractive gains also for plywood, veneer, particleboard, and hardwood products. China, a nation of one billion people embarking on a major reindustrialization program, represents huge potential. Shipments were near zero in 1979, over $200 million in 1982, and could exceed $600 million by 1990. So far most exports have been logs, but sales of lumber are beginning. Other small but potentially attractive Asian markets include Australia and Korea.

Europe: The timber industry expects exports to reach $1.6 billion by 1990--compared with about $700 million in 1981. Our best customers are and will continue to be West Germany, the United Kingdom, and Italy. We now ship primarily high-grade products and have a 1 or 2 percent market share. The real potential is in construction lumber, plywood, and panels as timber frame construction gains favor. Hardwoods are also likely to benefit as U.S. oak and other temperate hardwoods gain greater acceptance relative to tropical species.

Middle East: Wood shipments to the Middle East have been small but growing in recent years. Growing demand for lumber, plywood, and such products as railroad ties and telephone poles could push 1990 exports to between $250 million and $300 million.

   Latin America: Exports could more than double, to about $330 million annually by 1990. Construction lumber and plywood are in the best position to benefit from the goal of several nations to sharply increase the availability of housing.

World Trade Problems

   The discussion thus far would suggest great days ahead for wood exports. But remember, I have talked only about potential. Your success in reaching that potential will depend upon success in overcoming a number of serious trade barriers and marketing problems. Solving these problems will require close coordination between industry and government, and an aggressive effort on the part of each.

Trade Barriers

   A discussion of official trade barriers alone would make a long paper. Let me highlight just a few problems.
   Japan's plywood tariffs of 15 to 20 percent are prohibitive. Veneer tariffs of 15 percent clearly do not make sense--even from the point of view of Japan's own best interest--in view of growing restrictions from Asian suppliers of logs and veneers. Duties on certain species of lumber are illogical considering Japan's commitment to buy more finished products. On the other hand, the U.S. wood products industry has made some progress in getting plywood standards accepted in Japan.
   Turning to Europe, we find a host of tariff and nontariff barriers. For example, softwood plywood faces high tariffs after filling an annual quota of 600,000 cubic meters on shipments into the European Community. Other restrictions such as phyto-sanitary certificates on oak logs and limitations on bark on lumber often seem to be applied largely as a trade irritant.
   The tariff system in Latin America is best characterized as high, vague, often changing, and sometimes applied, sometimes not. Published rates range from near zero to 160 percent. Frequently, requirements such as import licenses and certificates impede trade. The strangest aspect of these trade restrictions is that they are continued in the face of efforts by the national

governments to provide more housing for their people, where imports are essential to that effort.

Marketing Problems

As difficult as it will be to reduce trade barriers substantially, we must also overcome a number of marketing problems if we are to reach our export goals. Many marketing problems are not clearly defined and no specific avenues are established for their removal. Many of the problems are overseas; some are within the United States and perhaps in your own company.

Let us consider some problems that are common worldwide. Potential customers and users may not be familiar with U.S. species, their characteristics, and suitability for intended end use. U.S. grades and standards may be different from those of their current supplier. Timber frame construction may not be widespread in the market you select. For example, five years ago, the United Kingdom built very few timber frame houses, the rate is now about 30 percent and may be 70 to 80 percent five years from now. In some potential markets, wood may not be a traditional construction material, as in Latin America.

Your potential customer may not be quite convinced that a U.S. supplier will continue to service him when U.S. housing is strong. Foreign buyers have long memories and have not forgotten that the U.S. industry lost interest in foreign markets after an effort at penetration nearly 20 years ago. In short, you want a dependable customer; that customer demands a dependable supplier!

Marketing problems in specific countries will run the entire gamut of possibilities. In one country, it may be credit; in another it may be exchange rate fluctuations. Sometimes, getting enough information to make a successful bid can be difficult. Domestic transportation must be worked out and your own foreign marketing techniques developed. Your ability to supply specific sizes and qualities is also critical. In summary, problems related to each transaction must be identified and solved.

Joint Industry/Government Efforts

Previous sections of the paper attempted to spell out some of the opportunities ahead for U.S. solid-wood producers, as well as some of the problems to be faced. For a long time, many industry leaders have been very much aware of both the opportunities and problems. The question to be answered was how to speed up their resolution. It was agreed in late 1979 that the Foreign Agricultural Service (within the U.S. Department of Agriculture) would work closely with the industry to increase exports. Congress has agreed to the inclusion of forest products among the agricultural commodities supported by this agency.

Because the Foreign Agricultural Service (FAS) may be new to many of you, I will describe briefly those functions of interest to the solid-wood products industry. FAS was established 30 years ago with the simple mission to maintain, develop, and expand export markets for U.S. agricultural commodities, now including forest products. During this 30 years, agricultural exports (excluding forest products) increased from $3 billion to $43 billion in 1981.

The activities of FAS can fit under three major headings--market access, market information, and market development. The approach used is unique in that industry experience and expertise is joined with government efforts in all of our endeavors. FAS initiates, directs, and coordinates the Department of Agriculture's formulation of agricultural trade policies and programs related to world markets for U.S. agricultural products. It monitors international compliance with bilateral and multilateral trade agreements. It identifies restrictive tariff and trade practices which act as barriers to U.S. agricultural commodities, then supports negotiations or other activities to remove them.

This agency maintains a worldwide agricultural intelligence and reporting service to provide U.S. producers and traders with information on world agricultural production and trade that they can use to adjust to changes in world demand for U.S. agricultural products. This is done through a continuous program of reporting by 74 posts located throughout the world, covering approximately 110 countries. Reporting includes information and/or data on foreign government policies, analysis of supply and demand conditions, commercial trade relationships, and market opportunities. Forest products will be included in the near future. FAS is now beginning to use advanced computer and telecommunications technology to improve and speed the flow of information between the posts and Washington.

FAS analyzes agricultural information essential to the assessment of foreign supply and demand conditions in order to provide estimates of the current situation and to forecast the export potential for specific U.S. agricultural commodities. The Service formulates policies and administers agricultural commodity export programs including export credit.

The Export Credit Guarantee Program (GSM-102) is designed to expand U.S. agricultural exports by stimulating U.S. bank financing of foreign purchases on credit terms of up to three years. The program operates in cases where credit is necessary to increase or maintain U.S. exports to a foreign market and where private financial institutions would be unwilling to provide financing without Commodity Credit Corporation (CCC) guarantee. The destination country must provide a reasonable expectation of repayment. In the past, the program was geared toward bulk commodities. Now, any agricultural commodity (including forest products) whose export furthers CCC's long-range market development objectives may be considered on a case-by-case basis.

All of these functions come together in the program called <u>Foreign Market Development</u>. Every agricultural commodity exported in a significant amount is promoted by one of the 56 associations participating under this program. The National Forest Products Association (NFPA) represents the wood products industry. NFPA serves as the industry coordinator. Industry member associations actually plan and conduct the activities overseas with appropriate financial and other assistance from FAS. Currently, 15 member associations are active in the program. These include: The Southern Forest Products Association (SFPA), American Plywood Association (APA), National Lumber Exporters Association (NLEA), Northeastern Lumber Manufacturers Association (NeLMA), Western Wood Products Association (WWPA), Maple Flooring Manufacturers Association (MFMA), and the Fine Hardwoods-American Walnut Association (FH-AWA).

The activities undertaken by the associations and by FAS are intended to offset problems such as those discussed in previous sections. For example, the industry and FAS have made some headway in convincing Japan to establish plywood standards more acceptable to the United States. The next and more difficult task is to convince Japan to reduce tariffs. At the same time, we will continue to encourage Japan to take more lumber and other manufactured wood products in their total purchases. It has been a long struggle, but accord is near on procedures that will allow shipments of oak lumber and logs to the European Community to continue. Our program provided the forum in which these procedures could be worked out. In the future, we will be working to increase or eliminate the EC quota on plywood imports. Both APA and NLEA have representatives stationed in Europe to help expand markets. Other representatives will be added.

To continue examples of market development activities, both APA and SFPA have been making major efforts in Latin America where the critical problems are high tariffs and the current use of masonry as the traditional building material. Techniques used include educational and technical seminars and on-the-spot assistance on construction of demonstration units. High tariffs continue, but Venezuela has allowed duty-free entry on certain panelized components. Some government credit also has been used in the Caribbean. The program has been extremely useful in providing for the publication of technical material in the foreign language needed, and in providing travel by technicians and key decision makers.

These are but a small sample of the kinds of activities undertaken by the industry working within this program. If it is likely to be effective, and we can afford it, we will try it!

In late 1982, Congress enacted a new tool to encourage exports: the export trading company legislation. This is one of the few pieces of legislation in over a decade aimed at giving American business major new tools to penetrate and expand export markets. It will encourage many small- and medium-sized firms--especially exporters of value-added and specialty products--to enter the export market for the first time, as well as provide new opportunities to existing exporters.

In general, the strength of the legislation is its reliance on the ingenuity, productivity, and efficiency of the American business and financial community. Instead of mandating a particular form of trading company or imposing an inappropriate foreign model on U.S. industry, it leaves it up to the U.S. private sector to develop what is likely to be a highly diverse group of trading companies. The legislation is designed to attract producers of goods and services, banks, export management companies, shipping companies, and other export service businesses into an effective joint effort to exploit foreign markets. They may do this either to export their own products or to act as a "one-stop" service for unrelated clients.

The legislation would: (1) permit bankers' banks, bank holding companies, and Edge Act corporations--subsidiaries of bank holding companies--to invest in export trading companies, and (2) allow export trading companies to apply for an antitrust certification preclearance, giving greater assurance that their activities would not make them liable to antitrust action. The new legislation also applies to the export of services, which was not covered by the Webb-Pomerene Act. The Commerce Department will administer the certification process in consultation with antitrust enforcement agencies. In addition, the Commerce Department has the promotional and match-making role in fostering the formation of export trading companies.

Conclusion

In this paper, I have suggested where export opportunities may exist for U.S. wood products, I have outlined some of the problems, and I have described efforts by industry and the USDA to open and expand markets. We are now to the $64,000 question: "Should you be in the export market?" Only you can make that decision. Let us hope that you decide to try exports only after you have recognized that development of sound export markets is not simple and requires a long-term commitment. Let us hope that you know your market, your own capabilities, limitations, and how to address all of the problems in between. If you decide to try the export market, I wish you every success.

ENVIRONMENTAL ISSUES AND
THEIR INFLUENCE ON WORLD TRADE IN WOOD

John Larsen

It is no news that United States industry is becoming increasingly dependent upon its ability to compete efficiently in the world economy. Fifty years ago, America exported relative little; today we export a great deal. U.S. exports as a share of GNP have doubled in the last decade and today about one-fifth of our industrial products are exported. I will leave the in-depth analysis of what this means to others more qualified than I. But in simple terms, it is clear that we are going to have a very hard time being prosperous in these United States if we do not sell our products successfully in international markets.

To do so, we must have competitive products in terms of customer needs, quality, price, and availability. This is particularly true in the wood products industry, where we are not only a minority supplier to the world (countries other than the United States supplied about 80 percent of the world's import requirements for both pulp and lumber last year), but are up against very sophisticated competitors both in terms of technology and marketing.

One determining factor in this international competition is, without a doubt, the political philosophy and economic structure of both those countries that are potential buyers of our products and those countries where our competitors are based. An equally important factor in determining U.S. competitiveness is productivity--the efficiency of our production versus that of other competing nations. I believe a crucial element in determining the overall productive efficiency of the U.S. wood products industry involves the costs and constraints placed on production by the current methods for regulating and enforcing environmental standards. It is on this topic that I would like to focus my attention. Environmental protection and regulation need not, in my judgment, be incompatible with efficient production. I do not believe that we have to sacrifice environmental resources or reduce our environmental values in the United States to complete effectively with other countries of the world.

It is quite clear that the majority of Americans are concerned and committed to protecting our environment. I applaud that concern. At the same time, I would like to suggest that our current approach to environmental protection in this country is not necessarily in the public interest--not when the public interest involves protecting both the environment and our nation's economy.

In theory, one of the purposes of environmental regulation is not only to provide protection for human health and the environment but to increase the efficiency of our economic output as well. The economic rationale supporting the increased efficiency theory goes something like this: environmental standards improve economic efficiency when the recovery and utilization of wastes result in a positive return from investment. Many would argue, however, that in reality these "returns on investment" seldom compete with alternative investment opportunities.

Regardless of that issue, it is obvious that environmental protection procedures are necessary to satisfy the public's desire for protecting air, water, and land resources. I include myself as a member of that public, and I imagine most of you do as well. Having stated that fact, however, does not answer the critical question: What is the level of protection desired by the public, taking into account that there is a very definite cost to them associated with such protection?

It would be nice if we could divorce environmental protection measures from cost--but we cannot. Therefore, after taking into account that certain levels are absolutely necessary for public health and safety, the question becomes one of economic efficiency. How do we satisfy public desires for protection in the most efficient manner?

In principle, most economists would have us seek the level of environmental protection that would be purchased by the public to meet their desires if there were some form of competitive marketplace for them to do so. Today environmental regulatory processes in the United States are not coming even close to meeting such efficiency criteria. This is particularly true for those industries that must compete in world markets--such as our wood products industry.

In this nation's well-intentioned efforts to assure that environmental goals are reached, we have established uniform, nationwide programs and standards that generally ignore economic efficiency--and, therefore, I would submit, the larger public interest. The reasons for this are largely political. Even the best of our legislators and regulators operate in an environment where the survivors are those who have performed best at maintaining and enhancing their political support. Economic efficiency is, therefore, often sacrificed to political expediency.

Because environmental issues are usually cast as "motherhood" issues in order to maximize their political usefulness, the natural legislative response has often tended to result in extreme standards--expressed in such terms as best control technology, zero discharge, lowest achievable emission rate. Regulatory agency strategies have also been politically opportunistic at the expense of economic productivity, focusing on strict, uniform control requirements rather than case-by-case determinations of the most cost-effective means of compliance.

Let me reiterate: I am not questioning the value or the legitimacy of achieving our society's needs for safe products, safe workplaces, or clean air and water. I am questioning the wisdom of the way we are going about it. And I do feel that many of the regulatory programs established during the 1970s have contributed directly to the problems we are having competing in international markets by imposing unnecessary constraints and costs on the production of U.S. goods.

The fact is that many of the nations competing with us have regulatory systems that have far less serious impact on economic efficiency than those of the U.S. While it is true that some of these countries have compromised safety, health, and the environment in the process, it is also true that many others have achieved satisfactory environmental results with far more cost-effective programs. I am not here to defend those who have poor environmental performance. As I have stated, I strongly support the need for adequate protection of the environment. What we must do, however, is improve the efficiency of regulatory programs directed at achieving our environmental goals.

Since the early 1970s, when the Environmental Protection Agency was established, U.S. forest products firms have been required to invest several billion dollars to comply with new standards and procedures. During that same period, the economic performance of our industry has declined in absolute terms. Obviously, if our regulatory programs have had a negative impact on productivity, they have also been a contributing factor to the decline in our competitiveness in world markets.

For a number of years, I have been a member of an international committee which meets routinely to exchange information on approaches to social regulation being taken by each country. As a result of these discussions, I have found that the U.S. has taken a significantly different approach in social regulation than most of the countries which seriously compete with us in international wood product markets--and that difference has resulted in a competitive disadvantage for the United States. This competitive disadvantage in the United States is the result of two major problems in the environmental area:

1. Regulatory standards in the U.S. tend to be more uniform and oriented to the maximum that is technically achievable. Most of the other nations, on the other hand, allow greater flexibility in finding the most cost-effective way to solve specific environmental problems.
2. The U.S. has put into place incredibly complex approval processes for obtaining construction and development permits. The processes are principally based on the adversarial system. They provide a mechanism for opponents of developments to delay decisions for many years--even indefinitely. Most of our foreign competitors, however, can expect final decisions on controversial projects in time frames of between 6 and 24 months--they are able to obtain a definite "yes" or a definite "no."

Let me give you a specific example which illustrates the problems associated with uniform, technically based standards. By July 1, 1977, U.S. pulp and paper mills were required to meet an effluent standard called Best Practicable Treatment. It was to be followed later by another level called Best Conventional Technology. The Best Practicable Treatment level generally required removal of pollutants which create demand for oxygen to a level of between 85 and 95 percent of that obtained by removing only solids.

This treatment is commonly referred to as "secondary treatment." All facilities were required to meet the standard. At many wood products facilities this standard was simply not needed to provide adequate oxygen levels in the receiving waters. In other instances, something less than 85 percent reduction would have satisfactorily met water quality needs.

Capital requirements to meet these standards came at a time when there was a severe downturn in international pulp and paper markets. During the same period, our Scandinavian competitors were also responding to increased public demand for protection of water quality. The regulatory agencies and industry in Norway, Sweden, and Finland, however, approached their respective environmental issues using regulatory processes which stressed collaboration and negotiation rather than confrontation.

In other words, the Scandinavian governmental agencies served as facilitators for working out individual differences between industry and environmental groups. They did not approach this task legalistically; therefore, the issues were not decided by legal criteria, complex court judgments, formal proceedings, appeals, and remands. They were instead settled on the basis of case-by-case judgments as to public environmental needs, economic conditions, and individual company situations.

In the period from 1976 to 1981 the Scandinavian governments exercised much greater flexibility in implementation of their water quality protection programs. Their progress has largely satisfied environmental interest groups, and the economic costs of their programs have been significantly less than those in the U.S.--thereby giving them a competitive edge over their U.S. competitors on this facet of production costs.

Let me give you an example of the problems associated with obtaining permits and approvals for new facilities in the U.S. This example is even closer to home. In 1975, my company, Weyerhaeuser, purchased a properly zoned, existing industrial and shipping site at DuPont, Washington. This followed an extensive internal process of assessing and identifying ways to make us more competitive in ocean transportation of finished forest products. In that same year, we applied for approval to replace an existing dock on this site and to construct a cargo system utilizing new, highly efficient, cargo handling technology. The proposed dock is approximately a mile to the north of a river delta having important environmental values. The projected capital cost of the facility in 1975 was $80 million. It cost in excess of $2 million to do just the environmental studies and to prepare the environmental impact statements.

Eight years have passed. Let me give you a progress report: We have obtained two permits, both of which are currently being appealed through several layers of the courts. We are in the process of an agency review and approval for a tidelands lease, which will probably also end up in court. To date, we have won every court case and every appeal. How long it will be, however, before we can consider getting on with the project or, if we should lose, exercise alternative options for the property, is anybody's guess. In the meantime, world markets for wood products are becoming more competitive.

At about the same time we purchased the DuPont property, the Canadian forest products industry also arrived at some of the same conclusions we did regarding the need to improve the efficiency of ocean transport for wood products. They planned a similar project at Nanaimo, British Columbia. It also had very similar environmental conflicts, and a number of environmental groups strongly opposed the project. The British Columbia government, however, worked with all parties to resolve those problems. In 1977, approval was sought to construct the facility. In early 1979, the project was approved and initiated. In late 1980, they began transporting wood products from the facility.

Meanwhile, back at the ranch, our company is still looking forward to several more years of delay. I don't use this example to try to persuade you we are right and those who oppose us are wrong. The more important question is whether the public is well served by processes that allow proposals like this to be left in limbo for such an extended period of time. I represent a school of thought that believes our current methods of resolving environmental disputes are failing badly. Moreover, I believe there is a growing public recognition that this failure is putting not only environmental quality but also the economic health of this country in jeopardy--particularly in relation to our competitive position in international markets.

In short, the public interest, which clearly involves both the need to maintain environmental quality and a strong economy, is not being well served by our present system of establishing and enforcing environmental policy--a system which usually begins with well-intentioned legislation, but too often ends up hopelessly mired in interminable litigation. Our judicial system is an admirable one, protecting the weak against the strong, presuming innocence, and guaranteeing the protection of the law to all citizens, rich or poor. But it is also founded on the adversary system; it involves a confrontational process. It postulates that, in any litigation, there will be a winner and a loser. This adversarial approach may be well suited to criminal and damage law, but when the parties to a dispute are not individuals seeking justice, but rather are diverse groups pursuing sometimes abstract concepts concerning economic health, energy, and environmental quality, I submit that as a nation, we simply cannot afford to see one party win and the other party lose.

Yet, in the process of trying to resolve our most important environmental disputes through the courts--disputes which involve the quality of life and welfare of thousands of people who may not even be a part of the proceedings--this is exactly what we do. Through the adversarial process we attempt to seek win-lose justice, an almost impossible task under the circumstances for either a judge or a jury. It is my experience that this approach creates bad law, perpetuates unresponsive government, and fails miserably to serve any close approximation of what is economically most efficient. It seems to me that it is time we seriously ask ourselves--as members of a free market economy in competition with other vigorous world economies, and as citizens in a culture striving to maintain and enhance valuable environmental qualities--can we afford to continue this adversarial, open-ended, win-lose system that makes the attempt, but seldom succeeds, at resolving complex environmental issues? I don't think we can. I think we owe it to ourselves--and to the public--to find better ways to reach the environmental goals of our society.

I am not so naive as to believe that we can resolve all issues without falling back on our court system. Our democracy guarantees access to the courts for those who cannot otherwise resolve their problems by less combative means. On the other hand, it seems to me we ought to be attempting to put new procedures into place that are directed at avoiding these legal battles. As a businessman, I suppose I can be accused of overemphasizing the need for efficiency and the timeliness of decisions. Yet the cost of a lack of efficiency and timeliness falls on everybody--taxpayers as well as those of us trying to compete in international markets.

I see the increasingly adversarial nature of American public life as a problem of our society--a problem which, when adopted into regulatory frameworks, is associated with the competitive position of industry in world markets. I am convinced that those of us in industry have a responsibility and an opportunity to alleviate at least a part of this problem by showing a willingness to explore, with interest groups who oppose us on social regulatory issues, new approaches for resolving our differences--differences which, if not resolved, will surely lead to a new generation of legislation, followed by new regulatory processes, followed by new rounds of litigation, resulting in a terrible waste of everybody's time and money.

The idea of using different methods to resolve environmental conflicts--methods which are styled to use more collaborative approaches--would clearly be more in line with the procedures followed by other major countries of the world, our competitors in world trade. If we could demonstrate that better solutions can be found by reasoning together, and that acceptable, more cost-effective agreements can result, then it seems to me that we may well take a major step toward returning civility and rationality to our democratic processes, not to mention enhancing our competitive position in world trade--a task which appears to me to be crucial for our future prosperity.

U.S.-JAPANESE TRADE RELATIONS AND THE PULP AND PAPER INDUSTRY IN JAPAN

Takashi Akutsu

Trade between the United States and Japan, both ways, reached $62 billion in 1981 by U.S. statistics and $64 billion by Japanese statistics. This is the largest overseas trade relationship and one of the most important economic relations in the world. The U.S. is the largest export and import market for Japan, and from the U.S. side, Japan is the second largest export and import market, surpassed only by Canada. Japan is the largest customer of U.S. agricultural products. In 1981, this market reached $6.6 billion, accounting for 15 percent of total U.S. farm products exported. According to Japanese statistics, the size of invisible trade in both directions is also great and reached $28.7 billion in 1981. Our direct investment in the U.S. has increased rather rapidly in recent years and our mutual direct investment amounted to $1.9 billion in 1981. When one takes into account these and other transactions, the total economic activity between our two countries far exceeds $100 billion.

West Coast-Japanese trade has expanded more rapidly than the total U.S.-Japanese trade. According to Department of Commerce statistics, U.S. exports to Japan through the Pacific Coast states--California, Oregon, Washington, Alaska, Hawaii, Montana, and Arizona--were 47 percent of total U.S. exports in 1970, and in 1980 accounted for 58 percent. U.S. imports from Japan through the Pacific Coast similarly expanded from 41 percent of total U.S. imports in 1970 to 49 percent in 1980. Japan's investment in the U.S. also is concentrated on the West Coast. According to the survey made by the Japan Trade Center (JETRO) in 1981, nearly 50 percent of the 238 manufacturing establishments in the U.S. covered by the survey were located in the Pacific Coast states (Alaska, Washington, Oregon, Idaho, Montana, Wyoming, California, Nevada, Utah, Colorado, Arizona, New Mexico, and Hawaii).

Today we have a number of problems between our two countries in economy and trade. Trade imbalance has become quite substantial and we hear a great deal of criticism from our American counterparts. We also hear complaints that Japanese exports are taking thousands of jobs away from Americans. Considering the size of our mutual economic capacities (about 30 percent of the total world GNP) and the extent of trade activities between us, it is no wonder that many problems arise between our two countries. Both countries have a great responsibility to develop the world economy, and from that point of view, we should make utmost efforts to solve our problems in an amicable and constructive manner. Recently, unfortunately, the situation has led to greater protectionist sentiment and campaigning among legislators, industrialists, and labor unions.

Trade balances are determined by a variety of factors, including the competitiveness of industries, industrial structure, exchange rate of currency, and the like. Properly, trade balances should be evaluated not on a country-by-country basis but on a global, multilateral basis. And also when we look at the economic relations between our two countries we have to take into account not only trade balances but also the invisible trade balance. It is widely known that the invisible trade balance is decidedly in favor of the U.S. ($2.2 billion in 1981 by Japanese statistics).

As I have mentioned, many factors are responsible for trade imbalances. First of all, it should be pointed out that, as the global point of view suggests and as the Economic Report of the President states, the main explanation of Japan's surplus in manufacturing trade and in trade with the U.S. is that Japan, with few natural resources, incurs huge deficits in its trade in primary products, especially oil, and with primary producers, especially OPEC. The surpluses in the rest of Japan's trade offset these deficits.

If we then look into the matter of our relations with the U.S., we can find several reasons for trade imbalances. First of all, there is the competitiveness of industries, which in itself reflects various internal and international conditions, such as the differences in productivity and the quality of products.

Second, there is the situation of a too strong dollar as a result of high interest rates in the U.S. Some have blamed Japan for manipulating a cheaper yen, but I understand that recently the U.S. government has admitted that we are not responsible for that. A cheaper yen is not beneficial to Japan. It will raise the price of our imports, and not only those industries directly affected but also many others that heavily depend on imports of raw materials will suffer from it. Also consumers will lose. A cheaper yen accelerates exports, thus aggravating the current trade friction problems. And a cheaper yen means Japan has to pay more in yen for the same amount of imports while it can receive only less foreign exchanges for Japanese exports.

Third, there is the sluggish growth of the Japanese domestic economy. Fourth, export efforts have been more earnestly pursued by the Japanese than by Americans. President Reagan stressed the need for greater export efforts on the part of American business in his speech at San Francisco in early May 1983. Currently, a number of public and private agencies are working hard to take advantage of the opportunities offered under the new Export Trading Company Act. We earnestly hope their efforts will be fruitful. We are also making efforts to help U.S. businessmen find export opportunities in Japan through the business consulting system, under which we, at our JETRO office, receive inquiries concerning their exports to Japan and provide U.S. businessmen with guidance and counsel.

Some point out that foreign products cannot penetrate Japanese markets because of artificial trade barriers. We don't think this is the main reason for our favorable balance of trade. But we are paying more attention to the criticism from abroad and making utmost efforts to solve these problems. On the export side, Japan has voluntarily exercised export restraint in some areas such as the automobile industry.

The trade imbalance and trade friction problems should be solved not by protectionist policy but by the balanced expansion of trade. This is the Japanese policy and this view is shared by the Reagan administration. The main reason the trade imbalance has developed into the current serious friction dispute is directly related to the prolonged world recession that hit not only the U.S. and Europe but also Japan following the two oil crises. Their impact on the economy of the major countries, along with some structural problems, is the underlying factor. We have to achieve global economic growth by the expansion of trade in a balanced manner and we have to revitalize the world economy with continuous dialogue and efforts among the nations. Protectionism is the worst answer to the problems facing us. This view is also explicitly expressed by President Reagan.

Japan has made utmost efforts to open its market, and we think it is now one of the most open in the world. We have accelerated the reduction of tariff rates well ahead of the schedule of the Tokyo round of negotiations, continuing to maintain the Japanese tariff level as the lowest among nations. We are facing severe criticism from abroad that our market is closed especially because of nontariff barriers. As for import restrictions by import quota, we retain some on agricultural products, but import restrictions on industrial goods are kept to only a small number of items.

Recently, many procedural matters, such as inspection and certification, have been criticized as trade barriers. They have been implemented not to create import restrictions but for various other reasons, such as maintaining safety, and we do not think that these procedural matters are the main reason for the current imbalances. But since they are perceived to be inhibitive to foreign exporters, we have initiated several import promotion measures. Two sets of import promotion measures were initiated during the first half of 1982 and Prime Minister Suzuki made a special statement on the opening of the Japanese market. Following Mr. Nakasone's election as prime minister, our government decided to implement additional import promotion measures in the middle of January 1983. These measures include:

- Reduction or elimination of tariff rates on certain products
- Relaxation of import restrictions on certain products
- Improvement of import testing procedures and the establishment and strengthening of the functions of the Office of Trade Ombudsman (OTO)
- Promotion of imports
- Other measures involving export policy, industrial cooperation, and government procurement

We still have some problems in the field of agriculture. Import restrictions on beef and citrus fruits are widely discussed. The real economic effect of liberalization of imports of these items is said to be not so substantial, but the restrictions are often asserted by the American side to symbolize the Japanese trade barriers. Every country has difficult political problems regarding agriculture and we are not immune to them. It is necessary to continue to talk about this matter, and we hope that some compromise can be reached without significant repercussions in both countries.

In a word, Japan's position is that we have made the utmost efforts to open and liberalize our market and we have taken various measures to promote imports. The attitude to welcome foreign products was a policy emphasized by our former Prime Minister Suzuki and has been reaffirmed by Prime Minister Nakasone. At the same time, we are endeavoring to revitalize the world economy and to solve the trade problems by mutual expansion of trade.

I would now like to turn to the subject of the pulp and paper industry in Japan, especially in its relation to the U.S. This industry in Japan has depended heavily on imports of raw material. Our wood chip imports account for about 45 percent of our total consumption, and this figure is up more than 20 percent in the last ten years. Nearly 50 percent of our wood chip imports come from the U.S. The U.S. share in the Japanese wood chip market is, therefore, more than 20 percent. Pulp imports have also increased and account for more than 15 percent of our pulp consumption. We import mostly from North America, with about 30 percent from the U.S., and this U.S. share has increased in recent years.

The international trade of our paper products had a rather small share in Japan, with limited exports and imports in the past. But recently paper imports have increased rather rapidly, and in the first half of 1982 our imports of paper surpassed our exports for the first time. Imports accounted for nearly 4 percent in 1982. Especially, imports from the U.S. have increased rapidly. The market share of U.S. paper products in Japan was 1.9 percent during the period from January to November 1982. It was less than 1 percent three years earlier. Newsprint and kraft linerboard are the two largest items from the U.S.

We decided to further reduce the tariff rate on some paper products including kraft linerboard from April 1983, even though the Japanese paper industry is suffering from the current recession.

The demand for paper is rather sluggish in Japan for several reasons: (1) the current economic recession, (2) change in transportation methods, (3) the users' conservation-oriented attitude, (4) substitution of petrochemical products, and (5) change in GNP structure (the shift of consumer demand from commodities to services, a shift to tertiary industry).

On the other hand, we have a number of problems on the supply side. We have excessive capacity and excessive competition among manufacturers, which have brought about a weaker position for these companies. The sharp rise in the price of energy and raw materials is also depressing the pulp and paper industry.

But the pulp and paper industry in Japan has been attempting to adapt itself to the new changing situation. The report "The Vision of Paper and Pulp Industry in 1980's" proposes: (1) consolidating the industry's production structure by shifting the investment from quantitative expansion to qualitative consolidation, structural transformation by disposing of excessive capacities, and so forth, (2) improving raw material supply by diversifying overseas supply sources, recycling more scrapped paper, and utilizing more such material (Japan's recycling rate is one of the highest in the world), and (3) encouraging technological development for the conservation and rationalization of energy and raw materials consumption.

The paper demand in Japan was forecast to increase 3.5 percent from 1979 to 1990 according to the "Vision" report. Quite recently, in January 1983, the projection was modified downward to 2.6 percent. The main reason for this change was the weaker forecast for GNP growth.

The pulp and paper industry in Japan is now preparing to adjust to the new situation by the Law on Temporary Measures for Structural Adjustment of Specific Industries. This is in accord with the ideas of the Positive Adjustment Policy (PAP) agreed upon by the OECD ministerial conference in 1982. It is hoped that these measures will prove successful.

EXPORT CREDIT INITIATIVES: EXPLAINING CREDIT PROGRAMS

Melvin E. Sims

The Import of Exports

At a time when nearly every budget's bottom line can only be described with phrases like "belt tightening," "budgetary restraint," or "fiscal responsibility," it might come as a surprise to hear that one item will be increasing by 930 percent. No, that's not the annual inflation rate for Argentina. That figure refers to projected United States Department of Agriculture (USDA) expenditures for the market development of forest products from fiscal 1981 to 1984.

Increases for market development, though not of the same magnitude, are being enjoyed by a number of commodities. The reason is simple. Farm exports, including forest products, continue to be a priority for this administration. Farmers depend upon exports for about one-fourth of their income. The public at large benefits, too, as each agricultural commodity exported helps to balance our nonfarm deficit. There are also the jobs made possible by a healthy and growing agricultural system--24 million people are employed today in agriculture and agriculture-related industries. In short, exports are vital to our economic well-being.

Unfortunately, the sources of strength in our agricultural exports have been eroded by a variety of factors: weak economic conditions throughout the world, financial instability in a number of countries, the strong U.S. dollar, losses related to the Soviet embargo, continued East-West tensions, unfair trade practices by some of our competitors, and restrictive market actions by some of our buyers.

In terms of dollars, these factors have caused our agricultural (excluding wood products) exports to drop $4.7 billion to about $39 billion in fiscal year 1982, falling from the year-earlier record of $43.8 billion. If the forecast for this year is correct, the pattern will continue, with exports dropping off to around $36 billion.

The Elements of Recovery

The Reagan administration has stepped up its efforts to increase American exports. On January 11, the President announced a second phase of the blended credit program. This program provides for interest rates below commercial levels to finance U.S. agricultural export sales. The second phase provides $250 million in direct, interest-free export credit under the GSM-5 program, and at least a billion in Commodity Credit Corporation (CCC) export credit guarantees.

A real standout in the second phase has been the $20.5 million credit package announced for Jamaica. For the first time, lumber--a value-added product--was included among those commodities approved for such a sale, amounting to 16.5 million board feet of lumber or about $6.5 million in blended credit for the sales of lumber.

The blended credit program has been a real success so far. The response from importing countries after it was announced last fall was immediate and positive. At that time, the program provided for $100 million in direct credits and $400 million in credit guarantees for use in the current fiscal year. Within a month, the total $500 million had been committed to specific export markets. This cleared the way for the sale by private U.S. exporters of more than two million tons of wheat and significant amounts of corn, vegetable oil, soybean meal, and cotton. The response to the announcement that additional funds would be available was equally positive, and we are processing applications for their use.

Just a week after that announcement, Secretary Block announced, on January 18, that we had struck an innovative wheat flour deal with Egypt. The arrangement calls for the commercial sale and delivery of one million metric tons of U.S. wheat flour to Egypt over the next 12 to 14 months. This single arrangement represents one-sixth of the annual world trade of wheat flour and will generate more than $150 million in U.S. export earnings.

The flour price negotiated with Egypt was at a level that would meet the subsidized competition. USDA will provide, on a competitive bid basis, enough wheat free from CCC stocks to enable U.S. suppliers to contract for sale and delivery to the Egyptian market at the agreed price. Financing of the sales will be through the GSM-102 Export Credit Guarantee Program of the CCC. Financing guarantees of up to $117.7 million will be provided. The arrangement will almost double total U.S. flour exports.

In addition to these programs, USDA is using other financing tools to increase U.S. exports. They can be summarized briefly:

- Public Law 480, dating from 1954, authorizes the programs through which the U.S. has provided more than $30 billion over the years in food aid to help developing nations and to meet emergency situations. Market development is one of the basic purposes of that act, and we are working to make those programs more workable, more realistic, and more effective in meeting export growth objectives.
- P.L. 480 is a food assistance program, but one of its primary aims is to stimulate economic growth in the recipient countries, ultimately to make them commercial customers for food and other products as well.
- P.L. 480 has helped many countries move from aid to trade. Primary among them is Japan, which is now our largest single-country agricultural customer and the most important market for the U.S. lumber industry.
- The GSM-102 credit guarantee program has been enlarged from $2.0 billion to $4.8 billion since the Reagan administration took office. We have sharpened the focus of that program to pinpoint market-building opportunities.
- Finally, existing programs include the effort to deal with the subsidies code of the General Agreement on Tariffs and Trade (GATT). We have maintained strong and consistent pressure on the European Community (EC) for almost two years to effect changes in its export subsidies. One of the reasons for the general lack of progress on that front is the weakness of the GATT subsidies code. We will continue to push for needed changes, both in the code and in EC policies. The flour sale to Egypt was a clear signal that the United States is not going to stand by and lose its markets to subsidized competition.

The Credit Element Closer Up

As you can see from this brief overview of our current methodology for expanding markets, credit programs have become crucial elements.

The rationale for U.S. credit programs--both concessional and commercial--has remained pretty much the same over the past 28 years. These programs were created to develop U.S. export markets, to combat hunger, and to foster economic development abroad.

As overseas markets have changed their needs for such things as short-term commercial credit or additional commodities, U.S. credit programs have been refined and tailored to reflect new international situations.

The potential for future growth in U.S. agricultural export sales depends to a large extent on the export credit tools available. Authorization exists for a broad range of U.S. programs, but only Public Law 480, GSM-5, and the GSM-102 Export Credit Guarantee Program of the CCC are covered under current government spending authority.

When P.L. 480 was introduced in the mid-1950s, major recipients were in Western Europe and Japan, recovering from World War II. By the 1960s, the focus of P.L. 480 shifted to developing countries, many of which were just gaining their independence.

The work of P.L. 480 in the 1970s continued the development work begun in the 1960s. Today in the 1980s, the goals of earlier years have become more crucial than ever.

Many developing countries still need the financial boost they get from being able to import on a long-term, concessional basis and to develop their own agricultural infrastructures.

Along with the development of concessional financing, there arose the need for credit programs to help former concessional recipients make the transition to commercial purchases. That need was answered by CCC's credit programs, for 25 years in the form of direct credits, and today as credit guarantees.

Credit guarantees have some important advantages. For one, they introduce countries to the use of commercial credit channels in importing agricultural products. More importantly, the system of credit guarantees introduced in 1979 has substantially reduced direct government intervention in export financing. The Reagan administration believes this is a more productive and realistic approach to credit programs.

The GSM-102 Export Credit Guarantee Program offers importing countries credit from U.S. banks at commercial interest rates. This credit, which is guaranteed by the CCC, is used to buy U.S. agricultural commodities or processed derivatives therefrom.

The GSM-102 program is designed to expand agricultural exports by stimulating U.S. bank financing of foreign purchases on credit terms of up to three years. In every transaction the foreign buyer's bank must issue an irrevocable letter of credit covering the port value (f.o.b. value) of the commodity exported. CCC's guarantee will cover most of the amount owed to the U.S. bank in case the foreign bank defaults.

The program operates in cases where credit is necessary to increase or maintain U.S. exports to a foreign market and where private financial institutions would be unwilling to provide financing without CCC's guarantee. That organization does not wish to provide guarantees where commodities would be purchased for cash in the absence of CCC's program.

Generally, the program is aimed toward countries where the guarantees are necessary to secure financing of the exports and where the destination country has the financial strength to provide a reasonable expectation that foreign exchange will be available to make payments as scheduled. Although in the past the GSM-102 export program has been geared mainly to bulk agricultural commodities, it is now placing increasing emphasis on higher value specialty crops and processed products. Any agricultural commodity or processed derivative which looks as if it will further USDA's long-range market development objectives will be considered.

As you know, forest products are relatively new to the program. It was not until fiscal 1981 that the Foreign Agricultural Service made available $4 million in credit guarantees under GSM-102 for exports of softwood lumber and plywood to Jamaica. In 1982, the amount was increased to $6.4 million. Thus far in the current fiscal year, we have approved $8.3 million. More recent announcements bring this amount to just under $30 million.

That's not bad for the new kid on the block. Forest products are drawing more attention from this administration because we are concerned not only with increasing exports in general but also with building our trade in the value-added products.

Thus far, U.S. credit programs have been very successful. The transformation of former concessional markets to full commercial trading partners proves well that aid does lead to trade. The list of countries that have made the switch, including Japan, Korea, and Taiwan, continues to grow. The budget for direct credit and concessional financing has dwindled substantially in recent years. And credit guarantees have emerged as a very important market development instrument.

It is easy to see why credit is such a valuable tool as we try to get our agricultural exports moving forward once again. Right now it is an uphill battle because we are competing during very difficult economic times. The ground we gain now will be tough going but well worth the effort. Demand really is just around the corner.

Where you are concerned, I do not think it is an idle pipe dream to project exports of solid-wood products in excess of $7 billion by the end of the decade. Of course, it would mean more than doubling the $2.9 billion figure for fiscal 1982. It may well be possible, particularly with the use of such effective tools as the GSM-102 program.

For USDA and this administration, exports are of the utmost importance. We encourage and support the unflagging efforts of industry to build America's share of the export market. And we will continue to do so. To a large extent, the fate of our farmers and foresters, as well as the general economy, depends upon our success.

EXPORT PROMOTION, IMPORT RESTRICTION, AND
ANTITRUST CONSIDERATIONS IN U.S. TRADE WITH JAPAN

John O. Haley

As the other topics of this symposium illustrate, most discussions of international trade policies tend to focus on economic and business considerations or direct governmental efforts either to promote exports or to restrict imports. Similarly, during the past few years the dispute over the United States-Japanese trade imbalance has tended to emphasize direct government involvement on both sides of the Pacific to reduce tariffs, quotas, and other legal restrictions on the export to Japan of products from the United States and corresponding arrangements to restrict importation of Japanese products to the United States, at least in such quantities that they adversely affect American producers and cost jobs. Having been the principal postwar advocate of free trade policies, the United States is torn by the current economic and political pull toward various means to protect American industry and American jobs from the effects of international competition, at least where, as in the Japanese case, it is perceived to be unfair.

Until recently, Japanese firms selling or buying in the American market have had the best of both worlds. They have been subject to protectionist, producer-welfare policies in Japan and yet free from the antitrust constraints imposed on their American competitors. At the same time, their American competitors were correspondingly disadvantaged in both markets. Like the United States and the European industrial nations, Japan has made use of an array of measures both to promote exports and to restrict competitive imports. These measures have included tariff barriers, quotas, and a variety of direct and indirect subsidies. What distinguishes Japan from other industrial nations, however, has been its emphasis on private arrangements by the industries concerned to advance in tandem into foreign markets and to exclude any serious competitive threat from abroad. For most of the postwar period, the promotion of cartel and cartel-like arrangements has been fundamental to Japanese international trade policy. We have our

Webb-Pomerene Act, and Europe--or at least West Germany--its antitrust exemption for export cartels. What is unique about Japan, therefore, is not its use of export and import cartels, but rather the centrality of their role.

Time does not permit me to explore here in any detail the full variety of cartel practices in Japan, their historical background--especially the pattern of government regulation adopted between the late 1930s and the war--or indeed to assess the extent to which procartel policies in Japan ultimately have had negative, long-term consequences for the Japanese economy. Perhaps we can deal with some of these questions during discussion. What I would like to do in this presentation is to describe the legal framework for foreign trade restrictions and the variety of patterns that characterize the Japanese approach, from direct legal constraints to purely private arrangements. Time permitting, I shall then attempt to outline the antitrust considerations under U.S. law as well as Japanese law, with brief mention of the most recent developments in U.S. law.

The Foreign Exchange and Foreign Trade Control Law is Japan's basic regulatory statute governing foreign trade. It grants the Japanese government--that is, the Ministry of International Trade and Industry (MITI)--the power to designate particular products whose export and import is restricted through licensing. Under the law and related administrative regulations, both prices and volume can be set. In 1964, at the height of Japanese protectionism, 211 items were formally restricted under the law, many of which (national treasures, narcotic drugs, explosives, military equipment, items on the CoCom list) were controlled for reasons that had nothing to do with commercial trade policy. Products with obvious competitive trade implications included automobiles, machine tools, ball bearings, and textiles.

The second statute is the Export and Import Transactions Law enacted 1952. A counterpart to our Webb-Pomerene Act and the export and import cartel exemptions of the German anticartel law on which it was modeled, the Export and Import Transactions Law is essentially permissive. It allows private manufacturers and exporters to enter into agreements or establish associations to restrict the quantity, price, quality, and other matters related to exports or imports. Such agreements must be filed with the Japanese Fair Trade Commission (FTC) and, until 1955, had to be approved by it. Since 1955, approval has been a MITI prerogative, a very significant change in that the FTC is Japan's exclusive antitrust enforcement agency, and MITI has been the primary institutional advocate of Japanese procartel policies. No approval is permitted under the statue, however, if--among other factors--the agreement is unduly restrictive on any party's right to enter or withdraw. Japanese export and import cartels are thus, as a matter of law, voluntary. There is provision in the Export and Import Transactions Law for direct governmental regulation, under which MITI can issue an order requiring parties to abide by industry-agreed-upon restrictions if the purposes of the cartel cannot be attained without such order. This is referred to as an "outsider regulation" but has been used only rarely to enforce export and import cartel agreements. Since 1955, over 130 products, including lumber products and particularly plywood, have been subject to export cartels directed at the U.S. market.

Cartel-like restrictions are also established under a variety of other statutes. For example, it is possible under the Medium and Small Enterprise Promotion Law, to engage in joint buying and selling arrangements, which obviously may restrict both imports and exports. There are also extralegal private arrangements outside of any statute, many of which involve informal governmental approval and supervision. Where Japanese officials believe import or export restrictions desirable, they may suggest or "guide" the parties to enter such an agreement, often at the instigation of some firms in an industry in order to assure effective agreement by the industry as a whole. Because of the ease with which export and import cartels have been formed under the Export and Import Transactions Law or other statutes to promote particular industries, usually such guidance has been coupled with a formal, legally authorized cartel arrangement, although cartels in these cases are reported to the Fair Trade Commission and are made public. Where publicity is considered undesirable for political or other reasons, a purely private arrangement with governmental guidance has been the common pattern.

Finally, there are purely private arrangements, which the parties hide from the government, including MITI or whatever ministry has jurisdiction over that particular industry. To the extent that agreements not formally exempted by statute from Japanese antitrust regulation involve a substantial restriction of competition in the Japanese market, they risk being challenged under the Japanese Antimonopoly and Fair Trade Law. This statute was a product of the American occupation, but has survived despite repeated attempts to weaken its substantive provisions as well as its enforcement. In formal terms, the statute is quite severe. Its scope and its sanctions are considerably more extensive than German law or the Treaty of Rome. In fact, believe it or not, next to our own, Japan has the strongest antitrust legislation in the world. From 1953 through the mid-1960s, however, its enforcement was weakened by personnel and budget cutbacks and the appointment of commissioners to the FTC who assured less than active enforcement. But as indicated by recent Japanese Supreme Court and Tokyo High Court decisions involving output restrictions and price-fixing in the oil industry successfully prosecuted as criminal violations of the Japanese antitrust law, anticompetitive arrangements that violate Japan's antitrust law are not exempt or otherwise permissible because of informal governmental acquiescence. Administrative guidance that promotes violation of the law, and the restrictive conduct itself, are illegal.

Some may ask how export cartels have injured American producers. After all, it can be argued, they have tended to fix minimum prices and thus, if effective, have kept the participating firms from dumping or at least selling at even lower prices. Indeed, the usual public justification for export cartels by Japanese officials has been that they are used to prevent dumping. Moreover, legally voluntary export cartels have been the principal device used by Japan to satisfy U.S. demands for export restraint. Some ask, therefore, what moral or legal standing producers have for complaining. In response, I would note two common features of Japanese export cartels.

First, they have reflected or permitted more extensive collusion than simply in connection with exports and imports. In almost every case I have seen in detail, such cartel arrangements have been tied to collusive attempts to stabilize prices in the domestic Japanese market. In conjunction with other barriers to imports they have thus permitted if not encouraged firms to keep prices in Japan above international market levels and thus supported (or even necessitated because of oversupply) expansion of overseas sales below price levels that would have prevailed otherwise.

Second, few cartel arrangements have involved prices alone. Most also involve customer allocations and other restrictions that limit competition among Japanese sellers in the American market. Such arrangements permit deeper discounts and price cuts to selective customers or for particular product lines where the Japanese are not as competitive as U.S. producers while maintaining higher prices for products or with customers for whom Japanese producers have had a competitive edge. Not having to compete among themselves, they have been able to concentrate their efforts to outbid their American competitors.

Of the injury caused American producers by Japanese import cartels, little need be said. Concerted buying arrangements have enabled Japanese traders to obtain the lowest possible price for American products and then to maintain artificially high prices at home or, in some cases, to limit purchases from abroad to ensure that the products they or related firms produce would not be subject to significant competitive pressures. This pattern has been especially prevalent, I believe, in sales through the general trading companies and cooperatives.

Even where the injury has been slight or difficult to prove, there is still the issue of fairness. Simply put, it has been unfair for Japanese firms to be able to collude in setting prices and restricting output or supply, while their American competitors have been subject to antitrust liability for much less overt or potentially harmful conduct. That such unfairness has had negative consequences is evident in the consequently intensified demand in the United States for protection, with all of its harms to international trade and the consumers.

Our responses to Japanese trading practices have not, however, adequately dealt with the problem. During the late 1950s through the early 1970s, with few exceptions neither Republican nor Democratic administrations displayed much concern. Where, as in textiles, the political influence of a particular industry was sufficiently strong to demand some response, the pattern was to press the Japanese to limit exports. Our efforts on the import side were centered on the threshold barriers of formal restrictions--both tariffs and quotas and demand for liberalization of equity investment restrictions. The cartel problem received scant attention. Yet once the formal barriers were removed, we found that the problem did not go away. We began to see Japan as an onion with new layers of restrictions each time an outer barrier was removed.

A complicating factor has been ignorance and misconception. Books like Vogel's Japan As Number One and the profusion of literature on Japanese management created or reinforced an impression of Japan as a well-run trading machine: efficient industry, efficient government, and industrial policies carried out by enlightened bureaucrats. If there was or is a problem, it seemed to be ours. The result has been predictable. On this side of the Pacific we hear first the increasingly strident calls for protection: If the Japanese are more efficient, we can only save ourselves and our jobs by erecting barriers. Second, there are the voices of those who would have us emulate what they believe to be the policies that have led to Japan's success. A couple of years ago, for example, the U.S. Federal Trade Commission sponsored a seminar at which one of the principal themes was an end to rigorous antitrust enforcement and a procartel policy like that of Japan.

The Japanese, hearing such views, react in predictable fashion. If they are successful because they are more efficient, then, according to all the American economic doctrine they have imbibed since the end of the war, they <u>deserve</u> access to our markets and we are wrong to deny them the benefits of the free trade we have preached for so long. Moreover, what right do we have to complain if we are simply protecting our inefficient industries or are otherwise copying what they have done?

The prevailing views about Japan are terribly flawed, however, as many if not most of you know. To debunk a few of the major myths about Japan that plague us:

1. As to efficiency, except for a few large-scale manufacturers—especially the automobile and the electronics industries—the Japanese economy is remarkably inefficient. Publicly run institutions from the national railroads to universities, the distribution system, and primary industry are far less efficient than in the United States.

2. Governmental industrial policy has not been a principal cause for Japanese trading success. There can be little doubt that protection of the automobile and electronics industry in the 1950s helped these industries gain valuable time to become internationally competitive. But there are few, if any examples, of governmental success in targeting an industry or a firm within an industry for subsidies, protection, or other forms of favorable treatment, except cases which were obvious to any reasonably knowledgeable observer. Had MITI officials had their way, in fact, there would not have been a Sony or a Honda.

3. Nor have restrictive practices, planned cartels, concentration, or the lack of antitrust enforcement over the short or long run contributed substantially to Japan's success. The fact is that the most successful Japanese industries are the most efficient and tellingly the most competitive domestically. There were only four motor vehicle manufacturers in Japan in 1945, but today there are eleven. In electronics the largest firm, Matsushita, is a postwar entrant

along with Sony. The largest retailer is Daiei--established in 1958. In the short run the ability of Japanese firms to cartelize may have promoted, as some very knowledgeable economists say, risk-taking as a safety net in that in the event of overexpansion a cartel would be allowed. But the industries where barriers to entry and price uniformity have prevailed over any length of time are, as elementary neoclassical economics teaches, among the least efficient and least able to meet competition from abroad. Japanese industry has been the least successful in these areas where governmental efforts to restrain competition have been the most successful.

4. Competition, not collusion or concentration, therefore, is the key to Japanese success. Competition for technology and a greater share of domestic and foreign markets has let to innovation and efficiency.

In Japan today, the call for competitive free trade is hardly uniform. As all of you know, agriculture, forestry, and fisheries are among the least competitive and therefore most protected areas of the Japanese economy. An approach to the Japanese trade problem that seeks to protect what we perceive to be less competitive American industries and adopt policies that permit collusion or governmental "guidance" with the same effect thus confirms the Japanese approach and provides justification for maintaining the existing restrictions.

As an alternative, I would like to suggest what may seem to be a rather radical approach: more vigorous international antitrust enforcement, particularly by government agencies, as to both export and import cartels. To date the federal government has not challenged a single export cartel directed at the U.S. market. And only recently has a federal enforcement agency been willing to act against an import cartel--in the seafood industry. As a result, private damage actions have been the only mechanism for enforcement. But private actions, while often effective, are difficult and far too costly to maintain.

In the one suit in which an export cartel has been challenged--the Zenith case--after millions of dollars were spent over a period of a decade, the trial court dismissed the action before trial, holding--among other points--that the manufacturer-plaintiffs did not have standing to bring an action for a price-fixing cartel. The decision on appeal is expected at any time now. What is desired is not a greater number of litigated cases--that would benefit only the lawyers--but rather a strong proenforcement stance by the FTC and the Justice Department.

Would greater international antitrust enforcement really be a solution or even beneficial? I believe so. To be sure, antitrust enforcement will not aid uncompetitive industries. Less efficient firms will not be able to maintain their markets against more efficient Japanese competitors. Nor should they. What greater antitrust enforcement would achieve is to ensure that there is competition and that it is fair. Since the late 1960s and early 1970s the number of authorized Japanese export and import cartels has decreased by near 30 percent overall, with even greater reduction in export to the U.S. This, I believe, was in large part because of publicity involving the Zenith case. There have been an increasing number of private antitrust actions challenging other practices. From my discussion with Japanese lawyers and in-house legal advisers, I am convinced that U.S. antitrust actions have had a major influence on Japanese business practices--far more than recent trade negotiations. Antitrust enforcement induces a private response rather than placing the Japanese government in the dilemma of undermining its sources of political support. Truly competitive industries can be expected to cease their collusion. Those that are not would be forced to rely on formal government measures for protection; but these are often difficult for the Japanese government to employ. They force unwelcome political choices--between injury to its efficient industrial producers or injury to the less competitive sectors of the economy. Formal restrictions, also, give us something to negotiate against. Informal cartel arrangements we more easily overlook than formal quotas and tariffs.

Nor need greater antitrust enforcement be limited to actions in the United States. American firms have too long ignored their remedies under Japanese law. The more complaints that are filed with the Japanese Fair Trade Commission, the stronger its hand in dealing with the problem of collusion in the domestic market.

Antitrust enforcement both here and in Japan provides no panacea, but it is surely a better approach than the protectionist warfare in which we are poised to engage.

FEDERAL TRADE LAWS AND THEIR IMPACT
ON INTERNATIONAL TRADE IN WOOD

James G. Yoho

Introduction

This paper focuses on recently enacted and pending federal legislation dealing with international trade, and is limited to that legislation which has now, or promises to have, significant impact on forest products. It is also limited by the thrust of this symposium to legislation of significance to exporters.

Before going into detail regarding recent or pending trade legislation, I would like to exercise my academic prerogative and take time to set a brief historical background to help us appreciate "where we are today" in respect to international trade legislation.

Down through World War II most of our trade laws were aimed at controlling the flow of imports, primarily through the use of tariff duties. Like most industrializing and industrialized nations, we were interested primarily in two things:
1. Protecting our processing industries from foreign competition so they could have an opportunity to grow and develop.
2. Protecting American workers from the competition of cheap foreign labor.

In order to achieve these objectives we embraced a system of variable rate tariff duties on manufactured goods, including many forest products, while raw materials were generally permitted to enter duty-free.

As to exports from the United States, there were virtually no restrictions. Similarly, there was little or no help for exporters. They were essentially left to shift for themselves. There were, of course, some exceptions to this, as there are to most generalizations. For example, we did give a little encouragement to exporters through the Webb-Pomerene Act, which dates back to 1918. This act, which might be viewed as the predecessor of the Export Trading Company Act of 1982, was intended to strengthen the hand of U.S. exporters by legally facilitating cooperation among them in exporting American goods. Unfortunately, the act gave exporters little or no assurance that they would be free of antitrust prosecution if they elected to cooperate under its aegis, and as a consequence the act was little used--particularly in the post-World War II period, when business felt continually threatened by government antitrust activity.

Another, and probably the major, exception to the above generalization was the maritime legislation, various forms of which date back many years. For example, there was the Jones Act, with which you are all too familiar. The main thrust of most of our maritime legislation was to regulate, protect, or subsidize.

After World War II, when we emerged as the dominant economic force among nations, we became champions of the worldwide free trade movement. If we did experience a little discomfort here and there in our economy from the importation of some items, the matter was dismissed as being cheaper than foreign aid which, in those days, was being doled out in large doses, to the developed world even more than to the developing world. As the leader of the free trade movement we encouraged the formation of the European Economic Community (EEC) as a means toward the economic recovery of Western Europe. We also promoted the General Agreement on Tariffs and Trade (GATT) and participated generously in its subsequent Kennedy round of tariff reduction negotiations in the 1960s, and in the further reductions of the Tokyo round of the 1970s. Most of the post-World War II trade legislation, down through the Trade Act of 1974, was in accordance with this political philosophy of free trade.

Incidentally, it was Section 124 of the Trade Act of 1974 which authorized the president to continue bilateral negotiations with individual signers of GATT. This authority expired in January of 1982 and, unfortunately, legislative action to extend such authority apparently died with the 97th Congress (H.R. 4761 and S. 1902). With it we lost the likelihood of any near-term solution to several issues of interest to the forest products industry on which talks had been under way for some time. These included Canadian tariffs on U.S. molding and millwork plus the quota which limits EEC duty-free softwood plywood imports from the U.S. to 600,000 cubic meters per year, subject to year-to-year _ad_ _hoc_ exceptions. Fortunately, extension of Section 124 authorization was not needed to permit the continuation of negotiations with Japan regarding that country's tariffs on the importation of U.S. softwood lumber and plywood.

Another turning point in our political philosophy, which is the basis for today's international trade legislation, took place after the Arab oil embargo of 1973 and 1974 and the subsequent worsening of our already chronic trade deficits. This gave rise to an aggressive trade policy which is reflected in the recent and pending legislation which will be developed here in some detail. At the same time there was a growing realization on the part of American industry, and particularly the forest products industry, that it must give serious attention to world markets as well as domestic markets if it wished to survive and thrive.

The legislative effort which followed our turn toward an aggressive trade policy is characterized by the removal of discouragements and outright obstacles to exporters, the establishment of new encouragements for exporting plus governmental assistance to industry in gaining access to foreign markets and competitive equality with foreign exporters. But there are many, including Professor Frederick Amling, an international trade expert at George Washington University, who think that our government is not doing enough to help American exporters gain equality in international markets. He was quoted recently in the Wall Street Journal as accusing Washington of a failure to recognize the realities of international trade; in Amling's words, "it's economic warfare and we aren't tuning up for it. We are a little bit naive."

Now, as a nation, we appear to have made another partial turn in our attitude toward international trade without having abandoned our old course. I refer to the growing protectionist attitude which has arisen over the past couple of years, and which is reflected in a number of proposed pieces of legislation, none of which has yet been passed by Congress. But perhaps it is wrong to characterize this latest attitudinal development toward international trade as protectionist--a more descriptive term might be "retaliatory." I say this, because the new attitude, which many have embraced in both government and industry, appears to stem as much from an impatience with our lack of progress in gaining equality in foreign markets as it does from any desire to shield us from unfair competition.

Recent Legislation

Federal Reorganization of Trade Responsibilities

A pacesetter among recent international trade legislation grew out of a long debate in the Congress and within the Executive Branch. The result was the 1980 reorganization of the federal government's international trade effort. The action taken represented a compromise between those who wanted to see the establishment of a full-fledged Department of International Trade and those who wanted to see the effort handled as an expanded trade function within the Department of Commerce. The compromise resulted in the establishment of the Office of the U.S. Trade Representative (USTR), headed by a person having the rank of a diplomat, within the Office of the President. The Office of the Trade Representative was given responsibility for international trade policy and negotiating authority, while administrative functions, on the other hand, were assigned to the Department of Commerce. There were large increases in staff to handle these new functions and responsibilities.

This legislative action and the resulting administrative structure are important milestones because they represent a dedication on the part of the government to a new aggressive foreign trade policy. Moreover, since its founding, the USTR has actively pursued negotiations on behalf of the forest products industry's exporting interests. And likewise, Commerce has been helpful to our industry to the full extent of its obligations.

FAS Responsible for Forest Products Export

The responsibility in the federal government for the promotion of the export of U.S. forest products, however, rests with the Foreign Agricultural Service (FAS) in the Department of Agriculture. The Agricultural Trade Expansion Act of 1978 was the first legislative step in the assignment of this duty to FAS; and the responsibility was fully mandated by the Agricultural Act of 1981.

Long active and extremely successful as a promoter of agricultural exports through cooperative programs with industry, FAS established a full-fledged forest products staff about a year ago, and by August of 1982 it had reached an accord with Commerce to avoid duplication of effort in the forest products area. Combined funding for the cooperative program which was established between FAS and

the forest products industry has grown from a modest few
thousand dollars in fiscal 1980 to about two million dollars
for fiscal 1983. The program is handled through the
National Forest Products Association (NFPA), with most of
the funds going to NFPA's member associations for specific
action programs. Funding, which comes partially from
government and partially from industry, is expected to
continue to grow in the years ahead.

Foreign Corrupt Practices Act

The Foreign Corrupt Practices Act (FCPA) of 1977 is a
good example of a piece of legislation which has had a
negative impact upon America's export activities. This act
was hastily passed with the noblest intentions as an
aftermath of the Watergate and Lockheed bribery scandals.
Unfortunately, during consideration of this legislation, the
business community was reluctant to speak up in a
constructive way lest such action be interpreted as a
defense of corruption. Now the act's lack of clarity on the
antibribery issue and its burdensome accounting requirements
serve to discourage many would-be exporters. Its
antibribery regulations in particular have steered many
forest products exporters away from the lucrative Middle
East market.

DISCs Establish Competitive Tax Equality

Domestic International Sales Corporations (DISCs) were
authorized in 1972 after long and careful study by the U.S.
Treasury. This legislation actually predates our turn
toward an aggressive trade policy, but it has had a very
positive impact on our ability to export competitively. In
addition, exploring it here serves to illustrate many of the
tax problems peculiar to American exporters.

DISCs were intended to give U.S. exporters competitive
equality with foreign exporters through a device consistent
with out extraterritorial tax concept. In contrast with our
approach, many countries permit and encourage their firms to
set up overseas subsidiaries in tax-haven countries, for
handling exports. Usually under these circumstances, the
home country of the parent company does not tax the
repatriated earnings of such subsidiaries and does not
bother to police the intercompany pricing of goods
transferred to the subsidiary.

Most of the major exporting countries place great reliance on the value-added tax. In contrast, we depend heavily on corporate income and excise taxes. According to GATT, exporting countries may rebate direct taxes but not indirect taxes on export items. And unfortunately for us, the value-added tax is considered direct whereas income and excise taxes are judged to be indirect.

The 1972 legislation permitted exporting companies to set up paper subsidiaries in the U.S. known as DISCs to handle their exports. Among other qualifying prerequisites, 95 percent of a DISC's gross receipts must come from exports, and the goods destined for export must be transferred into the DISC at arm's length value. Initially, the tax benefits derivable from using the DISC arrangement were quite substantial, but in the government's quest for revenues the advantages have been severely watered down. Under the reduced benefits of the Tax Reform Act of 1977, some 75 percent of a DISC's profits were taxable, either to the parent or to the DISC, while taxes on the remaining 25 percent could be deferred indefinitely. The Tax Equality and Fiscal Responsibility Act of 1982 further reduced these tax deferral benefits.

DISCs have indeed been an important stimulus for U.S. exporters, as companies of all sizes have moved to take advantage of the possibilities. As early as 1976, some 75 percent of the U.S. exports were being funneled through DISCs; and tax returns for 1980 indicate that there were approximately 8,000 DISCs in use that year, doubtless many of them by forest products exporters. On the negative side, Treasury has estimated that DISCs cost our government about $1.4 billion annually in lost revenues, but they also admit that DISCs probably led to an increase in annual exports ranging from $6.2 billion to $9.4 billion.

Double Taxation of Americans Working Abroad

For many years the double tax burden on the earned income of American nationals working abroad constituted a serious deterrent to firms interested in staffing overseas offices with U.S. citizens. Prior to the current tax year, the earned income of such persons could be subject to full taxation by both the United States and the host country--our government's unique tax position in this instance again being based on its extraterritorial concept of authority over our individual and corporate citizens. This taxation policy has resulted in the loss of many overseas contracts by U.S. engineering and construction firms that could not afford to locate large numbers of Americans overseas. This in turn has resulted in lost markets for large quantities of American goods--including wood products, which might well have been specified on many projects handled by American firms.

The Economic Recovery Act of 1981 alleviated the most serious aspect of this double tax burden by permitting exclusion from U.S. taxes of the first $75,000 of income earned abroad, and also by allowing a nontaxable living allowance. The exclusion will increase in an annual step-wise progression from $75,000 for the tax year 1982 to a maximum of $95,000 in 1996.

Federal Control of Exports

The Export Administration Act (EAA) of 1979 provides for government regulation of U.S. exports. It authorizes export controls, primarily for national security and foreign policy purposes, and also provides the means for meeting antiboycott requirements as well as permitting export restrictions on goods in short supply. This is the legislative authority under which the administration recently moved to restrict grain exports to the Soviet Union and, more recently, acting through our extraterritorial concept of authority, attempted enforcement of sanctions on technology transfer and equipment sales from U.S. subsidiaries abroad for the big European pipeline project.

Export Administration Act authority has been applied mainly to high technology items and licensing, and has not been applied administratively to the export of forest products. At the time of the passage of the act, however, an amendment was added restricting western redcedar log exports from state-owned lands. At the same time, there was a threat to restrict all raw material exports because of congressional concern regarding the effect American scrap iron exports were having on domestic supply and prices.

Export Trading Company Act

The final piece of recent international trade legislation which I would like to mention briefly is the Export Trading Company Act of 1982. I say briefly, because later in the program substantial presentations are scheduled on the subject. Such a session is certainly well deserved because this legislation may, indeed, turn out to be very beneficial and useful to American forest products exporters.

It should suffice here to point out that the ETC Act was passed and signed into law last October 8 after what must be described as a long and tortuous journey through Congress, particularly considering the urgency of doing something to encourage U.S. exports plus the fact that the bill had broad bipartisan support. In my judgment, Congress had four basic objectives in mind in passing this legislation:

1. To legalize an American business structure for the international arena comparable to the highly successful Japanese trading companies.
2. To give small- and medium-size businesses the advantage of banding together to achieve critical economies of scale.
3. To dampen the long-standing threat of antitrust action against American business firms cooperating in the sale of U.S. products abroad.
4. To create more opportunities for financing U.S. exports by enabling greater involvement by banks.

After Commerce and the Federal Reserve Board have finalized their regulations pertaining to the formation and operation of an ETC, business will have a better idea regarding the attractiveness of this new device and how to proceed specifically to take advantage of it. It is going to require a much longer time, however, and a great deal more experience, before we will be able to evaluate the opportunities this legislation holds for our fragmented forest products industry.

Pending Legislation

A great many bills have already been introduced into the 98th Congress which bear, either directly or indirectly, on foreign trade. Not all of this legislation appears likely significantly to affect the Forest Products Industry, but in some cases the hidden opportunities or indirect impacts could be greater than anticipated. Accordingly, it behooves forest products exporters to monitor all pending international trade legislation.

This task is simplified by the fact that most of the bills which have been introduced this session first saw the light of day in the 97th Congress, or have been seriously proposed before. Likewise, many of the bills are simply aimed at revising, renewing, or replacing existing legislation. Thus, on the surface, it would seem that few if any surprise sleepers in the form of new bills are likely. There is always the danger, however, that an export-disruptive amendment, particularly for a given industry, will creep into the picture. Moreover, there is always the possibility of an international backlash reaction to a piece of trade legislation of little direct concern to forest products exporters, but which could affect them seriously.

Federal Reorganization of Trade Activity

The issue of federal organization for handling international trade matters has surfaced again in the 98th Congress. Senator Daniel Patrick Moynihan (S. 21) and Senator William V. Roth (S. 121) have introduced bills which would lead to a consolidation of the federal government's international trade efforts by establishing a full-fledged Department of Trade and Commerce. Roth's bill would create a new cabinet-level department by consolidating USTR with the International Trade Administration (ITA), now a part of Commerce. The new department would also include the Export-Import Bank (Eximbank) and the Overseas Private Investment Corporation (OPIC). Moynihan's bill would also consolidate all of the trade functions into a single department, but his bill would include all of the present Department of Commerce.

If enacted, reorganization legislation might lead to a more efficient and effective handling of international trade matters. Depending on the shape of the new organization that emerges, however, it is conceivable that there could be adverse impacts upon the interests of the forest products industry. At this juncture the industry needs a strong negotiating authority to help it gain better access to foreign markets, particularly the Japanese market, and to a lesser extent the EEC. To date, however, all of the proposals put forth which call for a merger of USTR activity could lead to a weakening of our negotiating clout. Likewise, it is possible--though not probable--that FAS might be transferred to the new department, thereby somewhat impairing the effectiveness of this helpful agency. In turn, this could lead to a weakening of the forest products industry's highly successful cooperative program with Agriculture.

Foreign Corrupt Practices Clarification and Simplification

A bill entitled The Business Accounting and Foreign Trade Simplification Act (S. 414) was introduced recently by Senator John Heinz of Pennsylvania to clarify certain bothersome provisions of the Foreign Corrupt Practices Act (FCPA) of 1977. This is essentially the same bill as the Foreign Corrupt Practices Clarification Act (S. 708) which died in the 97th Congress despite strong sponsorship by Senator Chafee and passage by the Senate.

The new bill would alleviate considerably the burdens placed on U.S. international trade activities by FCPA on a number of counts:

1. It changes the ambiguous "reason to know" standard for determining the liability of U.S. concerns for third party actions to a bright-line standard of direction or authorization of an illegal payment.
2. It reiterates Congress's original recognition that "facilitating" payments to foreign officials may be necessary in order to do business in certain countries and defines the types of payments which would be permissible.
3. It simplifies the accounting standards requirements of FCPA by requiring only a single system of internal accounting controls sufficient to provide "reasonable assurances" that they would achieve certain desired objectives, such as management accountability and record accuracy. It also seeks to diminish the confusion about the parent company's liability for the accounting practices of its foreign-based subsidiaries.
4. It eliminates the concern of business over the multitude of statutes which conceivably could be used to prosecute companies for the corrupt payment violations described in the FCPA. The new act would be the exclusive provision for prosecuting violations.
5. It provides for consolidation under the Justice Department for the enforcement of the law's antibribery provisions.
6. It provides for more expeditious and useful guidance to the business community through Justice's business review procedure.
7. Finally, it recognizes that only multilateral strictures on corruption can effectively check such practices in international business.

Clarification and simplification of FCPA has long been of high priority to our industry. It would seem that passage of the recently introduced Business Accounting and Foreign Trade Simplification Act could have only a favorable impact upon the export interests of the forest products industry.

Domestic Content Legislation

Strictly speaking, domestic content legislation would have to be viewed as import rather than export directed. Moreover, it is essentially aimed at highly manufactured goods, such as automobiles, rather than commodities. For these reasons, such legislation would seem to be of little concern to forest products exporters except for the fact that, if enacted, it could result in a severe backlash reaction toward the importation of all American goods and services from some of our most important customers.

As the debate over such legislation is about to begin again, a worldwide survey of foreign trade product requirements, recently completed by the Office of the U.S. Trade Representative, has provided new fuel for the proponents of such legislation. This survey shows that local content requirement regulations are rampant, there being some 222 such regulations in 50 different countries.

Two domestic content bills (H.R. 5133 and S. 2300), which would have required foreign-built automobiles to have as high as 90 percent U.S.-made content, died in the recent November-December lame duck session of Congress, but a new domestic content bill has been introduced in the House by Representative Richard L. Ottinger of New York, amid growing public support for such a move.

The Generalized System of Preferences

The Generalized System of Preferences (GSP), which offers preferential tariff rates for developing countries, is essentially another piece of import legislation of interest to the forest products industry and forest products exporters. We have a large imported-hardwood-products industry in this country which is largely dependent upon the developing world as the original source of raw material for the products they distribute. Exporters, on the other hand, are interested in such legislation because developing countries often offer important and growing markets for forest products; and it is quite possible that in renewing GSP authority, the administration will seek to make the program contingent upon our gaining improved access to such markets.

GSP expires at the end of 1984 and the administration has indicated that it will seek extension in some fashion beyond that date. This has spurred the Trade Policy Staff Committee of the USTR to schedule hearings next month in

Washington, San Francisco, and New York to begin exploring alternatives.

Reciprocity Legislation

Today's reciprocity legislation might be described more accurately as "market access" legislation, since it seeks to clarify the president's power to retaliate against restrictive and basically GATT-illegal trade practices. As Senator Heinz has described it, the purpose of reciprocity legislation is very simple: "...to give our Administration more leverage in opening other nation's doors and in persuading them to accept GATT disciplines."

Senator John C. Danforth of Missouri was the champion of such legislation in the 97th Congress and his original bill (S. 2094) was known as the Reciprocal Trade and Investment Act. A companion bill (H.R. 6773) was sponsored by Representative Bill Frenzel of Minnesota. Both bills died in the lame duck session of the 97th Congress. A new bill (S. 144), introduced in the 98th Congress by Senator Danforth and his colleague Senator Heinz, is essentially the same as Danforth's earlier bill. Senator Heinz, who is rapidly becoming the main congressional spokesman for this legislation, points out that during the multilateral trade negotiations (MTNs) in the late 1970s, we made substantial tariff concessions, largely in return for agreements from others to accept increased discipline over their behavior in the international marketplace. Since then, according to Heinz, we have witnessed a world increasingly beset by violations of the very standards which had been agreed upon. Now many believe it is necessary to clarify present law with respect to the president's authority to seek improved market access and to provide for retaliation if that access is denied.

It would definitely appear to be in the interests of forest products exporters to strengthen the president's negotiating hand to seek improved market access for American exports. Industry strongly supported extension of the president's tariff negotiating authority under Section 124 of the 1974 Trade Act which expired last year; but that proposal now lacks sufficient support to be revived. The reciprocal trade bills are not exact substitutes for extension of Section 124 negotiating authority, but they could very well lead to achieving similar objectives, perhaps at lower trade-off costs.

Export Administration

There is general agreement among exporters, economists, and policy makers that the Export Administration Act of 1979, which is due to expire at the end of September, has resulted in some devastating blows against America's export interests without giving rise to the successful achievement of intended foreign policy objectives. As a result, strong moves are now under way in Congress either to replace the act entirely or to change it substantially. (It should be noted that none of those pressing for EAA revision dispute the necessity of controlling the export of American technology which is critical for national defense purposes.)

Senator Heinz, Chairman of the International Finance Subcommittee of the Senate's Committee on Finance, has introduced a bill (S. 397) which would provide sanctity for contracts which may be in existence at the time trade controls are imposed, and strengthens the criteria for imposing controls for foreign policy purposes. In addition, the Heinz bill would shift enforcement of the act back to the Customs Service; however, the export licensing authority would remain in Commerce's International Trade Administration. Another EAA bill (S. 434) has been reintroduced by Senate Banking Committee Chairman Jake Garn of Utah which would entirely replace the 1979 act by creating an independent federal authority to be known as the Office of Strategic Trade. Garn's bill would completely remove the export control function from the Department of Commerce.

Another new bill (H.R. 1566), which would extend and amend the Export Administration Act for another two years, was introduced in the House late in February by Representative Don Bonker of Washington, Chairman of both the International Economic Subcommittee and the House Export Task Force. Apparently, Bonker's bill would also result in the curtailment of the president's authority under the act and, in specified instances, make such action subject to congressional approval.

Given the general concern in Congress over the negative impacts the old Export Administration Act has had on U.S. exports, and given the renewed interest by the administration in bolstering exports, it seems certain that EAA will be replaced with something which would better serve the interests of the total American export community.

Eximbank Charter Renewal

The Export-Import Bank's charter expires September 30 and therefore must be renewed very soon. The Eximbank, as you probably know, promotes exports by extending loans, loan guarantees, and insurance to U.S. exporters in tight competition with foreign government-supported competition. Exim's efforts have usually been in support of big-ticket or high-technology items, hence its fate would not seem to be of great interest to forest products commodity exporters. The Exim, however, is an important weapon in our arsenal for pressing competitor nations to agree to end all export-credit subsidies. Unfortunately, Congress appears to be dragging its feet on taking up this important trade issue, with preliminary hearings expected to begin only in the near future.

DISC Replacement Alternatives

After long resisting the GATT Commission's ruling that the DISC approach violates its agreed-upon international rules, both Congress and the administration appear to have resolved to find a GATT-compatible substitute for this highly successful, competitive tax equalizer for American exporters. Representatives Bill Frenzel of Minnesota and Sam Gibbons of Florida first introduced legislation (H.R. 5179) in 1981 which would replace DISCs with an International Sales Corporation (ISC). This approach would have been incorporated into the U.S. tax system, the same territorial concept used by our GATT critics. Frenzel and Gibbons introduced a somewhat modified version of their bill (H.R. 7311) late in 1982 which added a proviso to permit the ISC to sell services as well as products, and change the name to International Sales and Service Corporation (ISSC). Meanwhile, early in the summer of 1982, Senator David L. Boren of Oklahoma introduced a somewhat similar bill (S. 2708) calling for the establishment of Export Sales Corporations to replace DISCs. These Export Sales Corporations would have to be estalished abroad (logically in tax-haven countries) as foreign subsidiaries subject to local taxes but not U.S. taxation. However, ESC dividends would be taxable to the parent company, except for an intercorporate dividend-received deduction.

A different approach was introduced in the House late last month by Guy Vander Jagt of Michigan. Vander Jagt's bill, called the DISC Revision Act of 1983 (H.R. 1673), would preserve the basic structure of the DISC and impose an interest charge of 4 percent on accumulated DISC income which had been subject to deferral. Despite the completely different approach employed in Vander Jagt's proposal, he claims it would be GATT-acceptable. Also, many observers have suggested it would probably be preferred by small businesses which would likely be reluctant to go through the expense and hassle of setting up overseas subsidiaries.

There has been considerable discord within the administration over what to do about replacing the DISC, with Treasury--probably concerned over potential losses in revenue--being the principal foot-dragger. Eventually, and consistent with his intense desire to improve America's competitive export ability, President Reagan gave Treasury orders to come up with a GATT-legal alternative. Treasury finally floated its proposal in mid-February to the Cabinet Counsel on Commerce and Trade, who gave it approval early this month. Like the Boren bill, and the earlier Frenzel/Gibbons bills, the Treasury proposal calls for the establishment of foreign subsidiaries that would actually take title to goods from the parent, at arm's-length pricing, for resale elsewhere abroad.

There is no doubt but what a GATT-acceptable substitute for maintaining the all-important tax equivalent advantages of DISCs is vital to all U.S. exporters, including the exporters of forest products.

Shipping is another area of interest to forest products exporters and at the moment several pieces of such legislation are pending which warrant your attention.

Maritime Regulatory Reform

Maritime regulatory bills, which would replace the Shipping Act of 1916, were introduced in the last Congress by Senator Slade Gorton of Washington (S. 1593) and by Representative Walter Jones of North Carolina (H.R. 4374). The House acted favorably upon its version but Gorton's bill was blocked from coming to a vote in the Senate. Gorton immediately reintroduced his bill in the 98th Congress and it passed the Senate on March 1. House passage is expected soon.

It is difficult to predict what impact, if any, such legislation, once passed, will have on the cost of shipping forest products. The bill, which proposes giving shipping conferences greater antitrust freedom, was motivated mainly by a desire to enable U.S. carriers to be more competitive. As passed by the Senate, the measure is expected to give comfort to shippers by clarifying their long-standing antitrust exemptions. Reform deregulation did not go as far as some had hoped, thanks mainly to a last minute amendment from Senator Ted Stevens of Alaska, which knocked out provisions for ending tariff filing requirements with the Federal Maritime Commission for ocean common carriers.

Port User Fee Legislation

Port user fee bills, which were intended to shift all port development costs from the government grant category to the shippers who actually use the ports, were introduced into the last Congress (S. 1692 and H.R. 4627). But this strict approach apparently died with the 97th Congress because clones of these bills have not been introduced this session of Congress. Rumor has it, however, that Senator Mark Hatfield of Oregon is trying to put together a new bill which would provide for a cost-sharing approach consistent with the administration's thinking by combining its desire for economy in government with the necessity to sharpen our export competitiveness. In his State of the Union Message, the president said: "It's time for us to get together and enact a port modernization bill." Only a few hours later administration sources leaked a draft copy of "The Deep Water Harbor Improvement and Cost Recovery Act of 1983," which contains new and reduced cost-sharing formulas for port development and maintenance.

As proposed, this new legislation would doubtless lead to increased costs for all shippers, but the increases would be lower than under legislation introduced in the 97th Congress. Moreover, the new proposal provides for a uniform port user fee thereby avoiding the disruptive possibility of large cost increases at some and no increases at other individual ports as would have been the case with earlier proposals.

Cargo Reservation Legislation

A bill entitled The Competitive Shipping and Ship Building Act of 1983 (H.R. 1242) designed to require greater use of U.S. flagships in the import and export of bulk commodities has been introduced by Representative Lindy Boggs of Louisiana; and a companion bill is reported ready for introduction in the Senate by Claiborne Pell of Rhode Island or Russell Long of Louisiana.

Mrs. Boggs points out that presently less than 4 percent of our bulk cargo imports and exports are carried in

U.S. flag vessels. Under her proposal, each importer or exporter of bulk cargoes would be required to ship and receive 5 percent of such cargoes in U.S. bottoms the first calender year following passage of this act; and that portion would have to increase by one percentage point yearly so as to reach at least 20 percent within 15 years. Boggs' bill would also make the Secretary of Transportation responsible for setting guideline-maximum rates in accordance with shippers' costs. The Secretary would also be responsible for requiring and helping shippers to reduce their operating costs, and U.S. shipyards to reduce their construction costs.

It is assumed that the term "bulk cargo" would include forest products, but this is not entirely clear from the Boggs bill, which defines it as "cargo transported in bulk without mark or count." Aside from this definitional uncertainty, the requirement to ship in U.S. bottoms would no doubt increase the costs of exporting U.S. forest products. One estimate which I have seen for forest products places the costs per ton carried of operating a U.S. vessel without subsidies at approximately twice that of a foreign flag of convenience vessel.

Another aspect of cargo reservation which I want to touch on today concerns the enacted or adopted legislation of other countries which could force us into bilateral or multilateral cargo reservation agreements with such countries. Already the United States has negotiated bilateral agreements with Brazil, Argentina, China, and the Soviet Union which reserve to each of these trading partners' vessels a portion (usually 40 to 50 percent) of the cargo shipped between them. Currently the U.S. is on the verge of, or already involved in, negotiating similar agreements with the Philippines, Indonesia, South Korea, and Venezuela.

The recent drive on the part of these nations to strike bilateral cargo reservation agreements with us probably stems from their impatience with the slow ratification of the so-called Liner Code developed by the United Nations Conference on Trade and Development (UNCTAD). This code would allow countries which ratify it to divide the shipping of their liner cargo trade, giving each trading partner 40 percent, with the remaining 20 percent going to cross traders. This code is intended to become a legally binding international treaty when shipping tonnage owned by ratifying partners exceeds 25 percent of the world total.

Both bilateral reservations and the Liner Code would, by their nature, reduce competition among ship operators, and thereby limit the options of shippers to pick and choose the cheapest and most efficient service. The end result would almost certainly be increased transportation costs, diminished quality of service, and higher prices which would have to be passed along to buyers. The ultimate result would probably be a loss of foreign market share by cost-sensitive exporters to those producers who are able to operate cheaper shipping lines.

THE CASE FOR ELIMINATING JAPANESE DUTIES ON WOOD PRODUCTS

John Ward

Trade relations between the United States and Japan are at a low point in 1983. Both economies are suffering the results of difficult recessions. The United States experienced a negative balance of almost 4.3 trillion yen in its trade with Japan in 1982.[1] Protectionist interests in America (and elsewhere) are threatening actions which could be negative for Japanese exports. And the United States, in spite of some responses by Japan, is still calling for that nation to open its markets wider to American exports.

Reactions in the United States are understandable. America's open markets make it Japan's largest export customer--9.0 trillion yen ($39 billion) in 1981, far above other customers. Japan's automobile producers have been supplying over 20 percent of the U.S. market while American auto workers have suffered 15 to 20 percent unemployment. Furthermore, efficient Japanese competition is affecting many other U.S. industries and workers.

The Opportunity

Within this context, Japan has much to gain and little to lose by taking the dramatic step of eliminating all duties on primary wood products. Japan is the largest importer of solid wood in the world: 1.8 trillion yen ($8 billion) in 1980. Its wood processing industries are dependent on imports for nearly two-thirds of their raw materials. And Japanese forecasts project a continued high dependence on imports into the 21st century. Yet, in spite of this need, wood products remain an example of a trade

---

1. Data used in this paper are from the Japanese Ministry of Finance, the U.S. Department of Commerce, and the Food and Agriculture Organization (FAO) of the United Nations. U.S. dollars have been converted to Japanese yen at a rate of 1:230.

practice for which Japan has often been criticized--that of bringing in raw material at zero tariff while protecting its own finished product manufacturers with high tariffs. Note the figures in table 1.

Table 1. Japanese Wood Imports (1980).

Import	Share of Total Consumption (percent)
Raw Materials:	
Softwood logs	50
Hardwood logs	83
Pulpwood and wood chips	59
Processed Products:	
Plywood	1
Veneer	2
Particleboard	7*
Hardwood lumber	8
Softwood lumber	14

*Overstated in this year because of domestic shortages.

Trade with the United States illustrates the problem. The United States is the largest supplier of wood to Japan, which in turn is its largest export customer. Yet U.S. shipments of raw materials substantially exceed those of finished products. In 1982, American exports to Japan totaled 292 billion yen ($1,269 million). Of this amount, logs, wood chips, and lumber, a processed product minimally affected by duties, equaled 288 billion yen ($1,251 million). However, U.S. exports of veneer, plywood, and particleboard amounted to only 1.3 billion yen ($5.8 million) or 0.5 percent of the total. These shipments included: veneer--0.6 billion yen ($2.6 million); plywood--0.5 billion yen ($2.2 million); and particleboard--0.2 billion yen ($1.0 million).

The most important reason for this difference is high tariffs. Japan's rates for major wood products are shown in table 2, along with comparable American and EEC rates. Another reason is a Japanese standard that precluded the use of American plywood for construction applications. However, through lengthy negotiations, a new standard has been proposed which should eliminate this problem.

Table 2. Tariff Rates.

Product and Japanese Tariff Code	Rates (in percent)*		
	Japanese	EEC	U.S.
Veneer (44.14.230, -.290)	15	4	0
Plywood† (44.15.111, -.119, -191)	15	0	8
Particleboard (44.18.100, -.200)	12	10	4

*Rates shown are the lowest rate to be reached (usually by 1987) under the 1979 Tokyo round of agreements, except the Japanese rates for veneer and standard softwood plywood, which were not decreased under those agreements.
†Rates shown are for the two product categories for which a comparison should be made. (1) Japanese rates are for the major product which the United States produces and for which it desires increased market access--standard softwood plywood and specially finished grades. (2) The European Economic Community rate is for the bulk of the softwood plywood which it imports. Plywood within certain thickness and other definitions and over a bound quota of 600,000 cubic meters annually is subject to a 10 percent duty. (3) The U.S. rate shown is for hardwood plywood, the primary product produced in Japan and exported by it to the United States. (The U.S. rate on softwood plywood, which is 20 percent, does not provide a true comparison with Japan. This rate, which affects virtually only Canada, was negotiated down to 8 percent during the Tokyo round, contingent upon a common standard for softwood plywood being agreed upon between Canada and the United States. Since no agreement has been reached, Canada continues to maintain its 15 percent and the United States its 20 percent rate on softwood plywood.)

It has been argued that U.S. market participation is low because American producers have been unwilling to meet Japan's size and quality requirements. In the case of veneer, American producers now supplying to Japanese specifications. U.S. plywood producers are still unable economically to produce the major standard Japanese sizes; but they cannot be expected to invest in the equipment necessary in the face of such high tariffs. As for particleboard and other products, the U.S. can and does meet Japanese requirements.

In sum, tariffs are the primary barriers remaining to increased Japanese imports of American wood products. Japanese tariffs are very high compared to U.S. and EEC rates, the average Japanese rate for all products imported of less than 3.0 percent, and the weighted average rate for all Japanese imports of forest products of only 0.2 percent. Moreover, they are extraordinarily high considering Japan's commitment to open its markets wider to U.S. products and its repeated statements that it will try to buy more processed wood products from the United States.

The elimination of these duties would create additional competition for Japanese manufacturers of plywood, particleboard, and lumber--producers that have already suffered deeply from the effects of a prolonged recession in housing construction. But any short-term negative effect on those industries would probably be minimal. Indeed, tariff elimination would provide certain positive benefits to these manufacturers.

Overall Benefits to Japan

Important benefits of eliminating tariffs would accrue to the Japanese economy itself. Dropping the nonproductive tariff tax on products so important to housing, construction, and other related industries will potentially benefit the Japanese consumer with lower final product costs. And it will help maintain the competitive position of wood relative to other materials, a benefit to both producers and consumers who have for centuries maintained a reverence for this product's aesthetic and structural qualities.

Japanese producers of other industrial products exported to the United States would also gain. Eliminating wood tariffs would build powerful support within, and positive publicity by, the U.S. forest industry, which has important influence in Washington, D.C., and throughout the United States. This support could assist in countering

America's growing protectionist movement. To illustrate: in December 1982, Japan announced small duty reductions on several pulp and paper items. Although this move was minimal, it nonetheless invoked forest industry news releases giving credit to Japan for improved market access. A major move by Japan to open its markets to wood products could build significant support within the industry.

The greatest beneficiary of tariff elimination might be the Japanese wood products industry itself. Dropping duties on finished products might be the most important step that could be taken to ensure long-term supply of raw material from the United States.

Although opposed by the U.S. government and the industry's national association, demands in America for increased restrictions on U.S. log exports have regularly recurred when U.S. housing demand is strong and Japan's need for logs is high. Continued maintenance of the Japanese tariff wall simply reinforces the arguments of some of those who wish to restrict U.S. log shipments. They maintain that the tariff is evidence that Japan is not sincere in its commitment to buy more American finished wood products. The elimination of the tariffs on these products would make the task of resisting such arguments much easier.

Effect on Specific Industry Sectors

Veneer Tariffs

Dropping veneer duties would be of direct and immediate benefit to the Japanese plywood industry. Japanese veneer tariffs are now 15 percent. During the Tokyo round of negotiations, Japan declined to make any reductions in these tariffs. The United States, in contrast, with strong encouragement from its own plywood industry, which wished to reduce raw material costs, moved to eliminate immediately (1980) all tariffs on veneer, then at 6 to 10 percent.

Japan is the second largest manufacturer of plywood in the world, producing hardwood plywood from raw materials imported mostly from Southeast Asia. But Japan's main suppliers are taking steps to reduce its raw material supply.

Japanese plywood manufacturers have been exploring alternatives. A most important one is softwood veneer from the United States, which would be used to produce softwood plywood or a combination panel with softwood in the inner plies and hardwood veneer on the exterior. Experience by Japanese plywood manufacturers to date has shown that softwood veneer offers important production advantages and product strength capabilities. In addition, plywood made from softwood veneer has been shown to have greater resistance to decay. It is also available from a dependable long-term supply source, the United States. Realistically, there should be no opposition in Japan to eliminating tariffs on veneer.

Softwood Plywood Tariffs

Japan did agree to reduce its tariffs on specialty softwood plywood during the Tokyo round--from 20 to 15 percent, still a very high level, over a three-year period starting in 1982 (construction grade plywood was not changed, remaining at 15 percent).

Japan argues that elimination of softwood plywood duties would be very damaging to its industry, now plagued with low returns and bankruptcies because of its housing recession. The American plywood industry understands the Japanese problem, since it has just been through the deepest depression in its own history. But the U.S. industry still desires the opportunity to compete on an equitable basis in Japan.

The Japanese plywood industry has been an important exporter of hardwood plywood to the United States, where it faces a tariff approximately half its own. Even though these shipments have declined substantially in recent years, they still totaled 7.8 billion yen ($34 million) in 1982, or 15 times the value of the softwood plywood the United States was able to export to Japan.

Even if the softwood plywood tariff were to be eliminated immediately, differences in panel size would preclude an immediate rush to U.S. products by Japanese end users. The standard panel size now produced in Japan is 90 by 180 centimeters (approximately 3 by 6 feet), while the United States produces only 122 by 244 centimeter panels (4 by 8 feet). Therefore, to compete for the great bulk of the Japanese market, U.S. producers must invest in new or modified equipment.

Increased market development efforts are also a primary requirement if the United States is to increase its share of the Japanese market. The result of such development efforts, not justified at present because of the high tariffs, could well be an expanded <u>total</u> market for plywood in Japan, thus benefiting both American and Japanese producers.

In summary, when market acceptance and U.S. investment requirements are considered, it would appear that any significant U.S. competition with the Japanese plywood industry would develop slowly. Thus, reduction in the tariff would provide equitable market access yet not cause major short-term dislocation in the domestic industry.

Particleboard Tariffs

Duties on particleboard (reconstituted wood panels which include standard and structural particleboard, waferboard type products, and other wood panels produced from wood chips and residues) received minor consideration by Japan during the Tokyo round. This tariff was reduced from 15 percent to 12 percent by 1987 compared to 4 percent in the United States.

Japan's particleboard demand, except for the last three years, has been growing rapidly. Growth in demand, primarily for the furniture industry, has even outstripped supply on occasion.

The Japanese particleboard industry needs no tariff protection. Rather, demand growth demonstrate a need for additional supply from external sources.

Summary

The elimination of tariffs on processed wood products is an essential step to be taken if Japan is to allow U.S. manufacturers to compete on an equitable basis within its marketplace. Moreover, Japan can gain by taking such a step:
- Potentially lower housing, construction, and other product costs and a strengthened competitive position for wood for its end users and consumers
- Greater stability of raw material supply for its wood products producers
- In the case of veneer, specific benefits for its plywood manufacturers
- A closer alliance with a major U.S. industry which supports free trade and open markets

In conclusion, elimination of high Japanese tariffs on processed wood products--particularly veneer, plywood, and particleboard--would provide major benefits for both Japan and the United States. The action would not only improve their trade partnership in wood products but would be a major step to strengthen the overall economic bond between the two nations.

TOWARD WORLD TRADE EXPANSION: A CANADIAN PERSPECTIVE

Eugene W. Smith

Many of the issues raised here are the subjects of intense debates in capitals around the world. The debates are intense because the stakes are high. We all face momentous choices with far-reaching consequences for the international economic and political system and for the values we attach to our relations with other sovereign nations.

The economic facts are obvious. The world is racked by the worst economic downturn since the great depression. It is true that inflation is down but so too is industrial production, the largest and steepest decline in decades. In the industrialized world, entire sectors such as the steel industry are undergoing massive restructuring as international shifts in comparative advantage work their way through our economies. In the developing world, many nations are carrying debt burdens which will impose austere, harsh conditions on their peoples for years. There is talk of an international financial crisis.

World unemployment has reached intolerable levels. Governments are being urged to create more employment opportunities and lead the way out of the morass, but governments complain their scope for action is strictly limited:
- Tight monetary policy does bring down inflation, but at the cost of jobs, at least in the short term.
- An expansionary fiscal policy does create jobs, but the resulting inflation hits those on fixed incomes and tends to cause interest rates to rise.
- Increased benefits for the needy are desirable, but they drain resources needed for investment in productive capacity.

Every positive domestic policy thrust contains within itself its own negative. We are frustrated.

It does not end here, for every government also insists that it cannot solve its economic problems on its own. The problems are international and require international solutions. It is difficult enough for us to accept the domestic constraints, the unavoidable trade-offs which restrict national policies. But when governments point to the limiting impact of only vaguely understood international abstractions such as global trade, foreign-exchange markets, and the international financial system, our frustrations only increase.

I would like to concentrate my remarks on the international situation, particularly on the trade system. I speak from a Canadian perspective but I know my views are shared by many others concerned with these issues.

If I were forced to suggest a single theme which threads its way through all our fears and frustrations, I would point to vulnerability. People everywhere feel exposed and defenseless, subject to forces outside their control. Our interests are directed toward world developments in ways we barely understand, and are buffeted by developments which appear uncontrollable. It is understandable that there are calls for dramatic and, if necessary, unilateral action; but, by their very nature, such calls may indicate a critical gap in our understanding of the modern world, and our traditional view of the authority and autonomy of the nation-state as it exists today.

To correct our understanding, we must examine how we got to where we are today. The international trade system provides the most readily available answer.

The General Agreement on Tariffs and Trade--GATT, as it is called--establishes a multilaterally agreed upon set of rules for the conduct of international trade. When GATT was first proposed in the 1940s, the industrial countries shared a set of assumptions about the importance of an open trade and payments system to economic recovery and growth. The postwar economic boom has vindicated that vision. Since the founding of GATT, world trade has expanded approximately 600 percent, an astounding figure. In the U.S., trade has expanded from 5 percent of total GNP to its current level over 12 percent, showing how important trade has become to the world's most powerful domestic economy. In forest products world trade amounted to $55 billion (U.S.) in 1980 and has been growing faster than world consumption. The roundwood equivalent of industrial forest products, traded internationally, now represents over one-third of total world consumption.

No one today would argue against the general proposition that domestic economic growth is positively related to the expansion of international trade.

The rationale for GATT, the political momentum behind its establishment, arose out of the economic chaos of the 1930s. The creators of GATT, led by the U.S., saw clearly that the destructive "beggar-thy-neighbor" policies of the great depression had to be replaced by an open trade and payments system. The International Monetary Fund was created to provide the necessary liquidity to fuel trade expansion in stable conditions.

The consensus in favor of trade expansion has held firm for over 30 years, despite the costs that certain industries have had to bear in order to adjust to a changing world environment. A complex pattern of international trade and finance relationships has developed to the point where the economic destinies of all countries are inextricably linked. This phenomenon, commonly referred to as <u>economic interdependence</u>, has appeared irreversible, at least until now.

We are, at this time, witnessing the strongest challenge yet to the postwar global economy--a challenge which one American scholar, with admirable foresight, described a few years ago as "revolt against the costs of interdependence."

Entire industries are calling for protection from foreign competition, even though that protection undermines the solid foundation upon which our interdependent world is built. Domestically, industrial sectors are pitted against each other in an effort to ensure that their interests are not traded off for a gain somewhere else. Questions of local and regional survival are raised as traditional industrial areas watch their economic strength trickle away to new areas--whether at home or abroad scarcely matters.

How quickly we forget the rapid and unprecedented economic growth which was the major benefit of interdependence.

Governments everywhere are besieged with appeals from hard-pressed domestic industries for protection from foreign competition. Of course, protectionist measures in one country invite countermeasures from other countries, as each sees export markets shrink and is forced to take steps to protect its domestic markets. Once the cycle of protectionism begins, it is very difficult to stop. It would not take long for the international trading system to unravel.

I do not for a moment underestimate the intensity of these pressures, nor do I want to leave the impression that the fears of unemployed people are groundless or ill-founded. On the contrary, it is the legitimacy of those concerns, the burden of economic hardship, which puts governments in such a dilemma.

The automobile industry is a case in point. In the U.S., it is estimated that as many as one in every seven jobs is linked to the automobile industry. The contraction of the industry results from many factors, including changing living styles, consumer preferences, and energy prices. It is also due in no small part to the rapid penetration of the domestic market by foreign imports.

The industry has done a remarkable job of adapting itself to the new environment. But tough times persist and so we now have pending legislation, known as the domestic automobile content bill, designed to protect American automotive jobs. The bill would require all car manufacturers, domestic and foreign, to incorporate a higher proportion of U.S. content in their automobiles as their penetration of the U.S. market increases. The measure is clearly protectionist but it enjoys wide support in Congress and throughout the country.

At the same time as the U.S. Congress is considering protectionist legislation, there are many who are also pushing for a new, aggressive American policy to open up foreign markets to U.S. products. One notable focus of this demand for aggressiveness is the services sector. This is not surprising. It is now estimated that the service sector accounts for over 65 percent of the U.S. gross national product. Services also account for over 30 percent of U.S. exports. The U.S. has developed a strong, highly competitive capability in such areas as financial services, insurance, and data processing. Many countries have restrictive policies in these areas so that U.S. services, although highly competitive, do not enjoy free access to those foreign markets.

Let's step back and look at these two policy initiatives in perspective. In an area where U.S. competitiveness has eroded, relative to some other countries, the call is for protection. Where U.S. products are competitive, the call is for expanded access to foreign markets. A foreign country, however, sees it in just the opposite way. Where its products are competitive, the U.S. is ready to restrict access; but in areas where it suffers a competitive disadvantage, the U.S. is demanding increased access. All countries are in a similar position. There are contradictory pressures to restrict imports and expand exports. I say "contradictory" because, as a general principle, both cannot be met. We can only expand exports if someone buys them, but if everyone is restricting imports then there will be fewer buyers overall, not more.

This takes us to the heart of a multilateral trade negotiation--and negotiation is the operative word here. A comprehensive multilateral trade agreement is a delicately balanced achievement. One country's comparative advantage is another country's comparative disadvantage, and it is at this point that trade-offs between countries take place. I will lower my trade barriers against your steel exports if you lower your barriers against my textiles. Each country seeks to maximize its advantages and limit its disadvantages. The goal is a balance of concessions based on mutual self-interest and, despite the limitations, it is remarkable how far we have moved in this direction. Over the last three decades we have seen develop an increasingly rational global allocation of productive resources based on comparative advantage.

I have referred to a hypothetical trade-off between the steel of one nation and the textiles of another. But look behind this international trade-off and we will see that there are also domestic trade-offs. For when one country achieves greater access for its steel exports in exchange for offering greater access to textile imports, it is also telling its domestic textile producers they must be prepared to face stiffer competition from abroad. The textile producer is losing some of his protection not for a direct benefit but in order to permit benefits to flow to steel producers in his own country. The trade negotiator makes the trade-off because he feels the nation as a whole will benefit, but tell that to the textile producer--especially when economic times become tough!

Trade policy experts like to talk of industrial "winners" and "losers." Back your winners and let the losers contract. Harsh language, but it reflects a harsh reality. Trade policy is a tough discipline. There is perhaps no other area of national life in which the diverse and disparate economic interests of a nation must be so reconciled and aggregated into what we term the "national interest."

Trade negotiators pursue the national interest. The postwar boom is directly and positively related to the growth in international trade and this has clearly served our respective national interests. We should be mindful of this since the opposite would also be true; a contraction of international trade will produce strong economic contraction at the nation level.

Let me give you an example. Canada last year imported $3 billion less in capital goods from the U.S. than it did the previous year. When our economy recovers, and assuming new trade barriers have not been erected, I expect we will resume our previous levels of capital goods imports. To us, this will mean an increase in our productive capacity; to the U.S. it will amount to a $3 billion new job package. In the same context, the domestic automobile content bill as currently drafted will seriously disrupt Canadian-U.S. automotive trade. The U.S. exports $12 billion worth of automotive products to Canada. Consider the number of U.S. jobs involved in producing those exports.

There is much talk these days of job bills, but how many of these are merely welfare in disguise? No amount of special legislation can create jobs equal to the number created by expanding trade.

A second point to remember is that nations will retaliate against the protectionism of other countries. This is important because there are many who believe that a unilateral withdrawal of trade concessions will not lead to retaliation in other countries. They are wrong. History is instructive on this point. In 1930, Congress passed the Smoot-Hawley Act which dramatically raised U.S. tariffs, often to prohibitive levels. Despite protests from many nations and petitioning by major American economists, President Hoover did not veto the bill. Retaliation by many nations was widespread. Canada alone imposed duty surcharges on 125 major U.S. products. Economic historians do not hesitate to point to the Smoot-Hawley Act as a significant factor in increasing and prolonging the severity of the great depression.

With this as background, we may ask why protectionism enjoys any support at all. The answer is not difficult. In general, protectionism does not emerge as a coherent policy articulated at the national level. It is always this industry or that region which seeks relief. In some cases, there is little appreciation or understanding of the wider policy concerns which argue against yielding to such pressures. But even when these wider concerns are known, we run up against the politically pressing need to make a special exception for this or that industry.

Democratic political systems are particularly susceptible to this type of pressure. Elected representatives, after all, are expected to be responsive to their constituents' concerns, and legislatures are under intense pressure to act. In the U.S., the management of this volatile situation poses a unique challenge. Individual congressmen and senators have greater independence than their colleagues in other countries, such as Canada, who are subject to strict party discipline on key votes. Also, the separation of powers in the U.S. gives Congress great freedom to initiate legislation independent of the administration. While many members of Congress have taken a strong stand for open trade, the Congress itself is now giving signals that it is willing to legislate to meet demands from local and special interest groups. Administrations, in our experience, tend to be more appreciative of the wider context in which economic policy is formulated and executed.

I do not wish to be branded as an alarmist, but people in Ottawa and around the world follow closely the events in Washington. In the last Congress, Canadian interests were directly harmed by protectionist legislation on trucking, specialty steel, and concrete and steel for highway construction. Mass transit, a sector where Canada is very competitive, has long been protected by "buy American" preferences. Our exports in other sectors such as telecommunications, uranium, and, of course, softwood lumber are threatened. In these cases, the sponsoring legislators are looking after special or constituent interests. Interestingly, a number of these legislators have, on the other votes, spoken out against protectionism. In other words, they are saying protectionism is bad--but my constituent is a special case and needs an exception to this general principle. We all understand this dilemma--the problem is that the cumulative effect of all these "special cases" is potentially very damaging.

It is not just damaging to us and I do not expect anyone to rise in indignation because Canadian interests are being harmed. But I do believe Americans should be concerned about the effect such legislation has on U.S. interests. Consider the consumer, who is deprived of cost-effective and attractive products from abroad. Consider also the U.S. workers who produce goods which are incorporated into Canadian products and then exported back to the U.S. On mass transit, for example, over 60 percent of the components of Canadian-built buses are manufactured in the U.S. Restrict access for our buses to your market and U.S. jobs as well as Canadian ones could be lost.

In addition to immediate costs to consumers and to a significant number of workers, there are larger and much more significant costs to an economy interdependent with a multilateral system. A proliferation of specific protectionist actions could harm such an economy beyond redemption.

Current world trends are worrisome. In addition to special protectionist legislation, there is also much talk of legislation to force other countries to trade fairly. But "fair" trade, like beauty, is in the eyes of the beholder. The unilateral approach and selectivity of the proposed U.S. reciprocity legislation could very well make things much worse for the U.S. as well as other countries.

The debates over trade in our respective legislatures and as reported by the media would lead one to conclude that only our own countries play fair, while the rest of the world is illegitimately restrictive. Dr. William Cline of the Institute for International Economics, however, recently published a paper in which he argues that the U.S. uses nontariff barriers to protect a far higher proportion of its domestic market for manufacturers than is often realized--a higher proportion than in Canada, I might add. His general conclusion states that there can be "no automatic presumption that the United States has a substantially more open market than those of its major trading partners."

The facts clearly show that all countries are guilty to a greater or lesser degree.

The U.S. Congress is critically important for Canada. We are each other's largest trading partner, with two-way trade at approximately $85 billion per year. We are so interdependent that there is virtually no U.S. economic policy, domestic or international, which does not have a direct impact on Canada. To an extraordinary extent, the domestic legislation of Congress virtually becomes U.S. foreign policy toward Canada.

We are caught in a vicious circle. Expanding trade is essential to economic recovery. Until that recovery comes, however, we are all being tempted to protect wounded industries, but this very protection will restrict trade and thereby delay recovery.

What do we do? Reference to recent management of the Canada-U.S. relationship may be helpful. Not too long ago, relations between our two countries were in danger of veering out of control. Your complaints about our energy and foreign investment policies and our complaints about your trade practices and environmental policies rose both in volume and visibility. After years of quiet diplomacy--so quiet that some Canadians regarded it as benign neglect--the parties went public. Congressional hearings were held, tough speeches given and warnings issued. The bilateral relationship became front-page news.

Fortunately, before any permanent damage was done, we managed to get things back under control. Our political leaders in both countries committed themselves to managing the inevitable problems which will arise between two different countries with so many transactions, in a spirit of good faith and mutual respect. Our respective foreign secretaries have decided to hold extensive bilateral discussions four times a year. We have modified certain policies to minimize their effects on each other. In areas where differences were irreconcilable we frankly acknowledged them. In some cases, we have resorted to the GATT dispute settlement mechanisms for approved multilateral treatment of our complaints.

I believe this is how we must handle the current global crisis. We must reject unilateralism in favor of multilateral solutions to our problems. We must exercise caution when considering legislative or regulatory remedies for our ills. We must demonstrate courage to resist the demands for protectionism. It is not easy to tell our peoples that there are no quick cures, that recovery will be slow and difficult.

Above all else, we must recognize we are in this together--the developing countries, on whom the burden of recession falls the hardest, as well as the industrialized world. The trade system is certainly imperfect and it contains many elements which appear unfair. But no country is entirely clean or virtuous.

There are some hopeful signs. The U.S., as the world's most powerful economy, is expected to provide leadership, and the current administration has taken a strong and important stand in favor of an open trading system. Also, the most recent economic news provides some basis for cautious optimism that recovery in the U.S. is under way.

I do not underestimate the dangers we face. They are there. But seeking a "quick fix" for politically expedient reasons today is to guarantee a much more ominous future. We have to work it out together. Those of us engaged in the forest products industry can help.

You may well wonder why I have been speaking for most of my allotted time without making more than passing reference to forest products; particularly when Canada accounts for almost one quarter of total world trade in forest products; is the world's largest exporter of lumber, pulp, and newsprint; and one Canadian job in ten is either directly or indirectly dependent on this trade. There are several reasons for this.

First, I wished to deal with this very important subject in the broadest possible context. Trade by definition is the exchange of goods and services between two parties and, as a consequence, we must give consideration to the vast number of goods and services that move in international trade if our objective is to increase our exports of forest products. While it is useful and necessary to examine such factors as supply/demand patterns, financing, transportation, and other related subjects, the fact remains that the basis of trade usually involves the exchange of quite unrelated products. As you all realize, tariffs and nontariff barriers are of paramount importance in determining whether or not this exchange takes place. Similarly, with or without tariff protection, nations must accept the need for constant change either to better serve foreign markets or to resist or accommodate imports, depending on the circumstances.

Canada, of course, wishes to expand its export markets for forest products of all kinds. In fact, over the past twenty-five years, we have expended a good deal of money, time, and effort to achieve greater market diversification. There is also every expectation that U.S. exports of forest products can and will increase. We welcome the participation by the U.S. government and industry in the promotion of U.S. forest products in various foreign markets. In our view, we would rather compete for a share of an expanding market than one which is static or contracting.

I would also like to explain that my frequent reference to our trade relations with the U.S. is related to the sheer size and importance of trade between our two countries. In 1981, the U.S. took 66 percent of our exports and provided 69 percent of our imports. Of our exports of forest products, the U.S. took some 70 percent of our total consisting mainly of newprint, lumber, and pulp having a value of $8.5 billion (CAN).

Over many years, trade between Canada and the U.S. has developed and flourished, based on interdependence and governed by the multilateral framework of GATT and through a series of bilateral trade agreements. For example, tariffs on newsprint were eliminated in the early 1900s to serve the growing needs of the U.S. northeast and central states. There was a comparative advantage in Canada provided by an abundant fiber supply and the availability of inexpensive hydroelectric power. We have, however, recently lost market share due in large measure to new lower-cost capacity in the U.S. South, which is a signal to us that we must adjust by improving our productivity or suffer the economic consequences.

You are all aware of the subsidy/countervail action initiated by a group of U.S. softwood producers against Canadian softwood imports. In our view, this action poses a far greater potential danger, including loss of trade benefits to both our countries, than most of us realize. Severe damage to the Canadian economy would be only part of the picture. Not everyone appreciates, for example, that on a national basis almost 50 percent of the raw material for

Canada's pulp and paper industry is derived from residue chips.

Similarly, a disruption of this trade would cause severe repercussions in Europe, Japan, and other countries, the extent of which is incalculable. We cannot believe that protectionist action in response to what we view to be short-term recessionary conditions could possibly be worth dismantling export market provisions which have evolved over many years and were negotiated in recognition of our mutual interest.

If Canada and the U.S., with the most extensive bilateral economic relationship in the world and a long tradition of cooperation, cannot successfully manage their relations, then we might well ask, "Who can?"

TIMBER MARKET AND TIMBER-PROCESSING INDUSTRIES IN JAPAN

Shizuo Shigesawa

Introduction

The Japanese people traditionally consume a large amount of timber and timber products. The annual per capita consumption of industrial wood has attained the level of one cubic meter.

Timber availability from domestic forests is expected to be considerably increased in the future as a result of aggressive efforts of afforestation work carried out since the postwar period. Under the current circumstances, however, the domestic supply of timber cannot meet the demand, and in 1982 some 65 percent of total demand was supplied by imported timber from all over the world, including North America.

Although Japan is one of the biggest timber importers in the world, the majority of imported timber is in the form of logs for raw material. This is because of the existence of wood-processing industries with large production capacity. In recent years, importing large quantities of logs has been more difficult, partly because of log export restrictions in producing countries, especially for tropical hardwood, which is the primary raw material for our plywood and sawn lumber industries.

Future timber imports must consist of more manufactured products, such as sawn lumber and plywood, instead of raw materials in the form of logs. This will become a critical problem for the Japanese wood-processing industry.

This report outlines the present situation of the timber market and discusses the potential for future adaptations of our wood-processing industry.

Consumption of Timber

For the last three years, domestic timber consumption in Japan has decreased. In 1982, total consumption of industrial wood is estimated to have been around 91 million cubic meters. A historically high record was set in 1979, when total consumption reached a level of 110 million cubic meters and per capita consumption was approximately one cubic meter.

This decrease in consumption has been due mainly to the worldwide economic recession, but more directly to the slowdown of new housing construction activity in Japan. This had a severe impact on wood-processing industries in Japan as well as other parts of the world, and not a few sawmills and plywood factories closed or curtailed their production.

In the long-term prospects, however, timber demand should recover with general economic development, and timber consumption is expected to increase beyond its previous record.

The housing construction market is most important, accounting for over 50 percent of total wood consumption. Traditionally, residential houses in our country have been made of wood. But in recent years, because of the urban development and steep upward trend of land prices, wooden-house construction has decreased considerably. Out of 1,150,000 new housing units constructed in 1981, 650,000 units were of wood--that is, about 57 percent in number and 61 percent of the total floor area, as shown in table 1.

Timber volume used per unit may vary considerably depending on construction system adopted. But according to the latest available survey results, in the case of the traditional Japanese wooden house the average timber requirements are 18.45 cubic meters of sawn lumber and 210 square meters of plywood per 100 square meters of floor area (that is, an average floor area of a Japanese wooden house).

The platform framing system, which was introduced to the Japanese market about ten years ago, is still not very popular in Japan, having been used in about 15,000 units per year in recent years. With this system of construction, requirements for timber and timber products may be somewhat higher than the traditional construction methods.

A relatively high level of housing activity has continued for more than ten years in Japan, compared with the international level, and the total housing stock has already surpassed the number of households by nearly three million units. Therefore, it is doubtful that the annual new housing construction will recover to the level of 1.5 million units in the near future.

Timber most in demand in Japanese housing construction is the extremely fine and delicate grades. The Japanese Agricultural Standard for sawn softwood lumber, for example,

Table 1. New Housing Starts in Japan.

Year	Total Units (1,000 units)	Floor Area (1,000 m^2)			Ratio of wooden(%)
		Total	Wooden	Non-wooden	
1975	1,356	112,422	77,587	34,836	69.0
1976	1,523	125,281	84,917	40,363	67.8
1977	1,508	126,818	83,559	43,259	65.9
1978	1,549	136,249	89,566	46,683	65.7
1979	1,493	136,515	88,621	47,894	64.9
1980	1,268	119,102	75,310	43,792	63.2
1981	1,151	107,853	66,145	41,700	61.3

Source: Ministry of Construction.

specifies 1,362 standard sizes, each item in one of three grades. As a result, the mechanism of distribution is rather complicated.

Other important timber uses are in nonresidential construction, furniture, joinery, industry, and packaging. But timber usage in these sectors does not seem to be specific to Japan.

Distribution of Timber Products

Timber products in the Japanese market are highly diversified in size and quality, so the mechanism for distributing timber products is rather complicated.

According to the report of the Ministry of Agriculture, Forestry and Fisheries (MAFF) published in 1982, the number of businesses that sell timber (excluding manufacturers) is over 18,000, categorized as follows:

```
Lumber auction market                              576
   (of which 74 deal primarily in sawn lumber)
Lumber distribution center                          69
Independent wholesaler                           4,830
Retailer                                        13,317
   Total                                        18,792
```

In lumber auction markets, registered bidders can buy the necessary quantities of qualified grades. Lumber auction markets play a fairly big role in the determination of general price levels of lumber and related products.

This wholesale system, developed after the war, allows members to procure all necessary building materials. Each wholesaler specializes in a particular product, and business is transacted by cross-trading.

The mechanism of distribution differs somewhat from that of such products as logs, sawn lumber, and plywood, and is, generally speaking, relatively complicated.

The analyses of distribution channels published by MAFF in 1982 are shown in figures 1, 2, and 3 for imported logs, sawn softwood lumber from imported logs, and plywood.

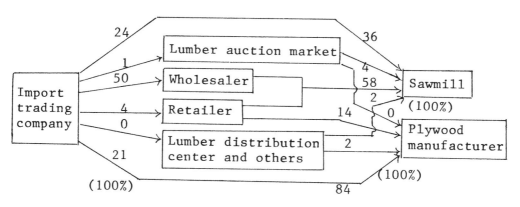

FIGURE 1. Distribution channel of imported logs.

In the case of imported logs, the trade flow is relatively simple, from the trading company to the sawmills, through wholesalers. The other important flow is direct sales of import trading companies to sawmills or plywood manufacturers. Almost all of the direct sales to plywood manufacturers are of hardwood logs from the countries of Southeast Asia.

In the case of sawn softwood lumber, a relatively high percentage is delivered from sawmillers to consumers directly, and auction markets are rarely used for this specific product. In contrast, about 20 percent of sawn lumber from domestically grown softwood species is sold through auction markets because the quality and appearance are quite different in each growing area.

In the case of plywood, transactions usually are made by cross-trade; auctions are not common. Manufacturers sell mainly through the trading companies which supplied the raw materials to them. The role of trading companies in

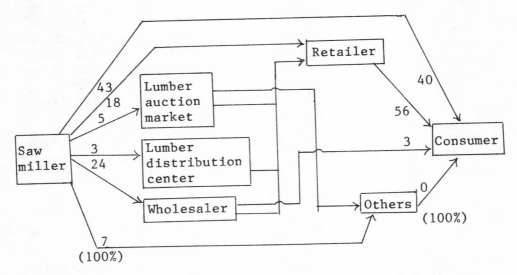

FIGURE 2. Distribution channel of sawn softwood lumber from imported logs.

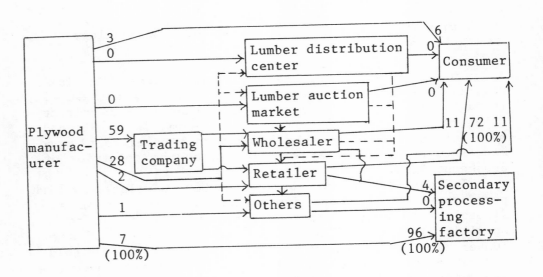

FIGURE 3. Distribution channel of plywood.

distribution of plywood has increased in recent years, reflecting the manufacturer's dependence on importing traders.

Recently, the possibility of establishing the forward market for a specific item of plywood has been investigated by some business circles, but a final decision has not been made.

Timber-processing Industries in Japan

General Aspects

In Japan, the development of timber-processing industries with very large production capacity has been supported by the high level of timber products consumption.

The majority of timber-processing enterprises are small- or medium-sized companies. The average production per mill is rather small compared with other countries, but most of the processing facilities are relatively modern and well equipped.

Since 1980, total demand for timber and timber products has decreased. Accordingly, almost all of the processing sectors have suffered severe financial difficulties in keeping up levels of operation, and many sawmills and plywood factories have been closed within the past year or two. In such sectors as sawn lumber and plywood, production capacity must be adjusted to cope with the actual and potential long-run demand for these specific products.

Sawmill Industry

In Japan, more than twenty-one thousand (21,000) sawmills were in operation as of 1981, as shown in table 2. The number of sawmills has decreased in recent years but the number is still very high in comparison with the consumption of sawn lumber. Total employees at that time numbered about 177,000 people, 17,000 less than the figure for 1980.

The number of mills in 1982 was even less and is estimated to have been a little more than 20,000, although final data were not available at the time of writing.

More than 90 percent of the sawmills are small- or medium-sized enterprises; the number of big mills with 150 kilowatts or more of output power is only a little over 2,000.

The details of raw material used in sawmills are shown in table 3, by type of logs. About 57 percent of the logs

Table 2. Number of Sawmills by Output Power.

Year	Total	Output Power (KW)				
		7.5–22.4	22.5–37.5	37.5–75.0	75.0–150	150 or more
1977	23,136	5,442	5,396	7,172	3,255	1,871
1978	22,794	5,077	5,302	7,215	3,255	1,945
1979	22,541	4,683	5,166	7,237	3,409	2,046
1980	22,241	4,359	5,055	7,273	3,444	2,110
1981	21,535	4,040	4,891	7,140	3,405	2,059

Source: MAFF 1982.

consumed in sawmills are imported, and among these imported logs the share of North American logs is approximately 50 percent. Therefore nearly 30 percent of sawn lumber produced in Japan is from U.S. logs.

Problems confronted by the sawmill industry in Japan are as follows:

- Number of sawmills. While there has been a decreasing trend in recent years, the number of sawmills is still too many in relation to the demand of sawn lumber. This is causing excessive competition in a limited market, and contributes to the unstable prices of the products.
- Improvement of products mix. In order to withstand severe competition from imported sawn products, sawmills must change their product mix to include semifinished products such as house components and planed parts for mill works.
- Adaptation to raw material supply. In the long run, the share of logs supplied from domestic forest lands will increase toward the end of this century. The production system should be changed structurally to meet the national requirement to utilize small diameter domestic softwood harvested in forest plantations.
- Rationalization of production. Considering the steep rise in cost of labor and raw materials, the system of production should be rationalized further.

Table 4 shows the breakdown of mill deliveries of sawn lumber by end uses. In the case of sawn lumber, nearly 77 percent of production is delivered to the construction sector, including house building.

Table 3. Raw Material Used in Sawmills by Type of Logs (1,000 m³).

Year	Total	Domestic Total	Total	Imported from:				
				Southeast Asia	North America	USSR	New Zealand	Other
1977	53,871	20,526	33,345	7,264	15,196	7,241	1,100	2,544
1978	54,976	20,482	34,494	7,469	15,497	7,836	1,132	2,560
1979	56,012	21,461	34,551	6,813	16,824	7,072	1,371	2,471
1980	52,074	20,953	31,121	5,547	15,870	5,937	1,368	2,399
1981	45,945	19,527	26,418	4,621	13,438	5,408	1,072	1,897

Source: MAFF 1982.

Table 4. Mill Deliveries of Sawn Lumber by End Uses (1,000 m³).

Year	Total	Construction	Civil Engineering	Boxes, Packages	Furniture, Joinery	Shipbuilding, Wagons	Other
1977	38,171	29,355	1,234	3,090	2,819	259	1,434
1978	38,846	30,023	1,275	3,041	2,803	251	1,453
1979	39,579	30,695	1,293	3,050	2,785	286	1,470
1980	36,858	28,260	1,239	3,156	2,512	252	1,439
1981	32,557	24,921	1,116	2,959	2,114	196	1,251

Source: MAFF 1982.

Plywood Industry

Commercial production of plywood in Japan originated in 1907. The Japan Plywood Manufacturers' Association (JPMA) celebrated the 75th anniversary of the plywood industry in 1982. Thus, it can be said that the Japanese plywood industry has a long history of production.

Most of the development and expansion in the plywood market occurred in the first postwar decade. Import laws were liberalized and large quantities of best-suited raw materials, mainly lauan, became available. Also, use of improved synthetic resins for adhesives contributed to the market improvement of plywood quality.

In 1953, quantity exports of Japanese plywood to the United States, the United Kingdom, and other industrial countries began and have since gained a favorable reputation in these overseas markets.

As a result, the decade of the 1950s was the most prosperous period for the industry. Some 30 percent of total production was exported, thus proving to be a major component in the Japanese economy.

Such a large quantity of exports to the United States caused undesirable friction with American producers, however, and a movement toward restriction spread rapidly. As a result, Japanese manufacturers and exporters decided to impose export control measures in order to avoid more serious friction between the two countries.

At the same time, plywood production and export by log-producing countries such as the Philippines, and newly industrialized nations such as South Korea and Taiwan, grew rapidly. From the beginning of the 1960s, exports of lauan plywood from Japan declined rapidly. Since then, these countries have produced almost all the lauan plywood formerly supplied to overseas markets by Japan. Today, most of our plywood exports are from domestic hardwood species grown in the northern part of Japan, mainly in Hokkaido.

Fortunately, with the rapid growth of the Japanese economy, domestic demand for plywood for construction expanded quite steadily, allowing the Japanese plywood industry to enlarge its production capacity and improve its manufacturing equipment. The production capacity of the industry continued to grow until just before the worldwide oil crisis in 1973; but since then, the demand for building materials has declined, along with the slowdown of housing construction. This has brought about a severe depression in the industry, with many mill closings and bankruptcies occurring.

Total plywood production in 1982 is estimated to be some 1.7 billion square meters (4 mm base). Although Japan is still ranked as the world's most important plywood producer next to the United States, the industry is today confronting various problems such as industrial structure, future raw material supply, increased costs, and competition in the marketplace against foreign products. Now seems to be an important time for the Japanese plywood manufacturers to adapt to the future conditions.

Number of factories. The number of plywood factories is shown in table 5, along with the numbers of independent veneer mills and prefinished plywood processors. Independent veneer mills mainly peel or slice domestic species for decorative purposes, and their production volumes are relatively small.

As the demand for prefinished plywood turned upward in the latter part of the 1940s, many processors--mostly small producers--began operation.

Table 5. Number of Factories.

Year	Independent Veneer Mills	Plywood Factories	Independent Prefinishers
1971	55	272	396
1972	52	254	410
1973	60	257	423
1974	58	265	446
1975	57	252	417
1976	56	244	411
1977	60	224	410
1978	55	217	394
1979	55	212	387
1980	53	199	392
1981	46	184	391

Note: Figures were taken at the end of calender years.
Source: MAFF 1982.

In the last decade, the number of plywood factories in Japan has decreased from 272 in 1971 to 184 as of the end of 1981. The final figure as of the end of 1982 is not available at the time of writing, but it seems to be around 170, because more than a dozen factories have been closed owing to financial difficulties. According to the government statistics for the end of 1981, total workers employed in veneer and plywood factories were 40,600 workers. This figure is 13 percent less than the number of the previous year.

Production. In recent years, production of plywood has not been very active; total production in 1981 was 1,770 million square meters (4 mm base). This figure is low compared to almost 2,150 million square meters in the peak year 1973, as shown in table 6. The volume of plywood and prefinished plywood cannot be added together because the figures for prefinished mean the volumes of products delivered to consumers after prefinished at factories, and are included in the figures for plywood.

The final figures for 1982 are not available yet, but they are expected to be 1,700 million and 333 million for plywood and prefinished products respectively.

Recent decreases in production reflect the slowdown of plywood demand due to the general recession and especially the low level of housing construction.

The breakdown of prefinished plywood by type of product is shown in table 7.

Table 6. Production of Plywood (million m^2; 4 mm base).

Year	Plywood	Prefinished Plywood
1971	1,799	503
1972	1,937	510
1973	2,149	561
1974	1,860	499
1975	1,542	417
1976	1,783	424
1977	1,869	420
1978	2,010	425
1979	2,132	444
1980	2,001	393
1981	1,774	337

Note: Prefinished plywood is not 4 mm base, but surface measure.
Source: MAFF 1982.

Table 7. Production of Prefinished Plywood by Type of Product (million m^2; surface measure).

Year	Total	Overlaid Plywood				Printed Plywood	Color Plywood	Other
		Polyester	Polyvinyl chrolide	Diallyl-phtalate	Fancy Veneer			
1971	503	30	26	--	58	207	30	149
1972	510	29	27	12	67	216	34	125
1973	561	30	27	9	85	221	40	148
1974	499	26	23	7	73	197	37	135
1975	417	21	20	4	67	166	34	105
1976	424	26	20	3	75	155	35	111
1977	420	25	18	3	86	161	29	97
1978	425	24	18	3	86	166	31	97
1979	444	23	18	3	83	181	32	103
1980	393	25	16	3	83	155	25	86
1981	337	24	16	3	65	130	24	76

Source: MAFF 1982.

Production System and Technologies. Most plywood factories in Japan are well equipped with modern machinery. Although the average production capacity per factory is rather small compared with factories in Korea and some other countries, the production equipment in Japanese plywood factories is considered to be among the most modernized in the world. The manufacturing system used in most Japanese factories is specially suitable for making plywood from large-diameter lauan logs. Veneer peeling is carried out under a reeling-unreeling system. Deck-system, generally used in most American softwood plywood factories, is not yet used in Japan.

The technology for manufacturing plywood has been substantially improved since the 1950s. The introduction of veneer-composers--which efficiently utilize narrow veneers, core repeeling lathes, and other modern machinery--has brought about striking improvements in recovery rate. The usual recovery rate of factories which produce a mix of thin plywood and concrete-form plywood exceeds 65 percent or even 70 percent in the most efficient factories. This high recovery rate seems to have no relation to the size of plywood produced.

The main reason for technical progress is considered to be the result of Japanese plywood manufacturers' efforts in developing more efficient machinery to compensate for the recent rise of timber prices. In addition, the worldwide upward trend of energy prices promoted efforts to increase energy self-sufficiency among these manufacturers.

Plywood with less free formaldehyde emission, anti-insect plywood, fire-retardant plywood, and other technical plywoods have been developed and marketed to meet the recent changes in plywood demand.

The specifications for these plywoods have been published under Japanese Agricultural Standards. Furthermore, the Japanese government revised the Standard for Structural Plywood last December in order to meet the future raw material supply of softwoods as well as hardwoods. This revision will also contribute to the acceptance of the North American softwood structural panels in the Japanese market. (See Appendix 2.)

International Trade in Plywood. The 1950s were the golden age of Japanese plywood exports, with a rapid growth rate about one-third of its total production mainly exported to the United States and European countries. Nevertheless, as previously mentioned, in the 1960s--due to increasing exports of plywood from the Philippines, South Korea, Taiwan, and other developing countries--the share of

Japanese products in these overseas markets decreased rapidly.

Table 8 shows plywood export figures from Japan. In 1981, the export share of total production was only 0.8 percent. The main component was domestic species plywood exported to the United States.

Table 8. Exports of Plywood (1,000 m²).

Year	Total	Lauan Plywood	Plywood of Domestic Species	Pre-finished Plywood	Percentage of Export in Total Production
1971	81,807	14,100	30,702	37,005	6.0
1972	67,603	2,280	33,313	32,010	4.7
1973	38,857	286	25,074	13,497	2.6
1974	30,841	209	20,590	10,042	2.2
1975	28,958	269	20,384	8,305	2.4
1976	33,448	307	25,324	7,817	2.5
1977	34,875	608	27,592	6,675	2.7
1978	26,265	480	21,332	4,453	1.9
1979	20,205	359	14,616	5,233	1.4
1980	13,831	565	9,748	3,517	0.7
1981	14,244	692	9,337	4,215	0.8

Source: Ministry of Finance.

The quantity of plywood imported to Japan is shown in table 9. In general, it is relatively small compared with the domestic production. In 1981, total imports were only a bit more than 5 million square meters--only one-third of the figure of the previous year, mainly because the slowdown of the Japanese domestic plywood market.

Imports of plywood recovered somewhat in 1982, with 4,090 thousand square meters imported in the first nine months of 1982 as shown in table 9.

Further increases of softwood plywood imported from the United States and Canada are expected to accompany an increase in market acceptances of these structural materials, especially for housing construction.

There are two prominent features to note in table 9. First, there are big year-to-year fluctuations in quantity;

Table 9. Imports of Plywood by Country of Origin (1,000 m^2).

Imports by Country of Origin

Year	Total	U.S.	Canada	South Korea	China	Singapore	Malaysia	Philippines	Indonesia	Other
1971	14,481	153	148	6,453	4,033	467	8	15	--	3,204
1972	33,991	126	132	18,150	14,559	8	308	257	--	451
1973	194,743	938	196	86,028	97,729	3,804	1,962	5,027	--	1,095
1974	104,914	247	86	54,094	45,857	2,392	233	1,005	--	1,000
1975	36,910	243	63	24,063	11,594	68	1	105	--	773
1976	20,879	263	173	18,022	1,961	5	--	1	--	454
1977	10,340	230	80	8,066	938	--	92	--	--	934
1978	11,940	296	98	9,132	700	--	6	777	--	931
1979	13,358	471	686	9,148	907	59	829	362	--	896
1980	15,060	665	2,309	7,367	382	--	1,556	51	2,600	130
1981	5,077	504	1,706	543	167	--	527	17	1,595	180
1982*	4,090	541	1,472	951	90	--	186	32	794	24

*January to September.
Note: Some imports from Indonesia before 1979 are included in "Other."
Source: Ministry of Finance.

second, there are also large fluctuations in the country of origin. In a normal economy, domestic production could fill domestic demand for plywood; but in years of high growth there is excess demand, rising plywood prices, and increasing imports. Whether this historical trend of imports will be continued in the future we are not sure, because of current changes in economical circumstances.

Problems confronted. Although production of plywood in Japan is still relatively high, the plywood industry is confronted with many serious problems to be solved. All these problems seem to require continuous efforts, huge amounts of money, and long periods of time for settlement. In fact, the industry is said to be near a turning point.

- Future supplies of raw materials, especially imports of tropical hardwood logs. In 1981, the plywood industry consumed some 11 million cubic meters of South Sea logs, excluding domestic logs. Considering the forest resources and timber export policies of tropical wood-producing countries, it is doubtful if Japan will be able to secure a stable supply of tropical hardwoods in the form of logs. If imports of tropical hardwood logs decrease drastically in the near future, plywood production in Japan will have to change some of its raw material input to other materials, including domestically grown or imported softwood species. To accomplish this change, it will be necessary to change a part of its production system to adapt to the potential raw material supply.
- Structural adaptation. Considering the predicted future level of housing construction and related plywood requirements, it will probably be necessary to reduce the total production capacity further, and the structure of industry must be changed.
- Competition with foreign products. As the production of plywood in log-producing countries such as Indonesia increases, the competition with these products will be amplified in the domestic market. On the other hand, the fact cannot be avoided that the softwood structural panels made in Canada and the United States will gradually make inroads into the market. The Japanese plywood industry has to exert all possible efforts to reduce production costs by use of technical improvements, and to develop new products to meet market requirements in order to withstand strong competition.
- Competition with other building materials. Particleboard and fiberboard will become increasingly

competitive as construction materials because their production requires lower material and labor costs than plywood. Medium density fiberboard (MDF), which is still in the early stages of popularity in Japan, is also taking a step forward in specific markets such as furniture manufacture. The production of plywood should be rationalized to secure existing markets for plywood against these wood-based panels.

Glued-Laminated Lumber Industry

The history of the commercial production of glued-laminated lumber in Japan is relatively short, starting in 1951. The industry has developed as a result of a short supply of high-quality softwood building materials—a special favorite of many Japanese people. Glued-laminated wood products have been well accepted in the housing material market since the beginning of their manufacture. The major increase in production has, however, occurred since the beginning of the 1970s.

The glued-laminated lumber industry in Japan developed in a quite peculiar way compared with those of Europe and North America. In those countries, production of laminated wood concentrates mainly on large-dimension, stress-bearing structural components. In contrast, Japanese production is concentrated on the production of house interior components such as pillars, thresholds, and stair-step boards.

The most popular products are laminated pillars, widely used for wooden houses and interior works of multifamily construction. These products are made of glued, sawn softwood laminae and are covered on four surfaces with clear softwood sliced veneers. Species of core laminae are primarily imported species such as hemlock, spruce, larch, and fir. Decorative surfaces are overlaid by domestic clear softwood such as Hinoki cypress (<u>Chamaecyparis</u> <u>obtusa</u>), a traditionally precious species in Japan.

According to the Japanese Agricultural Standard for Glued-laminated Timber, laminated lumber is divided into two categories, decorative and structural. Even structural laminated timbers have decorative surfaces, however, and are mainly used in interior work, although they were glued together with exterior-type adhesives.

Since the Japanese people are very fond of softwood surfaces in residential interiors, these products seem to have a promising future, considering the shortage of high quality softwood lumber.

There were 211 laminated timber factories in Japan as

of the end of 1981. The greater part of these are relatively small-scale factories, mainly depending upon a local market. Seventy-four percent of these factories come under the category of less than 30 employees.

Production figures of glued-laminated timber are shown in table 10.

Table 10. Production of Glued-laminated Timber (1,000 m^3).

Year	Decorative	Structural	Total
1975	156.9	61.4	218.3
1976	166.8	72.9	239.7
1977	171.0	76.8	247.8
1978	175.0	86.2	261.2
1979	176.3	117.5	293.8
1980	169.2	114.8	284.0
1981	159.9	91.2	251.1

Source: Japan Laminated Timber Manufacturers' Association.

As shown in table 10, more than half of the production is decorative, but the gradual increase in structural laminated timber has been a recent production trend.

The manufacturing process of glued-laminated timber is somewhat labor intensive compared to wood-based panels, but in recent years some manufacturers constructed mechanized factories for fairly large-scale production. Mini-finger jointers are widely used for the longitudinal joint of lamina. High frequency heating for curing adhesives is also used by some manufacturers. Stress grading of sawn lumber strips is not yet so popular in Japan.

In recent years, interest has been rising among some manufacturers for the production of large-dimension, stress-bearing structural laminated timbers for building other than residential construction.

Particleboard and Fiberboard Industries

Particleboard. The commercial production of particleboard started in Japan in 1953. Small-scale production had been practiced even before that time, but the output was very small because particleboard manufacturing

Table 11. Production and Breakdown of Deliveries by End Uses of Particleboard.

Year	Production (1,000 m²)	Deliveries by End Use (%)						
		Furniture, joinery	Electric equipment	Construction	Sewing machine	Ship-building	Musical Instruments	Other
1975	45,395	38.1	22.9	29.8	1.3	0.9	3.6	3.4
1976	58,494	37.2	24.6	30.2	0.3	1.0	2.9	3.3
1977	60,085	43.3	21.8	27.9	0.4	0.9	2.6	3.1
1978	61,190	49.8	22.9	20.8	0.4	0.6	1.9	1.8
1979	81,769	51.9	22.9	21.1	0.3	0.4	1.6	1.8
1980	85,177	49.1	27.1	18.6	0.3	0.4	2.5	2.0
1981	74,105	51.1	25.8	16.8	0.3	0.4	2.9	2.7

Note: Breakdown of deliveries excluding thin particleboard.
Source: Japan Fiberboard and Particleboard Manufacturers' Association.

was only a sideline of plywood mills for processing their own mill residues.

Practical commercial production started in 1953, when two Japanese enterprises constructed medium-scale particleboard plants with technology and equipment imported from West Germany.

There were 19 particleboard factories (excluding four thin particleboard plants) in operation as of the end of 1981, many of them equipped with imported machinery from West Germany and other European countries.

Table 11 shows the figures for production and breakdown of deliveries of ordinary particleboard by end uses. Production increased steadily until 1980, but since then output has remained somewhat lower, reflecting the general economic situation.

The most important sector of usage is the manufacture of furniture and joinery. The construction sector--such as floor underlayment and roof decking--has been affording a stable market also. Adhesives used for binding are usually urea- or melamine-formaldehyde resins.

One of the remarkable characteristics of the Japanese particleboard industry is the fact that almost all of the raw materials consumed are mill residues from the industry's own mills or purchased from other mills. Utilization of small logs is negligible, as shown in table 12. Accordingly, the owners of particleboard plants are usually plywood manufacturers. Only two companies are operating independently by using only purchased wood chips for raw material.

Fiberboard. The commercial production of fiberboard started in 1953, using--as with the production of particleboard--imported technology and equipment.

As of the end of 1981, 14 factories were in operation (7 hardboard, 4 insulating board, and 3 producing both types of board). Among hardboard plants, eight factories adopted a wet system and the remaining two installed a dry system.

Production figures are shown in table 13. Output in recent years has been at almost the same levels. Delivery ratio of hardboard by end use is shown in table 14. The automobile industry is the most important sector of consumption, followed by the construction sector, which takes about 20 percent of total production.

As with the particleboard industry, raw materials consist almost exclusively of mill residues. The greater part of the raw material is purchased by mills in the form of wood chips from primary processors of timber.

Table 12. Raw Material Consumption of the Particleboard Industry (1981).*

Item	Own Mill Residues		Purchased Wood Chips		Logs		Total
	N (softwood)	L (hardwood)	N (softwood)	L (hardwood)	N (softwood)	L (hardwood)	
Domestic species	17	99	142	79	27	23	387
Imported species	4	196	86	721	--	--	1,007
Total	21	295	228	800	27	23	1,394

*Excluding thin particleboard.
Source: Japan Fiberboard and Particleboard Manufacturers' Association.

Table 13. Production of Fiberboard (1,000 m^2).

Year	Hardboard	Insulating Board	Total
1975	70,628	22,614	93,241
1976	75,634	26,984	102,617
1977	74,510	29,567	104,077
1978	75,748	27,576	103,324
1979	78,460	33,957	112,417
1980	82,051	34,732	116,783
1981	70,673	27,000	96,673

Source: Ministry of Trade and Industry.

Table 14. Breakdown of Deliveries of Hardboard by End Uses (%).

Year	Auto-mobiles	Con-struction	Furni-ture	Electric Equip-ment	Pack-ages	Ship-building, Wagons	Other
1975	32.2	28.7	13.2	11.6	6.8	1.5	6.0
1976	34.7	28.5	11.6	11.3	7.9	1.1	4.9
1977	38.6	24.1	11.2	11.3	8.4	0.8	5.6
1978	38.2	20.2	12.8	11.0	9.7	0.7	4.7
1979	34.9	21.7	14.0	11.4	9.5	0.7	5.0
1980	38.5	20.1	12.6	12.9	8.1	1.1	6.7
1981	41.5	18.2	10.7	13.6	7.4	2.1	6.5

Source: Japan Fiberboard and Particleboard Manufacturers' Association.

An MDF factory in operation is so far the only one. Its nominal production capacity is sixty thousand (60,000) tons per annum. Statistical figures for the MDF are included in the item of hardboard.

APPENDIX 1: JAPANESE INDUSTRIAL ASSOCIATIONS

Name	Address	Phone
Federation of Timber-cooperative Societies in Japan	Nagata-cho Bldg. 2-4-3, Nagato-cho Chiyoda-ku, Tokyo 100	(03)-580-3215
Japan Lumber Importer's Association	Ushi Kogyo Kaikan Bldg. 3-13-11, Nihonbashi, Chuo-ku, Tokyo 103	(03)-271-0926
Japan Plywood Distributors' Association	Miyaji Bldg. 4-30-10, Taito, Taito-ku, Tokyo 110	(03)-832-2976
Japan Timber Products Storage Corporation	Rinyu Bldg. 1-7-12 Koraku, Bunkyo-ku, Tokyo 112	(03)-816-5595
Wood Technological Association of Japan	Ogi Bldg. 1-3-10 Shibadaimon, Minato-ku, Tokyo 105	(03)-432-3053
Japan Plywood Manufacturers' Association	Meisan Bldg. 1-18-17, Nishi Shinbashi, Minato-ku, Tokyo 105	(03)-591-9246
Japan Pre-finished Plywood Manufacturers' Association	Sasaki Bldg. 2-13-7, Nishi Shinbashi, Minato-ku, Tokyo 105	(03)-501-3684
Japan Plywood Inspection Corporation	Meisan Bldg. 1-18-17 Nishi Shinbashi, Minato-ku, Tokyo 105	(03)-591-7438
Japan Fiberboard and Particleboard Manufacturers' Association	Tanaka Yaesu Bldg. 1-5-15, Yaesu, Chuo-ku, Tokyo 103	(03)-271-6883
Japan Wood-flooring Manufacturers' Association	Mokuzai Kaikan Bldg. 2-5-11, Fukagawa, Koto-ku, Tokyo 135	(03)-643-2948
Japan Laminated Timber Manufacturers' Association	Takane Daini Bldg. 2-22-4, Nishi Shinbashi, Minato-ku, Tokyo 105	(03)-424-6527
Japan Wood Preservers' Association	Daisan Matsuzaka Bldg. 4-2-5, Toranomon, Minato-ku, Tokyo 105	(03)-436-4486
Japan Wood Processing Machinery Manufacturers' Association	Kikai Shinko Kaikan Bldg. 3-5-8, Shiba Koen, Minato-ku, Tokyo 105	(03)-433-6511
Japan Housing and Wood Technology Center	The 21st Mori Bldg. Annex 2-2-5, Roppongi, Minato-ku, Tokyo 106	(03)-583-8831

APPENDIX 2: JAPANESE AGRICULTURAL STANDARD FOR STRUCTURAL PLYWOOD

Effective Date
January 14, 1983

THE JAPAN PLYWOOD INSPECTION CORPORATION

Japanese Agricultural Standard for Structural Plywood

(Notification No. 2047 of the Ministry of Agriculture,
Forestry and Fisheries. December 14, 1982)

(Scope of Application)
Article 1. This standard shall be applied to structural plywood.
(Definition)
Article 2. The terms in the left column of the following table shall be as used herein as defined respectively in the right column:

Term	Definition
Structural plywood	Of the ordinary plywood (as prescribed by Article 2 of the Notification No. 383 of the Ministry of Agriculture, Forestry and Fisheries, April 11, 1964, establishing the Japanese Agricultural Standard for Ordinary Plywood), other than the plywood for fire-retardant shutters (as prescribed by Article 1 of the Notification No. 29 of the Ministry of Agriculture, Forestry and Fisheries, January 16, 1961, establishing the Japanese Agricultural Standard for Plywood for Fire-retardant Shutter), the plywood for concrete forms (as prescribed by Article 2 of the Notification No. 932 of the Ministry of Agriculture, Forestry and Fisheries, June 30, 1967, establishing the Japanese Agricultural Standard for Concrete Form Plywood), fire-retardant plywood (as prescribed by Article 2 of the Notification No. 1869 of the Ministry of Agriculture, Forestry and Fisheries, December 5, 1969, establishing the Japanese Agricultural Standard for Fire Retardant Plywood, plywood for scaffolds (as prescribed by Article 2 of the Notification No. 771 of the Minitry of Agriculture, Forestry and Fisheries, May 17, 1972, establishing the Japanese Agricultural Standard for Plywood for Scaffold), nonflammable plywood (as prescribed by Article 2 of the Notification No. 1650 of the Ministry of Agriculture, Forestry and Fisheries, September 11, 1972, establishing the Japanese Agricultural Standard for Nonflammable Plywood) and plywood for pallets (as prescribed by Article 2 of the Notification No. 6 of the Ministry of Agriculture, Forestry and Fisheries, January 7, 1977, establishing the Japanese Agricultural Standard for Plywood for Pallet), those plywood to be used for important parts from structural resistance in building constructions.
Type special	Structural plywood those are used outdoors, or at constantly wet places, and have passed the continuous boiling test as prescribed in Supplementary provisions 3-(1).
Type I	Structural plywood those are used indoors, and have passed the repeated boiling test as prescribed in Supplementary provisions 3-(2).

(Standards)
Article 3. The standards for structural plywood shall be specified as follows:

Item	Standard Specification	
	Class 1	Class 2
Degree of adhesion	Must come under either type special or type I	Ditto
Moisture content	Must pass the moisture content test as prescribed in Supplementary provisions 3-(3)	Ditto
Bending strength and Young's modulus	Must pass the bending test for Class 1 as prescribed in Supplementary provisions 3-(4)	Must pass the bending test for Class 2 as prescribed in Supplementary provisions 3-(4)
Compression strength	Must pass the compression test for Class 1 as prescribed in Supplementary provisions 3-(5)	
Quality of face and back	The grades given in the following table must conform to their respective quality standards for layer surfaces as specified in the following paragraph:	The grades given in the following table must conform to their respective quality standards for layer surfaces as specified in the following paragraph:
	Grade / Face / Back A-1 a b A-2 b b B-1 a c B-2 b c B-3 c c C-1 a d C-2 b d C-3 c d	Grade / Face / Back A-B a b A-C a c A-D a d B-B b b B-C b c B-D b d C-C c c C-D c d
Quality of core or cross band	Shall be conformed to the requirements for core and cross band for Class 1 as specified under paragraph 3	Shall be conformed to the requirements for core and cross band for Class 2 as specified under paragraph 3
Materials		Materials of face, back, core and cross band shall be those of Engelmann spruce (Picea engelmannii Parry) or better in strength
Composition of veneers	The number of plies, thickness and composition of veneers must conform to the	The number of plies, thickness and composition of veneers must conform to the standards as

	(Column 1)	(Column 2)
	standards as specified in the following table. In this case, the core or cross band made of veneers glued over each other in parallel grain directions should be regarded as one	specified in the following table:

(Unit: mm)

Thickness of panel	No. of plies	Thickness of veneers and composition
5.0	3	1.5-2.0-1.5
6.0	3	1.5-3.0-1.5
7.5	5	1.5-1.5-1.5-1.5-1.5
9.0	5	1.5-2.25-1.5-1.25-1.5
12.0	5	1.5-3.0-3.0-3.0-3.0-1.5
15.0	7	1.5-3.0-1.5-3.0-1.5-3.0-1.5
18.0	7	1.5-3.0-3.0-3.0-3.0-3.0-1.5
21.0	7	2.25-3.5-3.0-3.5-3.0-3.5-2.25
21.0	9	2.4-3.0-2.4-3.0-2.4-3.0-2.4-3.0-2.4

(Unit: mm)

Thickness of panel	No. of plies	Thickness of veneers	Composition
Less than 15.0	3 or more	1.5 or more	Total thickness of veneers parallel to the face veneer shall be not less than 40% and not more than 70% of panel thickness
Not less than 15.0, less than 18.0	4 or more		
Not less than 18.0, less than 24.0	5 or more	5.5 or less	
24.0 or more	7 or more		

	Column 1	Column 2
Warping or testing	Shall not severely injure the serviceability of the panel	Ditto
Dimensions	1. The standard width and length shall be as specified below:	1. The standard width and length shall be as specified below:

(Unit: mm)

Width	Length
910	1,820
910	2,130
910	2,440
910	2,730
955	1,820
1,000	2,000
1,220	2,440
1,220	2,730

2. The specified thickness shall be as follows

(Unit: mm)
5.0, 6.0, 7.5, 9.0, 12.0, 15.0, 18.0, 21.0, 24.0

(Unit: mm)

Width	Length
900	1,800
900	1,818
910	1,820
910	2,130
910	2,440
910	2,730
955	1,820
1,000	2,000
1,220	2,440
1,220	2,730

2. The recommended standard thickness shall be as follows:
(Unit: mm)
5.0, 6.0, 7.5, 9.0, 12.0, 15.0, 18.0, 21.0, 24.0

3. The discrepancies between the marked and measured dimensions for the items given in the left column of the following table shall be as specified in the right column respectively:

(Unit: mm)

Marked dimension		Discrepancy from measured dimension	
Thickness	5.0	+0.3	-0
	6.0	+0.3	-0
	7.5	+0.4	-0
	9.0	+0.5	-0
	12.0	+0.6	-0
	15.0	+0.8	-0
	18.0	+0.9	-0
	21.0	+1.1	-0
	24.0	+1.2	-0
Width and length		+0	-3

3. The discrepancies between the marked and measured dimensions for the items given in the left column of the following table shall be as specified in the right column respective:

(Unit: mm)

Marked dimension		Discrepancy from measured dimension	
Thickness	7.5 or less	+0.5	-0.3
	More than 7.5	+0.8	-0.5
Width and length		+0	-0.3

4. Diagonals shall not be differed by more than 3 mm in length

4. Diagonals shall not be differed by more than 3 mm in length

Marking	Marking	The following matters shall be marked in one block: (1) Name of product (2) Dimensions (3) Name or company name of the manufacturer or distributor	Ditto
	Method of marking	1. The matters for marking (1) and (2) shall be given in the following ways: (1) Name of product Mark as "Structural Plywood" (2) Dimensions Mark the thickness, width and length of the panel specifying they are in milimeter, centimeter, or meter 2. The matters specified for marking shall be given at proper position of each panel or package	Ditto
	Prohibited marking	The following matters shall not be marked: (1) Any term that contradicts the markings as required above (2) Any wording or other indication that tends to cause misunderstanding of the quality of the product	Ditto

2. The standards for the quality of veneers as prescribed in the preceding paragraph shall be as follows:

Item	Standards			
	a	b	c	d
Total of the widthwise dimension of sound knot, dead knot, loose knot, knot hole, open split, chip, joint gap, cross check, linear worm hole and patch	Not more than 1/20 of panel width	Not more than 1/15 of panel width	Not more than 1/10 of panel width (in case of Class 2, not more than 1/5 of panel width, but if thickness of face and back of the panel is not less than the thickness given in appendix table, up to 1/2 of panel width will be permitted)	Not more than 1/7 of panel width (in case of Class 2, not more than 1/5 of panel width, but if the maximum widthwise diameter of sound knot, dead knot, loose knot or knot hole is less than 65 mm, and the thickness of face and back of the panel is not less than the thickness given in appendix table, up to 1/2 of panel width will be permitted)
Sound knot or dead knot	Not more than 25 mm in diameter along the width of the panel	Not more than 40 mm in diameter along the width of the panel	Not more than 50 mm in diameter along the width of the panel	Not more than 75 mm in diameter along the width of the panel
Loose knot or knot hole	Not more than 3 mm in diameter along the width of the panel	Not more than 5 mm in diameter along the width of the panel	Not more than 40 mm in diameter along the width of the panel	Not more than 75 mm in diameter along the width of the panel
Patch	Not more than 50 mm in diameter along the width of the panel	Not more than 100 mm in diameter along the width of the panel	Ditto	Ditto
Bark pocket or resin pocket	Not more than 30 mm in long span	Not conspicuous	Not severely injure the serviceability of the panel	Ditto
Burl or vine streak	Slight	Not conspicuous	Not severely injure the serviceability of the panel	Ditto
Decay	Not permitted	Ditto	Ditto	Ditto
Open split (including chip or joint gap)	Not more than 20% of panel length in length, not more than 1.5 mm in width and not more than 2 in number	Not more than 40% of panel length in length, not more than 6 mm in width and not more than 3 in number; or not more than 20% of panel length in length, not more than	1. Split on panel shall not exceed 6 mm in width which located within 25 mm of parallel panel edge 2. In case of split which located other than the area written above,	1. Split on panel shall not exceed 6 mm in width which located within 25 mm of parallel panel edge 2. In case of split which located other than the area written above, (1) Split on panel

			6 in number	(1) the length of split which does not exceed 10 mm in width at a point 200 mm from the end of panel shall not be limited, provided that end of the split is tapered, or the length of split which does not exceed 15 mm in width at a point 200 mm from the end of the panel shall not exceed 50% of the panel length (2) Split on panel shall not exceed 50 mm in width when located within 200 mm from the end of the panel	shall not exceed 25 mm in width at a point 200 mm from the end of the panel and the end of split shall be tapered (2) Split on panel shall not exceed 75 mm in width when located 200 mm from the end of the panel
Cross check	Not permitted	Ditto		Very slight	Ditto
Worm hole	1. In case of round bores, not more than 1.5 mm in long span and not gathered 2. In case of linear bores, not more than 10 mm in long span and not more than 4 times the number of the panel are in square meter (rounding up any fraction to the unit) in number	Not conspicuous		Not severely injure the serviceability of the panel	Ditto
Other defects	Slight	Not conspicuous		Ditto	Ditto

(Note) "Total of the widthwise dimension of sound knot, dead knot, loose knot, knot hole, open split, chip, joint gap, cross check, linear worm hole and patch" is obtained by adding up all the diameter widthwise, width or length of these defects in the panel portion of 300 mm wide at right angle to its length where these defects are most numerous.

3. The standards for the core and cross band as prescribed in the preceding paragraph shall be as follows:

Item	Standard	
	Class 1	Class 2
Total of the widthwise dimension of sound knot, dead knot, loose knot, knot hole, open split, chip, joint gap, cross check, linear worm hole and patch	Not more than 1/5 of the panel width	
Sound knot	Not more than 75 mm in the widthwise diameter	
Dead knot, loose knot or knot hole	Not more than 75 mm in the widthwise diameter	Not more than 75 mm in the widthwise diameter. In case of veneer at least 2 plies removed from face or back, not more than 90 mm in the widthwise diameter
Bark pocket or resin pocket	Not severely injure the serviceability of the panel	Ditto
Burl or vine streak	Nor severely injure the serviceability of the panel	Ditto
Decay	Not permitted	Ditto
Open split (including chip or joint gap)	1. Shall not exceed 6 mm in width when located within 25 mm from parallel panel edge 2. In case of split which located other than the area written in the above item 1, (1) The width shall not exceed 25 mm at a point 200 mm from the end of the panel and the end of the split shall be tapered (2) The width shall not exceed 75 mm at a point 200 mm from the end of the panel	Ditto
Cross check	Very slight	Ditto
Worm hole	Not severely injure the serviceability of the panel	Ditto
Other defects	Not conspicuous	Ditto

(Note) Same as the "Note" for the preceding paragraph.

Supplementary Provisions

1. Sampling of Test Pieces

The panels from which the test specimens should be cut off (hereafter called "Sample panels") for the continuous boiling test, cyclic boiling test, moisture content test, bending test, and compression test shall be taken at random in numbers specified in the right column of the following table according to the size of the lot as given in the left column:

No. of panels of a lot	No. of sample panels	
1,000 or less	4 or less	In case of retesting, these number should be doubled
1,001 – 2,000	6	
2,001 – 3,000	8	
3,001 or more	10	

2. Judgment of Test Results

Of the test pieces collected from each sample panel as specified in the preceding article, the panels of lot concerned shall be judged to have passed the particular test when 90 percent or more of the test specimens from the lot conform to the standards for the test, but they shall be judged to have failed in the test when the percentage is less than 70 percent. In case the percentage is more than 70 percent but less than 90 percent, a retest procedure shall be carried out with the number of sample panel specified for retesting. In this case, if the percentage is 90 percent of more, the panels of the lot concerned shall be judged to have passed the particular test, but if it is less than 90 percent, they shall be judged to have failed in the test.

3. Methods of Testing

　(1) Continuous Boiling Test

　　a) Preparation of test specimen

　　　The test specimen shall be prepared out of sample panel as follows:

　　　(a) As for the 3-ply plywood, 4 test specimens of the shape as indicated in figure 1 shall be made out of each sample panel (type A in case of the plywood with face veneer measuring more than 1.6 mm in thickness, and type B in case of the plywood with its face veneer measuring 1.6 mm or less in thickness or of the plywood whose failures occur in face veneer when type A specimens are used). In this case, the kerfing shall be cut in such a way that an equal number of test specimens out of each sample panel have the lathe check of core veneer in the closed or open direction with that of the load.

　　　(b) As for the 5-ply, 7-ply, and 9-ply plywood, they shall be made into 3-ply by stripping off veneers in such a way that all the glued joints of the original sample panel are included in the test specimens. The test specimens shall be prepared by above (a).

Figure 1

Type A
Directions of lathe check of
core veneer and load are same
(closed type)

(Unit: mm)
Type B
Directions of lathe check of
core veneer and load are same
(closed type)

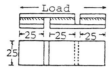

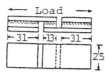

Directions of lathe check of
core veneer and load are
reverse (open type)

Directions of lathe check of
core veneer and load are
reverse (open type)

(Note) In case of softwood veneer is included in the composition of plywood, kerfing shall be extended two-thirds of the core ply.

b) Test procedure
Test specimens shall be submerged in boiling water for 72 hours, then cooled off in water at room temperature. Adhesive strength test shall be carried out while wet (by gripping both ends of the specimen, tensile loading at a rate of not greater than 600 kg/minute shall be applied to failure occurs, and measure the maximum breaking load; hereafter the same) to obtain the shear strength and average wood failure. In case of composed by hardwood species exclusively, measure the shear strength only.

(Note) The shear strength is calculated by the following formula. In case the thickness of ratio of core to face is 1.5 or more, multiply the factors obtained by the right column of the following table by the core/face thickness ratio.

Shear strength (kg/cm^2) = $\dfrac{Ps}{b \times h}$

where,
Ps = maximum load (kg)
b = width of test specimen (cm)
h = length between kerfing (cm)

Core/face thickness ratio	Factor
1.50 or ±2.00	1.1
2.00 or ±2.50	1.2
2.50 or ±3.00	1.3
3.00 or ±3.50	1.4
3.50 or ±4.00	1.5
4.00 or ±4.50	1.7
4.50 or more	2.0

c) Requirement for passing of specimen
Shear strength and average wood failure (%) shall be equal to or greater than the following standard values:

Wood species used in test specimen		Average wood failure (%)	Shear strength (kg/cm^2)
Hardwoods	"Kaba"		10
	"Buna," "Nara,"		9
	"Itayakaede," "Akadamo," "Shioji," "Yachidamo"		
	"Sen," "Hoo," "Katsura," "Tabu"		8
	"Lauan," "Shina," and other hardwood species		7
Softwoods			7
		50	6
		65	5
		80	4

(Note) In case of a test specimen made up of a combination of veneers of different species, the lowest standard value of shear strength of species involved shall be applied.

(2) Cyclic Boiling Test
   a) Preparation of test specimen
      The provisions of (1)-a) shall be applied mutatis mutandis.
   b) Test procedure
      Test specimens shall be submerged in boiling water for 4 hours, and then dried at a temperature of 60 ± 30°C for 30 hours again submerged in boiling water for 4 hours, and then cooled off in water at room temperature. Shear strength test with the specimens while wet to obtain shear strength and average wood failure.
   c) Requirement for passing of specimen
      The provisions of (1)-c) shall be applied mutatis mutandis.

(3) Moisture Content Test
   a) Preparation of test specimen
      Prepare two test specimens of an appropriate size out of each sample panel.
   b) Test procedure
      Determine the percentage moisture content by bone dry weight method. Some other methods than bone dry weight method may be applied in case such methods can clearly determine whether the specimen meets the required standards or not.

(Note) The bone dry weight is the weight at the time when the test specimen, being dried in a drier at a temperature between 100°C to 105°C is recognized to have reached to a constant weight; the moisture content (%) is to be calculated by the following formula:

$$\text{Moisture content (\%)} = \frac{W_1 - W_2}{W_2} \times 100$$

where,
W_1 = weight before drying (g)
W_2 = bone dry weight (g)

   c) Requirement for passing of specimen
      The average moisture content of the test specimens prepared from same sample panel shall be not more than 14 percent.

(4) Bending Test
    a. Class 1
    a) Preparation of test specimen

Prepare out of each sample panel two each of two different rectangular test specimens. As indicated in figure 2, one should be 50 mm in width at right angle with the main grain direction of face veneer, by 24 times of the plywood thickness plus 50 mm in length in parallel with the main grain direction; the other should be 50 mm in width in parallel with the main grain direction of face veneer, by 24 times of the plywood thickness plus 50 mm in length at right angle to the main grain direction of face veneer.

Figure 2

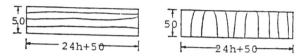

    b) Test procedure

As indicated in figure 3, carry out the test with the test specimens, placing the main grain direction of face veneer either in parallel or at right angle with the span direction. Measure the upper and lower limit loads within the proportionality range, corresponding strain and maximum load, and work out the bending strength and Young's modulus. The average loading rate applied should be not more than 150 kg/cm² per minute.

(Note) The bending strength and Young's modulus are to be calculated by the following formula:

$$\text{Bending strength (kg/cm}^2) = \frac{3 P_b l}{2 b h^2}$$

$$\text{Young's modulus (kg/cm}^2) = \frac{\Delta p \, l^3}{4 b h^3 \Delta y}$$

where,
P_b = maximum load (kg)
l = length of span (cm)
b = width of test specimen (cm)
h = thickness of test specimen (cm)
Δp = difference between upper and lower limit loads within proportionality range (kg)
Δy = strain at mid-span corresponding to Δp (cm)

Figure 3

(Unit: mm)

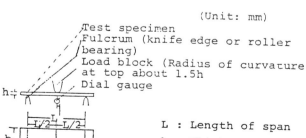

L : Length of span
h : Thickness of test specimen
b : Width of test specimen

c) Requirement for passing of specimen
The bending strength and Young's modulus shall be not less than the values given in the following table:

Thickness (mm)	Bending strength (kg/cm²)						Young's modulus (10³ kg/cm²)	
	0°			90°			0°	90°
	A	B	C	A	B	C		
5.0	420	380	340	80	80	80	85	5
6.0	380	360	320	140	140	140	80	10
7.5	340	320	280	120	120	120	70	20
9.0	320	280	260	160	160	160	65	25
12.0	260	240	220	200	200	200	55	35
15.0	240	220	200	200	200	200	50	40
18.0	240	220	200	200	200	200	50	40
21.0	260	240	220	180	180	180	55	35
24.0	260	240	220	180	180	180	55	35

(Note) "0°" and "90°" of the table mean the angle of span direction to the main grain direction of face veneer of the specimen, prescribed in Supplementary Provisions 3-(4)-b) "A," "B," and "C" refer to those standards of Article 3 that are largely concerned with strength.

b. Class 2
a) Test procedure
As indicated in figure 4, full-size test panel should be placed with face upside on the effective length (width of the panel) of loading piece, which is placed in the middle of span at a right angle to the face grain. Appropriate load should be applied according to the thickness, width and legnth of the panel. Young's modulus should be calculated by measuring the strain.

(Note) Young's modulus is to be calculated by the following formula:

$$\text{Young's modulus in bending (kg/cm}^2\text{)} = \frac{\Delta P \; l^3}{4 \; b \; h^3 \; \Delta y}$$

where
l = length of span (cm)
b = width of test panel (cm)
h = thickness of test panel (cm)
ΔP = difference between upper and lower limit loads within proportionality range (kg)
Δy = strain at mid-span corresponding to ΔP (cm)

b) Requirement for passing of specimen
Young's modulus in bending should not be less than the values listed in the following table:

Thickness (mm)	Young's modulus in bending (10³ kg/cm²)
5.0	65
6.0	60
7.5	55
9.0	50
12.0, 15.0, 18.0, 21.0, 24.0	40

(Note) As for the specimens of which thickness are between the figures of thickness listed above, Young's modulus in bending should be not less than the values calculated in proportion of thickness.

Figure 4

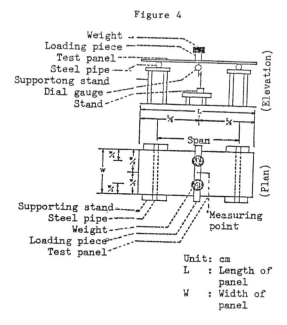

Unit: cm
L : Length of panel
W : Width of panel

(5) Compression Test
   a) Preparation of test specimen
      Prepare out of each sample panel two each of three different rectangular test specimens. As indicated in figure 5, one should be 50 mm in parallel by 25 mm at right angle to the main grain direction of face veneer; the other should be 50 mm at right angle by 25 mm in parallel to the main grain of direction of face veneer; and the third should be 50 mm for one side at 45° to the main grain direction of face veneer by 25 mm for the other side. In case of sample panels of 5.0 mm, 6.0 mm, and 7.5 mm in thickness, however, four each of these test specimens should be prepared, and two pieces having the same main direction be glued over each other.

Figure 5

(Unit: mm)

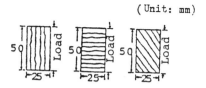

   b) Test procedure
      As indicated in figure 5, carry out the test with the test specimens for the three cases of load direction either in parallel, at right angle or at 45° with the main grain direction of face veneer of the test specimen and measure the maximum load and compression strength. The average loading rate applied should be not more than 100 kg/cm² per minute.

(Note) The compression strength is to be worked out by the following formula:

$$\text{Compression strength (kg/cm}^2\text{)} = \frac{P_c}{A}$$

where
P_c = maximum load (kg)
A^c = area of cross section of test specimen (cm²)

c) Requirement for passing of specimen
The compression strength of specimen shall not be less than the values given in the following table (kg/cm²):

Thickness (mm)	0°			90°			45°
	A	B	C	A	B	C	
5.0	160	140	140	90	90	90	80
6.0	140	120	120	120	120	120	80
7.5	140	140	120	90	90	90	80
9.0	120	120	110	120	120	120	80
12.0	120	120	110	120	120	120	80
15.0	110	90	90	140	140	140	80
18.0	120	120	110	120	120	120	80
21.0	120	120	110	120	120	120	80
24.0	120	120	110	120	120	120	80

(Note) "0°," "90°," and "45°" given in the above table show the angles of the main grain direction of face veneer of the specimen to the load direction as defined in Supplementary Provisions 3-(5)-b). "A," "B," and "C" refer to those standards of Article 3 which are largely concerned with strength.

Appendix table (concerning Articles 3-2, standards for veneers c and d)
(Unit: mm).

Thickness of panel	Thickness of face and back veneer			
	3 or 4-ply	5 or 6-ply	7 or 8-ply	9-ply or more
7.5 or ±9.0	2.5	-	-	-
9.0 or ±12.0	2.5	-	-	-
12.0 or ±15.0	2.5	2.0	-	-
15.0 or ±18.0	3.0	2.5	-	-
18.0 or ±21.0	-	3.0	-	-
21.0 or ±24.0	-	4.0	3.0	-
24.0 or more	-	-	3.5	3.0

Additional Provision
1. This notification becomes effective on January 14, 1983.

A MARKET FOR U.S. WOOD

John B. Crowell

This symposium offers the chance to discuss the U.S. Department of Agriculture's views and look ahead toward opportunities for timber growers and for the forest products industry to increase U.S. exports of wood to world markets. World demographics alone demonstrate an enormous future potential market for wood and wood products. Already there is a great discrepancy between the location of the world's most utilizable and productive forest resources and the areas of both greatest population and highest population growth. The future promises to intensify this discrepancy, meaning that forest resource-rich nations, such as the United States, are likely to have vast world markets for wood and wood products. At the same time, the United States clearly has the potential to produce substantially more softwood and hardwood than our domestic markets will require.

This points toward vast opportunities for greater exports to world markets. But there are four IFs we must come to grips with in order to realize those opportunities: (1) if some major policy decisions about managing the public forests are made and implemented so as to realize their full productive potential, (2) if investment opportunities on nonindustrial private U.S. forest ownerships are utilized, (3) if U.S. producers are willing to make permanent commitments to supplying manufactured wood products to foreign markets, and, last but not least, (4) if trade barriers barring or complicating access to major foreign markets can be overcome.

A Role for the Federal Government

The U.S. government can take three major roles to help increase U.S. export of forest products:
1. The federal government can and should make an important contribution toward ensuring a reliable timber supply. We can establish and carry out policies which

encourage the investment climate essential to improving the productivity of all forest lands.

2. We can directly improve management of public forest lands to increase their productivity within economically reasonable and environmentally sound limits.

3. We can help meet the substantial need for research on international trade, since there are several areas which we need to understand better if we are to increase exports effectively.

First Step--A Reliable Supply

Obviously, if we are to meet and expand markets for U.S.-grown wood and wood products, the first requirement is to ensure a reliable supply. The United States has the third largest softwood forest acreage in the world, behind Canada and the Soviet Union, but because our forests generally enjoy better growing sites and longer growing seasons, U.S. annual growth exceeds that of Canada, and approaches that of the Soviet Union. Utilization of the Soviet Union's forests is handicapped by inaccessibility and by limited transportation systems. Consequently, the U.S. is the world's largest producer of wood products. We are the second largest wood exporter--only Canada exceeds us, and that is simply because of its exports to the United States.

The U.S. government in general, and the Forest Service in particular, can make an important contribution toward ensuring a reliable timber supply. The government can set regulatory, investment, and tax policies which encourage productivity of privately owned forest lands. Many of these have been discussed by other speakers. One of the things that hasn't been discussed thoroughly is the role of the federal government as a forest landowner.

Marked Acceleration of Harvests

Especially in the western states, the federal government is a major landowner that can play a very significant role. The federal government controls a significant amount of the available resource in the West. At present the Forest Service controls about 55 percent of the standing softwood sawtimber in Oregon, Washington, and California, compared to less than 10 percent of the softwood sawtimber in the South. State by state on the Pacific Coast, the figures for national forest ownership of softwood sawtimber, and the percentages of the total for each state, are:

Oregon: 252.8 billion board feet, 61 percent
Washington: 133.8 billion board feet, 43 percent
California: 158.0 billion board feet, 62 percent

The Pacific Northwest national forest inventories reflect, primarily, vast old-growth timber stands. Much of the old growth has already been harvested on industry lands, but not on the national forests. While we don't have figures which define old growth, per se, we do know that in Oregon, Washington, and California, 717.2 billion board feet of sawtimber are from trees of 19-inch or more diameter, and equal about 73 percent of the softwood sawtimber inventory in that area.

There can be no doubt that we must increase national forest harvests if we are substantially to increase softwood sawtimber production over the next several decades. We must look to the national forests which have large inventories of old-growth softwood sawtimber. We are presently losing unacceptably large volumes of that inventory to mortality and disease, estimated at 4 billion board feet annually. Further, because of growth characteristics of these stands, their net annual growth is well below the biological and economic potential of the lands they occupy. By regenerating nonstocked areas, by harvesting and regenerating mature stands, and then by applying intensive management measures, such as spacing control and use of genetically improved trees, we can greatly increase current growth and can offer the prospect of continued high harvest levels from the western national forests for the indefinite future. And we can do so while managing the national forests for all of the various multiple uses, such as recreation, wildlife, grazing, water, and wilderness.

The Department of Agriculture has been making a major effort to make the national forests a more reliable supplier of timber and to put the sale of timber on a more economic and businesslike basis. We are also making efforts to improve the efficiency with which national forest lands are managed. We have adopted national forest planning regulations that emphasize economic efficiency and require the evaluation of economically optimal forest management regimes. These regulations also require the evaluation of departures to nondeclining yield harvest schedules. Such departures can result in increased harvest levels and also reduce the time it takes to bring a forest area under management.

Nonindustrial Private Ownerships

Just as national forests are the focus of increased production in the Pacific Northwest, nonindustrial private ownerships are essential to increasing forest productivity in the South. Long-term domestic demands and possibilities for export can be met by increasing the amount of investment on nonindustrial private forest lands, which make up 59 percent of this nation's commercial forest land. Much of that ownership is concentrated in the South.

Two recent Forest Service studies, however, cast considerable doubt on the accuracy of the conventional wisdom about these lands. This conventional wisdom, while recognizing the potential opportunities represented by nonindustrial private lands, has generally been pessimistic about overcoming the obstacles to realizing those opportunities. The studies are entitled: "Private Forestland Owners of the United States" and "Forest Management Decisions on Harvested Southern Forestlands: Why Landowners Do or Do Not Reforest Their Lands to Southern Pine." The studies found that 74 percent of the nonindustrial private lands is in ownerships of more than 100 acres, and is, therefore, in units which are reasonably cost effective to manage; that timber production becomes increasingly important as a landowner objective on the larger holdings; that nonindustrial private lands appear to be better managed and have more growing stock than has generally been assumed; that owners are making more effort than is generally realized to regenerate their harvested lands; and that forestry investments, in general, are financially more attractive than has often been assumed.

So there is great potential here. The federal government has a role in helping tap that potential. The Department of Agriculture has recently initiated a major effort whose objective is to forge a consensus among the various groups interested in nonindustrial private lands on what should be done to increase investments on these lands.

Industrial Ownerships

I would be remiss if I neglected to mention the potential of industrial lands. They make up 14 percent of U.S. commercial forest land, but account for 36 percent of the U.S. softwood production, making them the most productive forest ownership. As evidenced by what has been said at this conference, many companies, both large and small, are already showing great interest in increasing exports, and are taking the initiative to ensure they are out in front in capturing export opportunities. It is my understanding that some companies that own forest land are making forest management investments for the primary purpose of serving future export markets.

The Case for Increased Exports

We have highly productive forests. We have the potential to increase timber production from those forests rather markedly if we have the will to do so. In addition, we have several other advantages in world trade:

. This country is situated advantageously for access to worldwide markets, with ports on two oceans and the Gulf of Mexico. Clearly the Pacific Coast states, including Alaska, can serve the potentially vast Far Eastern markets, and have the standing forest inventory to do so. The East and South, in addition to serving the vast Eastern domestic market, can readily serve European and Middle Eastern overseas markets.

. We have available the technology and knowledge vastly to increase the growth and production of our forests. Existing data show the potential for more than doubling U.S. forest growth and harvest if economically efficient investments are made to this end.

. We have a well-developed infrastructure for production, manufacture, and distribution.

. Foreign demand is growing at the same time as Canadian and Scandinavian forests appear to be approaching, or have achieved, their foreseeable productive capacities. Our forests have an innate productive capability much greater than that of Canadian, Scandinavian, or Soviet forests because, as I said earlier, ours have longer growing seasons, occupy better sites, and are much more accessible.

Wood Exports Already Important

U.S. exports of forest products already are extremely important: they reached almost $8 billion in 1981, or 3.5 percent of all U.S. exports in 1981. U.S. exports equal about 15 percent of the timber products produced in this country, in terms of roundwood equivalent.

Our principal export markets at present are Japan for softwood logs and lumber, pulp chips, woodpulp, and paper and board products; and Western Europe for woodpulp, paper and board products, lumber, and plywood.

Industry Benefits from Exports

Development of overseas markets can dampen the impacts of a cyclical domestic market. It is worth noting, for example, that a number of mills on the West Coast which have been willing to adapt their production to sizes, dimensions, and grades acceptable in the Japanese market have been successful in maintaining operations during the depressed markets of the last several years. It has not been a boom period for them, because the Japanese are also suffering through a housing slump.

These mills, however, have maintained operations at levels which have enabled them to retain their crews and keep their firms viable during the disastrous period for forest products and home building from which we are just recovering as a result of finally successful but very painful efforts to control inflation.

During the last several years many midwestern hardwood veneer mills have adapted to produce veneers to specific European thicknesses. Now, many such mills are committed to exporting more than 50 percent of their products.

Acceptance of Manufactured Products

As the U.S. increasingly endeavors to expand markets for wood products by entering the world markets on a larger scale, U.S. companies must make a concerted effort to gain acceptance for manufactured products rather than for logs alone. The advantages are obvious: we create more jobs for Americans, and we reduce the domestic political effects of the prejudice against exporting raw materials. To do this, we need to be more cognizant of and willing to meet specifications required by other countries, rather than attempting to urge our specifications on the rest of the world. For instance, rather than trying to persuade the Japanese to build with two-by-fours, we need to accommodate their needs by milling to their preferred specifications.

Traditional Japanese building techniques use exposed structural members. Appearance is, therefore, much more important in Japan than in America. American producers who are willing to sort within grades to eliminate visual flaws will have better access to the Japanese market. They must also be willing to produce lumber to dimensions which best meet the traditional needs of Japanese builders.

USDA Export Efforts

As has been mentioned in this symposium (Harness; Sims), last year the Department of Agriculture established a small staff within the Foreign Agricultural Service to help the forest products industry seek increased exports. The staff has three primary functions: to provide analyses of international trade matters; to continue supporting the market development program already under way; and to help industry overcome the trade obstacles in foreign markets.

In addition to these efforts, the federal government can, through research, develop an increased understanding of the potential role of the U.S. in world trade of forest products. The Forest Service, even now, is developing a research initiative which will expand our understanding of international trade. This initiative will include the development of models for projecting and analyzing wood products trade and the evaluation of factors affecting this trade. Special emphasis will be placed on identifying key export restrictions and how to overcome them.

Other areas of study will focus on the transmission of pests and diseases, the degradation that can occur during shipment, and the comparison of our standards and product specifications with foreign counterparts to determine production opportunities. We hope these research initiatives will lead to improved understanding and to better-informed public and private policies and practices to enhance the long-term opportunity for the United States in world markets.

Need to Overcome Internal Barriers

In addition to industry's standardization of specifications that are not used outside of North America, there are several other trade and institutional barriers which must be overcome, if we are to reach more world markets. For instance, U.S. industry appears to be rather reluctant to seek world markets aggressively. In general, U.S. firms are most comfortable doing business with other U.S. firms. This is understandable, but it does not make for a fast-growing industry. The tremendous number and diversity of U.S. wood products companies also complicates efforts to enter world trade, compared with countries that have only a handful of producers.

Need to Overcome External Barriers

Overcoming trade entry barriers imposed by other countries is a very formidable obstacle, made even more difficult by the current world economic climate. One of its commonest forms is the imposition of tariffs designed to discourage imports of manufactured products. Although raw materials such as logs, chips, cants, and fuelwood are usually duty free, processed materials, such as planed lumber, veneer, plywood, and paper usually have tariffs ranging from 4 to 40 percent. Current efforts to negotiate reductions in these tariffs have been only modestly successful through the General Agreement on Tariffs and Trade. As John Ward mentioned, Japan imposes a range of tariffs on veneer, plywood, and particleboard. Japan is not the only country which imposes such tariffs. The European Community has a duty of 11 percent on imported softwood plywood. Even Australia recently introduced a new set of tariffs for wood products, ranging from 5 percent for rough-sawn timber and veneers to 37 percent for plywood.

Other External Barriers

Another form of barrier is an import quota. The European Community, for instance, has an annual quota on softwood plywood of 600,000 cubic meters. Although most European countries either favor a quota expansion or are neutral, France is trying to protect its one softwood plywood manufacturer, which is also the only one in Europe, by maintaining the quota.

Still another barrier is getting acceptance of our wood products in the whole spectrum of building codes, which vary greatly throughout the world, as can be imagined. Often these codes are designed in part to impose trade barriers, just as they frequently are used in this country to discriminate among various competing building products.

In some foreign markets, particularly in Europe, there is concern about diseases such as oak wilt. It is sometimes difficult to distinguish between the legitimate desire of countries to prevent the introduction of exotic diseases and the not-so-noble desire to protect markets for domestic producers. Better understanding of some of the diseases can help to develop appropriate responses to these issues.

Other external barriers which often are encountered are subsidies, voluntary restraints, competing export cartels, customs procedures aside from tariffs, restrictive product standards and grades, exchange rate controls, requirements for import licenses, and the necessity of passing import inspections.

Keeping an Open Mind for New Markets

While trying to break down barriers between U.S. industry and traditional markets, we also must recognize that world market opportunities may expand or contract drastically. For instance, China could become a major new world market for wood imports. That country has been buying large volumes of kraft paper for linerboard and sack manufacture.

To be realistic, one must also recognize the risk of international economic and political developments overseas which can drastically reduce or cut off markets. Such uncertainties must somehow be factored into investment decisions. The American forest products industry also needs to become a reliable supplier to overseas markets over time. Overseas markets should not be viewed as an option only when the domestic market is sour.

Conclusion

We know that the opportunities for increasing U.S. exports to the world markets are great, but government and industry must work together as partners, rather than as adversaries, to overcome some very major barriers. We must make our forests more productive to ensure a dependable long-term supply of wood for both domestic and world markets. We must be ready to meet foreign specifications for finished products, rather than trying to get the rest of the world to change to our specifications. We must make more progress in breaking down numerous barriers to foreign trade--some internal, but more of them external.

This is a tall order, but the potential rewards are great--both to the nation and to the forest products industry. They more than warrant the concentrated effort of industry and government to open up new trade opportunities and to rethink some of our traditional ideas in order to meet the requirements of a world drawn ever closer together.

GOVERNMENT AS A WOOD PRODUCER--ITS INTEREST IN WORLD TRADE:
BRITISH COLUMBIA

David Haley

Introduction

Canada is a nation richly endowed with forest resources, and the forest products industry is one of Canada's leading industrial sectors. On a world basis, in terms of timber supply and forest products production and trade, Canada plays an immensely important role. Twelve percent of the world's closed forest area and 15 percent of global coniferous timber reserves lie within Canada's boundaries (Persson 1974). In 1980, Canada produced 15 percent of the world's industrial coniferous roundwood, 13 percent of the total world supply of coniferous sawnwood, 16 percent of the woodpulp, and 33 percent of the newsprint (FAO 1982).

The Canadian forest products industry relies very heavily on export markets. In 1980, wood products valued at $12.8 billion (Canadian dollars), representing about 50 percent of all forest industry shipments, were exported. The most important commodities entering international markets were softwood lumber (25 percent), woodpulp (30 percent), and newsprint (29 percent).[1] Canadian exports in 1980 accounted for 19 percent of the total value of world trade in forest products and Canada contributed 62 percent of the newsprint, 44 percent of the softwood lumber, and 34 percent of the woodpulp traded on a world basis (FAO 1982).

---

1. Throughout this paper, the most recent data used for comparative purposes are for 1980. This is the most recent year for which a complete set of national, provincial, and world forest products statistics are available. 1981 data, where available, tend to be non-representative owing to the impact of the worldwide recession on forest products markets.

In Canada, forest resources are mainly in the public domain. Ownership rests not with the federal government but largely with the provinces. Of Canada's 342 million hectares of forest land, 77 percent is owned by the provinces, 17 percent by the federal government, and the remaining 6 percent is in private ownership (Environment Canada 1982).

While constitutionally the federal government of Canada has jurisdiction over international trade, the provincial governments, as major forest land owners, have a vital interest in forest products markets and marketing and, by virtue of their control over a high proportion of available timber supplies, are in a position to exert a significant influence over the nature and development of these markets. In addition to being involved in the development and promotion of export markets for forest products--both unilaterally and in cooperation with the federal government--the provinces can influence markets through resource policies which have an impact on the structure and location of forest industries, the product mix, the forms in which forest products can be exported, and, perhaps most significantly, the availability, location, and price of wood supplies.

In the remainder of this paper, I will direct my comments to the province of British Columbia--its markets for forest products and the ways in which the provincial government influences the development of these markets. British Columbia is not only the most important province in Canada in terms of timber inventories and the production and shipments of forest products, but also is the province in which I have made my home for many years and the region of Canada with which I am most familiar.

Markets for British Columbia Forest Products

According to Canada's most recent national forest inventory (Bonner 1982), British Columbia has 20 percent (51.5 million hectares) of the nation's productive forest land and 42 percent (9.7 billion cubic meters) of the total timber inventory. In 1980, British Columbia accounted for 35 percent of the total value of shipments of Canadian forest-based industries and 44 percent (Canadian $5.7 billion) of Canada's earnings from forest products exports.

The forest industry is by far the most important sector in the economy of British Columbia, consistently accounting for about 50 percent of the value of shipments of all manufacturing industries and more than 50 percent, by value, of British Columbia's foreign trade. Almost 10 percent of the provincial labor force is directly employed in the forest industry, and it is estimated there is an additional 15 percent if indirect and induced employment are considered (Ministry of Forests 1980). The forest industry is a major source of provincial revenues in the form of direct charges and fees and provincial taxes paid by forest companies and their employees. On a regional basis, the forestry sector is far more important than aggregate provincial figures would suggest, and many communities throughout British Columbia are dependent on the forest resource and the activities it supports as their major--in some cases only--economic base.

The economic health of British Columbia, as the current recession has demonstrated, is closely dependent (many would say alarmingly dependent) on the vicissitudes of world forest products markets and the ability of the province's forest industries to compete successfully in these markets.

British Columbia's forest products exports during the period 1976-80 included: lumber (49 percent); woodpulp (mainly bleached and semibleached sulphate) (29 percent); newsprint (9 percent); and softwood plywood (2 percent). The United States was, by far, British Columbia's most important market for forest products, accounting for 56 percent of the value of all exports during the period 1976-80. Japan was second with 13 percent, while countries of the European Economic Community jointly accounted for 14 percent.

Trends in British Columbia Forest Products Exports

The last 25 years have seen many changes in British Columbia's forest products industry in terms of the volumes and mix of products produced, the location of manufacturing capacity, and the markets served.

Between 1960 and 1980 the volume of timber harvested in British Columbia increased by 121 percent from 34 million cubic meters to 75 million cubic meters. This increase is considerably larger than the increases in industrial roundwood production experienced in either the United States (86 percent) or on a worldwide basis (73 percent) during the same period. Much of the increase in British Columbia was achieved by pushing the extensive logging margin further north into lower quality, less accessible timber stands, and by harvesting trees and logs of smaller dimensions. During the period, lumber production increased by 126 percent from 12.5 million cubic meters to 28.6 million cubic meters, woodpulp production by 195 percent from 1.9 million tonnes to 5.7 million tonnes, newsprint production by 75 percent from 0.8 million tonnes to 1.4 million tonnes, and softwood plywood production by 90 percent from 1.0 million cubic meters to 1.9 million cubic meters.

Lumber

A notable feature of British Columbia's lumber industry over the last 25 to 30 years has been the increasing importance of the province's interior as a lumber-producing region. In addition, during this period, there have been major shifts in the species and grades of lumber produced, and the importance of the United States as an outlet for British Columbia lumber shipments has increased considerably.

In 1950, over 70 percent of British Columbia lumber was produced on the coast. This figure had decreased to 54 percent by 1960, and by 1970 interior production had overtaken production on the coast and accounted for 51 percent of the total. By 1980, the interior produced 63 percent of all lumber manufactured in British Columbia.

In 1960, Douglas-fir accounted for 40 percent of the lumber manufactured in British Columbia, Hem-Fir 22 percent, and SPF (spruce/pine/fir) 23 percent. By 1980, SPF, at 55 percent of total production, was by far the most important lumber species group manufactured in the province. Hem-Fir was second, accounting for 27 percent of total production, and Douglas-fir had fallen into third place, accounting for only 11 percent of total lumber output. Between 1960 and 1980, SPF lumber production increased by 421 percent and Hem-Fir by 127 percent, while the volume of Douglas-fir lumber manufactured declined by 41 percent. SPF is the major lumber species group produced in the interior and accounted for most of that region's increase in lumber output during the 1960s and 1970s. Hem-Fir is the major lumber species group produced on the coast.

Expansion of the lumber industry into the interior of British Columbia witnessed increasing emphasis on the production of dimension lumber (2" x 4" to 2" x 12") and studs. Not only can these products be manufactured most efficiently in high-speed mills designed to maximize volume production from a diet of small logs, but they are also the products which have been in greatest demand in the burgeoning United States market. On the coast too, new mills have been designed, and some old mills upgraded, to produce construction-grade dimension lumber from small logs.

By the late 1970s, about 80 percent of the lumber produced in British Columbia was in dimension grades, about 8 percent was in select and shop grades, and the remainder was in construction-grade boards and timbers.

Since World War II, the United States has become increasingly dependent on softwood lumber imports in order to meet increasing demands in the face of flagging domestic production. This demand has been met almost exclusively by Canada and has provided an outlet for a high proportion of British Columbia's increased lumber production since 1960. The proportion of British Columbia lumber exported to the United States increased from 48 percent in 1960 to 65 percent in 1978, but in 1981--due to severely depressed United States markets--it dropped to 49 percent.

British Columbia's share of U.S. softwood lumber consumption increased from 8.6 percent in 1960 to 19.8 percent in 1980, having attained a peak of 20.4 percent in 1978. Canada's total share of the U.S. softwood lumber market increased steadily from 11.6 percent in 1960 to 29.0 percent in 1980.

The U.S. has provided the major market for the massive increases in lumber production which have taken place in the British Columbia interior since the mid-1960s. In 1965, the interior accounted for 44 percent of British Columbia's shipments to United States markets. By 1980, this figure had risen to 78 percent, and, in that year, 64 percent of interior production was shipped to the United States market.

Since the early 1970s, coast lumber producers have actively promoted offshore markets, particularly in Japan and the European Economic Community, in a deliberate policy of market diversification designed to reduce dependence on highly cyclical United States markets. This strategy has clearly met with some success. Shipments to Japan increased from about 14 percent of total coast waterborne trade between 1970 and 1973 to 29 percent in 1979 and 32 percent in 1980. Shipments to the countries of the European Economic Community increased from about 15 percent of coast shipments by water in the early 1970s to 26 percent in 1979 and 30 percent in 1980.

Several coast companies have upgraded mills and are improving log-making and sawing practices in an effort to increase the proportion of select and shop grades in their output mix. Price premiums for the higher grades, particularly for clears, are increasing, and the demand in offshore markets for these products is expected to remain strong. Maximum grade production is seen as the key to profitability in many coastal mills.

Woodpulp

Pulp production in British Columbia increased dramatically by almost 200 percent between 1960 and 1980. As in the case of lumber, much of the additional capacity put into place during this period was located in the interior of the province. The first interior pulp mill commenced production in 1962, yet, by the mid-1970s, the interior accounted for over 30 percent of total provincial production.

Bleached and semibleached sulphate pulps account for about 65 percent of British Columbia's total production. Most of this pulp is shipped as market pulp, although some is used as the chemical furnish in newsprint production. Mechanical pulp accounts for about 22 percent of total pulp production, and is used almost entirely in the manufacture of newsprint. The remaining pulp production is composed mainly of unbleached sulphate, used in the manufacture of kraft linerboard and sack kraft, and some sulphite pulp.

British Columbia's markets for pulp are more diverse than for lumber. About 96 percent of bleached sulphate pulp shipments are exported. The European market grew faster than the United States market during the last decade and in 1980 accounted for 39 percent, by value, of British Columbia's exports. In the same year, shipments of pulp to the United States amounted to 24 percent of total exports, with 20 percent going to Japan.

Newsprint

British Columbia newsprint production remained relatively stable during the 1970s at about 1.3 million tonnes per annum. In the early 1980s, as in other producing regions of the world, a substantial increase in capacity took place to bring total capacity in 1982 up to 1.7 million tonnes (CPPA 1982).

Close to 90 percent of British Columbia's newsprint production is exported. The major market is the western United States, which accounted for 80 percent of total exports between 1976 and 1980. Other markets are quite widely distributed in Asia, Oceania, and South America, and are of increasing importance. Europe purchases negligible amounts of newsprint from British Columbia, and shipments to Japan account for less than one percent of total exports.

Softwood Plywood

About 85 percent of Canada's softwood plywood capacity is in British Columbia. In 1970, 80 percent of the province's plywood capacity was on the coast, but, as in the case of pulp and lumber, much of the expansion during the 1970s took place in the interior, which now accounts for close to 50 percent of total production.

The interior industry mainly produces a utility or sheathing grade of plywood known as Canadian Softwood Plywood (CSP). The coast industry is producing increasing amounts of CSP (over 60 percent) but in addition produces "Douglas-fir plywood," which is manufactured in higher quality, sanded, and specialty grades. All British Columbia plywood is produced with an exterior glue line. The coast industry is faced with obsolete equipment, a shortage of suitable logs, and relatively low labor productivity. The interior plywood sector is modern and more efficient. Rebuilding and modernization of some coastal mills has taken place in recent years.

Softwood plywood faces high tariffs and/or quotas in export markets, and, as a result, less than 20 percent of British Columbia's plywood shipments are exported. The most important export market is the United Kingdom, followed by other countries of the European Economic Community, where softwood plywood enters under a duty-free quota of 600,000 cubic meters per annum. Since the late 1960s, the United Kingdom market has declined in importance while promotional efforts by both industry and government have increased market opportunities in other countries of Western Europe. In recent years, British Columbia has faced increasing competition in European markets from United States plywood producers in the South and Pacific Northwest.

The United States imposes a 20 percent tariff on softwood plywood imports while Canada maintains an import levy of 15 percent. These tariffs were scheduled to fall to 8 percent by 1984 under the Tokyo round of GATT negotiations following agreement on common standards between the two countries. Such agreement has not been reached, and it is unlikely that there will be a lowering of tariffs within the foreseeable future.

Role of the British Columbia Government

In British Columbia, where 94 percent of the forest land is provincially owned, the production and export of forest products are of such strategic importance to the economy of the province that any provincial government must regard the continued health of the forest industry and its ability to compete in international markets as important components of economic policy. Such policies are recognized in the programs of the Ministry of Industry and Small Business Development and the Ministry of Forests.

Ministry of Industry and Small Business Development

For a number of years, the British Columbia Ministry of Industry and Small Business Development (formerly the Department of Industrial Development, Trade and Commerce) has been involved in programs designed to develop and promote markets for British Columbia forest products overseas. These initiations have been largely directed at the lumber and plywood sectors. Important priorities are market diversification, in order to reduce dependence on shipments of lumber to the United States, and product diversification, particularly to promote innovations which will increase the values added to products before they leave the province.

The most notable of the Ministry's programs is the Cooperative Overseas Market Development Program (COMDP), which is jointly funded by the Province of British Columbia, the Federal Department of Industry, Trade and Commerce, and the Council of Forest Industries of British Columbia ($10.5 million Canadian each for the period 1980 to 1985). This program, which is administered by the Council of Forest Industries, was initiated in 1971. Successes of the program include the promotion of the platform frame method of house construction in the United Kingdom, Holland, and some other European countries, and the creation of an expanding market for CLS dimension lumber in Japan. Priorities of the program include diversification of markets for interior SPF shipments, and the development of larger export markets for plywood. In addition to consolidating British Columbia's position in the European and Japanese markets, new markets are being sought in areas such as Korea, China, North Africa, and the Middle East.

Ministry of Forests

The British Columbia Ministry of Forests Act, in setting out the purposes and functions of the Ministry, requires that--among other things--it "encourage a vigorous, efficient and world competitive timber processing industry in the Province." This mandate is fulfilled through measures designed to ensure that timber is manufactured within the province by controlling the export of logs and wood residues, and by forest policies concerning the disposition of public timber. The future development of the industry and its competitive ability are also profoundly affected by public forest policies concerning land use, the rate at which timber can be harvested, and investments in reforestation and timber management.

Log and Chip Export Controls

Control over the export of unmanufactured forest products has been an important feature of British Columbia's forest policy since the early part of this century. These restrictions do not take the form of a total prohibition on exports but, rather, require prospective exporters to acquire an export permit. A permit may be obtained if it can be demonstrated that the logs or wood residues in question are surplus to the requirements of timber processing facilities in the province. Applications for permits are reviewed by a Log Export Advisory Committee and a Chip Export Advisory Committee. Provincial restrictions apply to all public forest land and certain classes of private land. Similar federal controls cover all exports of unmanufactured forest products from the province and, therefore, restrict exports of timber from those private lands not falling under provincial jurisdiction.

Once a permit has been granted, exports that fall under provincial controls are subject to an export levy which is revised monthly and varies by species and log grade. At present, for example, Grade A Douglas-fir logs attract a levy of $22.04 per cubic meter while for Grade I western redcedar logs the levy is $1.90 per cubic meter.

The permit system has been successful in severely limiting log exports from British Columbia. Between 1971 and 1980 the volume of logs exported was 641,411 cubic meters per annum, or 0.97 percent of the provincial harvest during that period. When domestic markets are depressed, the volume of logs permitted to leave the province increases. For example, in 1982 logs exports rose to 1.27 million cubic meters or close to 2 percent of the provincial harvest.

In 1977, in response to the increasing surplus of sawmill residues in the interior, longer term guaranteed supply contracts for chips were allowed for the first time. A cooperative marketing organization--Fibreco--was formed to export surplus chips to overseas pulp mills, mainly in Japan, on behalf of 42 interior sawmills.

From time to time the wisdom of British Columbia's log export restrictions has been questioned. Could the province not benefit, it is argued, by taking advantage of the substantially higher prices which prevail at times for certain species and grades of logs in international markets compared to the domestic log market (Davies 1977)? The report of the most recent provincial Royal Commission on Forest Resources (Pearse 1976) suggested that exports be controlled, not through permits, but more directly through export levies which would be gradually reduced, "to enhance timber values and promote more efficient use of timber resources." It is most unlikely, however, given the strong opposition to the "export of jobs" which exists within the provincial labor movement, that log export restrictions will be relaxed.

The Disposition of Public Timber

While the British Columbia government clearly has efficient use of timber resources as one of its objectives, it also has goals of "balanced" regional growth, regional equity in the distribution of income, and stability of regional employment and incomes. In pursuit of these objectives, arrangements for granting cutting rights to public timber have evolved over the last 35 years under which private companies are guaranteed long-term timber supplies at appraised stumpage prices which are recalculated annually and revised monthly in response to changes in base selling prices (logs on the coast; lumber and pulp chips in the interior). Private sector obligations in these contractual agreements include annual cut control and timber utilization requirements, the construction and operation of designated timber processing facilities, reforestation and other silvicultural treatments, and varying levels of responsibility for other aspects of forest resource management.

In the absence of substantial private forest land holdings, such licensing arrangements--which protect licensees from the uncertainties of timber markets and the potential threat of governmental monopoly power--are said to be necessary to encourage capital investments in timber processing plants of a type and scale necessary to compete successfully in world markets. In addition, it is claimed, they strengthen and help stabilize regional economies by encouraging the establishment of permanent forest-based industries together with their accompanying infrastructure. Certainly, the spectacular growth of the British Columbia forest industry during the past 25 years, based almost entirely on export markets, and the development by the industry of remote areas of the province, seem to attest to the success of provincial government policies.

Of particular significance was the initiation, in 1962, of Pulpwood Harvesting Area Agreements. These Agreements granted exclusive long-term rights to the purchase of pulpwood, including sawmill residues, from a designated area in return for the licensee's guarantee to build and operate a pulp mill. These arrangements resulted in a flood of capital into British Columbia, much of it from abroad, and the doubling of the province's pulp capacity between 1963 and 1970.

In response to the expanding market for chips, sawmills in the British Columbia interior began to install log barking and chipping equipment. This trend was encouraged by government policy which offered sawmillers incentives in the form of reduced stumpages and increased allowable cuts if they agreed to harvest trees and logs of smaller dimensions and produce by-product chips suitable for pulp manufacture. Within a relatively short time period, the interior lumber industry underwent a transformation. Mills built to handle large logs gave way to fewer, larger, more sophisticated, high-speed mills designed to process logs down to smaller diameter limits into dimension lumber and chips. Cost efficiency was greatly improved and production increased substantially from 9.0 million cubic meters in 1965 to 14.2 million cubic meters in 1973.

Future Development of the British Columbia Forest Industry

The future development of the British Columbia forest industry and its ability to expand, or even maintain, its share of world forest products markets is, to a large degree, dependent on public policy.

In 1980, the Ministry of Forests, as part of the Forest and Range Resource Analysis, estimated that the British Columbia forest industry, in order to realize its potential market opportunities to the year 2000, must increase provincial log production to 91.1 million cubic meters,[2] 20

---

2. This was a median projection. Low and high projections ranged from 84.2 million cubic meters to 98.1 million cubic meters.

percent above the record 1978 level of 76.2 million cubic meters (Ministry of Forests 1980).

In the same document, however, an analysis of the province's timber supply portrayed a gloomy picture. As a result of land conversion to other uses, environmental protection requirements, and the removal of areas which are economically inaccessible, the productive forest land base of the province in public ownership was forecast to shrink by 25 percent to 35.4 million hectares by the year 2000.

Based on regional timber supply analyses, the Ministry of Forests adopted a goal of holding harvest levels at 75 million cubic meters per annum from both public and private sources; that is, approximately the level of harvest achieved in 1978 (Young 1981). Timber supply analyses also suggested, however, that an annual harvest at this level cannot be sustained and that "falldowns," or reductions in the cut below planned harvesting levels, will begin in some parts of all regions within 5 to 20 years. Maintenance of present cuts can only be achieved, it was said, by improved timber utilization, both in the forest and at the mill, more intensive forest management, and by a reduction in the present rate of forest land alienation to nontimber uses.

If the Ministry of Forests' assessment of the timber supply situation is correct and the proposed harvesting constraints are followed, the situation is serious indeed. British Columbia will not significantly increase its current contribution to world markets in solid-wood products, although further expansion of the pulp and paper sector will be possible through the use of mill residues and small dimension material. The consequences for the province as a whole of zero or--at best--slow growth in the forest industry sector present a problem which no one has seriously addressed.

It must be remembered, on the other hand, that the rate at which the old-growth resource is liquidated on public land is a policy decision. While the British Columbia Ministry of Forests does not subscribe to the U.S. Forest Service ideal of nondeclining even flow, current harvests are constrained by a general policy that the current harvesting level adopted for a particular unit will not cause future harvests to fall below the long-run sustainable yield of the unit.[3] In setting current harvest levels, no

---

3. Long-run, sustainable yield is defined by the British Columbia Forest Service as, "the volume of harvest that can be sustained from the productive forest sites with only minimal or basic silvicultural treatments. It is

allowances are made for anticipated future changes in technology, real timber prices, or investments in timber production. Yet in most cases an increased cutting rate over the next 20 years would not cause harvests to fall below today's level of sustainable yield for many decades hence. In short, the current harvesting ceiling adopted by the Ministry of Forests is probably conservative.

If it is assumed that the average rotation for public timber in British Columbia is 100 years and for private timber 60 years, then--assuming 35 million hectares of productive public forest land and 2.6 million hectares of productive private forest land--the area which could be cut over annually is approximately 393,000 hectares. While such a calculation is clearly simplistic, the difference between this area and the average area logged between 1972 and 1981 of 164,000 hectares per annum is difficult to reconcile.

At 75 million cubic meters per annum, about 0.8 percent of British Columbia's standing timber inventory of 9,731 million cubic meters (Bonner 1982) would be harvested annually; a very modest rate of inventory draw-down by any standard. Assuming an average rotation for the province of 100 years and an area of immature timber of 18.4 million hectares (Bonner 1982), with an average mean annual increment of 2.3 cubic meters per hectare (Bickerstaff et al. 1981), a current allowable annual cut for the province of about 140 million cubic meters is indicated by the Hanzlik formula. If the standing timber inventory is reduced by a generous 25 percent to reflect nonoperable areas, an allowable annual cut of about 115 million cubic meters, or 50 percent above the currently proposed level, would seem to be feasible.

In summary, given that markets are available in which British Columbia can compete successfully, an annual rate of harvest by the year 2000 of up to 100 million cubic meters--that is, about 33 percent above current levels--would seem to be realistic considering the province's existing timber resources and natural forest growth potential.

The magnitude of British Columbia's timber supply over the long run will be contingent on how effectively the timber resource can be renewed and its growth potential improved. The scope and success of such programs will depend to a large extent on public policy.

---

calculated from the volume available at the culmination of the mean annual increment for all sites" (Ministry of Forests 1980).

Reforestation and stand improvement projects carried out on public land by private companies holding long-term cutting rights are financed through credits against stumpage charges payable. In addition, companies which undertake silvicultural programs over and above their contractual obligations may be rewarded by an increase in their allowable cuts.

As more intensive forest management programs increase in magnitude, the flow of direct revenue to the provincial government in the form of stumpage charges will be significantly reduced. On the other hand, as the transition from old-growth to second-growth harvesting proceeds, average stumpage prices can be expected to rise. Second-growth stands will tend to be more uniform than old growth in terms of species composition and stem diameter distribution. Better accessibility and less rugged terrain will favor lower logging costs, but merchantable volumes per hectare will be less, average log sizes will be smaller, and there will be a lower proportion of clear wood. Stumpage prices will have to reflect the costs of timber growing (reforestation and stand management), which could be considerable.

Growth rates are modest in British Columbia, although they are among the highest in Canada. The average mean annual increment for the province is only 2.3 cubic meters per hectare, ranging from 9.1 cubic meters per hectare for good sites in the Vancouver Forest Region to 1.1 cubic meters per hectare for poor sites in the interior (Bickerstaff et al. 1981). For the province as a whole, good sites account for only 9 percent of the productive forest land area, while poor and lower sites account for over 50 percent.

Conclusions

The last 25 years have constituted a period of remarkable expansion for British Columbia's forest industry. This growth has been made possible by expanding world markets for forest products, public forest policy which has provided a favorable and stable environment for investments in manufacturing capacity, and a legacy of natural coniferous forests of high quality compared to most other areas of virgin timber remaining in the Northern Hemisphere. Further expansion to the year 2000, of about one percent per annum, is likely, but will call for modifications of the provincial government's cut control policies and for improvements in recovery standards by the forest industries.

Growth in British Columbia's shipments to traditional markets for forest products in Europe and North American is forecast to be modest during the next two decades in comparison with the previous 25-year period. Increasing emphasis will continue to be placed on maintaining or expanding market shares in Japan, and developing new markets in other countries--particularly around the Pacific Rim. Competition in this region will be intense, however, as huge volumes of radiata pine become available during the 1990s in Chile and New Zealand.

The old-growth forest resource which, to a large extent, has given British Columbia its comparative advantage in world markets, is already approaching depletion in some of the more developed areas of the province. The forest industry in the future will rely increasingly on artificially regenerated, managed stands which have been produced at considerable expense. British Columbia's ability to maintain or increase its level of timber production and compete successfully in world markets with both the traditional and the newly emerging wood producing regions will depend on how effectively reforestation and stand improvement programs can be implemented.

Given the pattern of land ownership in British Columbia, the success of silvicultural programs will depend heavily on public involvement and will be greatly influenced by public policy. British Columbia has barely tapped the potential productive capacity of its forest lands. Realizing this potential is the major challenge for British Columbia foresters in the years ahead. Programs are urgently needed to ensure that the limited resources available for timber management are used with optimum efficiency and that the better-quality forest land in all regions of the province, but particularly on the south coast and Vancouver Island, is adequately protected for intensive timber production.

References

Bickerstaff, A., W. L. Wallace, and F. Evert. 1981. Growth of forests in Canada. Part 2: A quantitative description of the land base and the mean annual increment. Environment Canada, Canadian Forestry Service. Information Report P1-X-1.

Bonner, G. M. 1982. Canada's forest inventory 1981. Environment Canada, Forestry Statistics and Systems Branch, Canadian Forestry Service, Chalk River, Ontario.

CPPA. 1982. Newsprint data 1981. Canadian Pulp and Paper Association, Montreal, Quebec.

Council of Forest Industries of British Columbia. Annual reports for the years 1970-1981.

Davies, D. W. 1977. Log export restrictions in British Columbia: An economic examination. M.S. Thesis, University of British Columbia, Vancouver, B.C.

Environment Canada. 1982. Canada's forests 1981. Environment Canada, Hull, Quebec.

FAO. 1982. Yearbook of forest products 1980. Food and Agriculture Organization of the United Nations, Rome.

Ministry of Forests. 1980. Forest and range resource analysis technical report. Vols. 1 and 2. Ministry of Forests, Province of British Columbia, Victoria, B.C.

Pearse, P. H. 1976. Timber rights and forest policy in British Columbia. Report of the Royal Commission on Forest Resources, Victoria, B.C.

Persson, R. 1974. World forest resources: Review of the world's forest resources in the early 1970s. Royal College of Forestry, Stockholm.

Young, W. 1981. Timber supply management in British Columbia--Past, present and future. Burgess-Lane Memorial Lecture, University of British Columbia, Vancouver, B.C.

THE ROLE OF JOINT VENTURES IN WORLD TRADE IN WOOD

Richard L. Atkins

Burlington Northern International Services is interested in opportunities for joint ventures, both in the United States and other countries. The new Export Trading Company Act, passed in 1982, is designed to encourage joint ventures leading to increased exports from the U.S. Since BN International Services is such a young organization, formed less than a year ago, and the implementing regulations under the new Export Trading Company Act have not yet been published, we have thus far not been involved in any joint ventures in international manufacture or trade. We have been investigating some possibilities and reviewing the results of joint ventures between other companies.

An example of a joint venture between two American forest products companies is a pulp mill near Eureka, California. It was built several years ago as a joint venture of Simpson Timber Company and Crown Zellerbach. Simpson had 300,000 acres of forest land and a large solid-wood manufacturing capability in the area. Crown Zellerbach had pulp mill operation and product sales experience. This was a successful partnership until recently when conditions in each company changed and Simpson bought out the Crown Zellerbach interest. Crown's strategy is now to sell off some assets and concentrate resources and capital on key domestic manufacturing and marketing operations. This seems to be a fairly common situation, where a joint venture works well for a while but one partner eventually assumes total ownership.

The same corporate strategy led Crown Zellerbach to sell its 50 percent interest in a paper mill at Laja, Chile. The purchaser was the co-owner, Chile's largest paper firm.

Crown Zellerbach has also made some joint venture investments that were not successful. In 1980 the company wrote off its $32 million 50 percent investment in a Netherlands affiliate. Competitive costs and productivity results indicated that the likelihood of future economic return was small.

Boise Cascade has a 50 percent interest in a company with two corrugated container plants in Austria and one plant in West Germany. Earnings from these operations have been depressed since 1980 because of the continuing recession in Europe, inflation, unusually high interest rates, and the higher exchange value of the dollar.

St. Regis Paper Company's principal joint venture affiliations are in Canada, England, Sweden, and Australia. These affiliates all registered a loss in 1982. Nevertheless, from 1973 through 1981 average annual equity earnings of $10.8 million from these joint ventures represented 9 percent of the $123 million average annual company net earnings. In 1982, the company wrote off its investment in a builders' supply chain in Great Britain because of continued business and financial problems. Problems in the French paper industry continued to affect the St. Regis affiliate, but St. Regis continued to benefit from the export of primary materials to this group. Its Austrian Paper and Packaging Company had an excellent year, while the Australian and New Zealand affiliates' earnings declined in the second half of 1982 due to the impact of world economic problems.

Champion International, another major U.S. forest products company, has major ownership of Weldwood of Canada Limited. In 1981, Weldwood had building product sales of $388 million, which amounted to 10 percent of total company sales. In pulp and paper the company's Brazilian subsidiary (of which Champion owns 99 percent) and Cariboo Pulp and Paper, a joint venture in which Weldwood and Daishowa and Marubeni are equal partners, contributed $153 million in sales--or 3.8 percent of total company sales. Income from these operations accounted for 28 percent of income from continuing operations of the company. Through Weldwood, the company owns or controls the timber and pulpwood on 4.5 million acres of timberlands in Canada, and through its Brazilian subsidiary owns or controls 97,000 acres in Brazil.

Scott Paper is an example of a U.S. forest products company that has made large investments in joint ventures with foreign companies. From 1977 through 1981, the earnings attributable to international affiliates accounted for 30 percent of the $119 million average annual net income of the company. A loss of $40 million in 1982, due to the worldwide recession, was the only loss during the period 1973 through 1982. In 1982, the $1.1 billion assets in international operations accounted for 45 percent of the total company assets.

Scott Paper's international affiliates are located in

21 countries throughout Europe, the Far East, Latin America, and Canada, with ownership percentages ranging from 33 to 100 percent.

With the exception of pulp, paper, and solid-wood products produced in Canada, the principal products made and sold by these operations are sanitary paper products similar to those sold in the domestic U.S. market. Market demand for these products is expected to grow faster in foreign markets than in U.S. markets. As a result, Scott Paper is devoting much of its research and development efforts to foreign needs and has accelerated the transfer of papermaking technology to them.

Georgia-Pacific is another forest products company making a positive commitment to joint ventures with foreign companies and/or governments. GP has exclusive cutting rights on 1.7 million acres of tropical hardwoods in Indonesia. It is currently operating a hardwood veneer mill and plywood plant in that country and has recently announced plans to enter the paper business in the Far East with an Indonesian kraft paper and linerboard mill to begin operation in 1986. This is a $410 million joint venture project in Northern Sumatra. Georgia-Pacific will provide equipment, technology, and management services in exchange for a 25 percent interest in the venture. The Indonesian government will provide $160 million in financing for a 50 percent interest, with the rest of the financing coming from the Japanese Import-Export Bank, Nichimen Company, and other commercial banks. An Indonesian timber company will contribute timber concessions for the remaining 25 percent interest in the venture.

In 1976, Weyerhaeuser formed a joint venture with Jujo Paper Company of Japan to build and operate a newsprint plant at Longview, Washington, based on studies started in 1973. The firm, called NORPAC, has been formed with 90 percent Weyerhaeuser investment and 10 percent Jujo Paper investment. Each firm takes 50 percent of the output, with Jujo's share going to Japan and Weyerhaeuser's share staying in the western U.S. The plant has the capacity to turn out 210,000 metric tons of newsprint per year. The biggest single customer is the Yomiuri Shimbun, which prints 13 million newspaper copies a day. Jujo is the largest pulp and paper company in Japan--the largest supplier of printing paper and the second largest newsprint supplier. Jujo is especially noted for its marketing experience; Weyerhaeuser has the resource base and experience in operating pulp and paper mills--although this was their first newsprint mill.

For many years, some of our major forest products firms have been involved in joint ventures in Southeast Asia, where there are vast reserves of hardwood timber valuable in the manufacture of hardwood plywood for use in this country and in Japan. These countries will no longer allow the export of their logs, as they prefer to manufacture the logs into lumber or plywood in their own countries, creating more jobs for their own people. As a result of restrictions and partnership requirements, some U.S. companies are withdrawing from Southeast Asia.

China passed a law in 1979 designed to increase foreign investment in that country through joint ventures involving major projects and large investments. China's economic policy reordering its priorities from heavy industry to consumer products has resulted in a drop in the number of joint ventures formed in China--from twenty each in 1980 and 1981 to only eight in 1982.

Future joint ventures in China will involve industrial projects such as the development of energy resources. Foreign businesses have complained about the difficulty of negotiating joint venture agreements in China. They assert that the time and resources needed to set up and operate a joint venture are not worthwhile for a small project.

China is drafting a second set of laws aimed at attracting foreign investment into Guangdong Province's three special economic zones. The proposed legislation would give legal protection to investors in such areas as technology transfer, patents, mortgages, foreign banking, and investment rights. Many foreign investors are hesitant about investing in a joint venture in China, however, because of a lack of legal protection for their investments. Regulations concerning economic contracts are particularly needed to protect investors. The lack of legal definitions has led to many arguments between investors and the Chinese over who is responsible for what.

Some alternatives to joint ventures in China are countertrade, in which foreign equipment is paid for with Chinese products, straight loans, leasing, and cooperative production. This last method is not governed by the joint venture law--the foreign party supplies capital and equipment, while the Chinese provide the physical plant, machinery, and labor. Profits are split according to an agreed-upon ratio.

Reports indicate that the Chinese are interested in the possibility of joint ventures in the forest products area, although we are not aware of any specific developments.

The Bank of China has recently agreed to take equity in joint ventures between Chinese and foreign firms, reflecting the bank's growing sophistication and liberalization, and possibly increasing foreign investors' confidence in Chinese projects. The chief aims of the bank's move are to help spur China's economic development and to use some of the bank's $9 billion in foreign exchange reserves.

Japan has been known as a country where joint ventures between Japanese and American interests have been relatively stable. As trade conflicts develop between Japanese and American interests, major changes are taking place in joint ventures also. Since the beginning of 1983, there have been eight cases where joint ventures were affected, including Komatsu Ltd. and International Harvester; Nippon Electric Company and GTE International; and Asahi Chemical Industry and Dow Chemical Company.

According to the Asian Wall Street Journal, a pattern seems to be emerging in which Japanese companies initially seek to form joint ventures in order to obtain foreign technology but eventually find that such ventures obstruct their efforts to build up their own strength. At the same time, foreign partners have become eager to liquidate joint ventures so that they can assume more active roles in profitable fields of business. As Japanese companies have grown, there has been less need for foreign capital to set up the joint ventures.

The number of sawmills in Japan is declining, and Japan will import an increasing volume of lumber. We have already seen where the Japanese have built or acquired sawmills in the Pacific Northwest, and we may see more joint ventures between Japanese and U.S. forest products firms.

Japanese forest products firms have been active in joint ventures in many other parts of the world. Oji Paper Co., Jujo Paper Co., and Daishowa Paper Mfg. Co. have been involved in a joint venture in Brazil to secure raw material. Cenibra and Flonibra, both joint ventures between Brazil's Rio Doce and Japan Brazil Paper & Pulp Resources Development Co. recently announced an agreement to merge. The partners invested 90 billion yen in Cenibra, which has been producing pulp from a 300,000 ton-per-year plant for six years. Flonibra purchased 160,000 hectares of land in the east section of Brazil for a pulp mill that would produce three times as much as the Cenibra plant.

Oji Paper Co. and Mitsui & Co. together acquired a 33 percent interest in New Brunswick International Paper Co. in Eastern Canada in 1980. The plant was to furnish 100,000 tons of paper a year to the Japanese. Because of market conditions, however, the Japanese asked the Canadian company to sell its products in the U.S. and European markets instead.

Alaska Pulp Co., of Sitka, is jointly owned by several Japanese companies, including Toray Industries, Inc., Teijin

Ltd., and Asahi Chemical Industry Co. The company has been affected by the poor market in Japan and the owners have agreed to increase the capital invested in the company, and to sell a lumber mill in Alaska.

The future prospects for joint ventures in South Korea appear somewhat doubtful at this time, as illustrated by Dow Chemical Company's pullout from the largest joint venture in South Korea. Dow had invested $145 million in its Korean operations and sold these operations to a group of Korean companies for $60 million. While reports indicate that both sides were at fault in the failure of this joint venture, Korea suffered a blow to its international image. Korea has been struggling to overcome a reputation as a difficult place for foreign investors.

Conclusion

As you can see by these examples, the joint venture business has had a mixed bag of results. There are many firms that have been successful partners in joint ventures and the profits have been substantial. There are also a number of firms that have not only not had profits but have lost some of their investment.

The United States must be considered one of the best places in the world for joint ventures in forest products because of the stable government, good raw materials, energy, and inland transportation in areas close to ports. Although there is adequate capital available within the United States for investment in new facilities, there may be better opportunities to attract capital from other countries which enjoy a positive balance of trade with the U.S. At the same time, the other country can tie up a steady source of wood or paper products, and the U.S. joint venture partner can secure a steady market for the product.

The new Export Trading Company Act will encourage joint ventures, especially among smaller companies. Joint venture partners who bring together access to raw materials, manufacturing capability, investment capital, and markets for products can expect to play a major role in the growing trend of forest products exports from the United States.

THE MARKET FOR WOOD-BASED PRODUCTS AS SEEN BY A SOUTHEAST ASIAN PRODUCER

Pedro M. Picornell

I have been asked to give you an overview of the market for wood-based products from the point of view of a Southeast Asian producer. To give you a summary of what has happened during the last few years would be to repeat a tale of woe that is not to anybody's liking. On the other hand, to give you detailed predictions of what will happen in the near future is, as you would say in your country, treading on very thin ice. In the course of my work between 1980 and the present, I have had to make a number of assumptions on what the wood products market would do within the next twelve to eighteen months. My record of success has been just about as good as that of the various boards of economic consultants to the leading U.S. business magazines which express their views on economic conditions and make such forecasts periodically.

When I talk of Southeast Asia, I am talking essentially of the countries that compose the ASEAN group--Indonesia, Thailand, Malaysia (which includes Sabah and Sarawak), the Philippines and Singapore, plus Papua New Guinea. Geographically, one may argue that Papua New Guinea is not part of Southeast Asia, but it is included here because much of its timber production ends up in the same markets as that of the ASEAN countries and thus it can be a factor affecting them. Of these countries, Malaysia, Indonesia, the Philippines, and Papua New Guinea are still actively engaged in the international trade of wood-based products. Thailand has not been an important factor in this trade lately, since their wood-based exports have been limited to relatively small volumes of very high-priced exotic hardwood lumber; and while Singapore once had a substantial plywood industry, its importance has been decreasing lately, mostly because of difficulties in getting the necessary raw materials and labor. Singapore appears to be shifting from the manufacture of plywood from imported logs to the further processing of plywood that is imported from other ASEAN countries. In the countries actively engaged in trading wood-based products, their capacities are much larger than their domestic demand; hence their wood-based industries (except for pulp and paper) are essentially geared to the export market. These represent the typical industries in developing countries supplying raw materials (logs) and semiprocessed goods (lumber and plywood) to developed countries. Our export-oriented wood products consist of logs, lumber and lumber products, plywood and plywood products. Those which are directed toward the domestic market are pulp and paper and composition board.

The present markets for our export-oriented manufactured wood-based products are in developed countries--the United States, Canada, Western Europe, and Japan. Logs are exported to Japan, Korea, and Taiwan, where they are usually manufactured into lumber and plywood. Korea and Taiwan re-export these to the developed countries mentioned above, while Japan uses most of them locally. It is presumptuous for me to tell you what the market for these products will be in your own countries--you should know this better than I do. Therefore, I will not be giving you sophisticated correlations between the rates of growth of your respective economies and the demand for wood products in your own countries. I will leave the risk of making such predictions to your own economists, who are certainly much more qualified to make them than I am. I will limit myself to presenting my views on how we in Southeast Asia expect to react to trends of the markets in your own countries, breaking down the discussion into the general product lines we have at present. I will also limit myself to the decade of the 1980s, as I expect to be retired by the end of the decade unless I am fired earlier for not having a better crystal ball.

In general, however, I will say, without having any specific growth figures or risking my neck on specific dates, that I agree that when the economy recovers in your countries, we will see a moderate upward swing in the demand for wood-based products for a rather limited period of time, as the pent-up demand for housing--which for various reasons has not been satisfied during the last two years of recession or depression (I leave it to you to choose what term you want to use)--comes into play. Then the demand will flatten out and settle to a steady but smaller growth rate as the effects of a limited population growth and slower rate of growth in personal incomes become apparent through the end of the decade of the 1980s.

Logs

Through the rest of the 1980s there will continue to be a respectable demand for good- and medium-quality logs from Southeast Asia in Japan, Korea, and Taiwan. In Japan, these are in demand for wood-based products for the local market, while Korea and Taiwan will try to keep their existing wood-based industries going as long as possible. The supply of logs from Indonesia and the Philippines, however, will continue to diminish and will eventually fade out. These countries already have announced their policies of discontinuing the export of logs, although we will see continuing pressures for the relaxation of log exports which will be at times difficult to ignore. The availability of logs from these countries will also decline as the sources of supply get further inland and more logs are used by local industry. Malaysia, specifically Sabah and Sarawak, will continue to export logs for some years to come, while Papua New Guinea will become a new, although different, source of logs. The timber stands in Papua New Guinea are different from those in other Southeast Asian countries in the very many unrelated species they contain, and the manufacturing philosophy in using them will have to be adjusted accordingly. Korea and Taiwan will increase their use of such logs, but will use them more and more for the domestic market than for export.

The one very interesting development in the availability of logs from Southeast Asia is the emergence of fast-growing plantation species which are capable of high yields on short rotation. During the last year and a half we have seen the first exports of Albizzia falcataria logs, and within the next two or three years we will see those of Eucalyptus deglupta. Albizzia falcataria has been shown to be a very promising wood for the production of mechanical pulp, although there are still some problems in its exportation in the form of chips or pulpwood: In sawtimber sizes, it shows promise in the manufacture of moldings, blockboard, and core veneer. Eucalyptus deglupta makes a good chemical pulp and has definite possibilities in the production of lumber and lumber products, and plywood. The planting of different varieties of ipil-ipil (Leucaena spp.) is also being looked into in a number of countries in Southeast Asia. This species makes good chemical pulp and composition board and is an excellent fuelwood, but its use in products such as lumber and plywood appears to be quite limited. While I see great promise for plantation-grown woods in Southeast Asia, substantial quantities will probably not be available until late in the 1980s.

Lumber and Lumber Products

The supply of sawn lumber and lumber products from Southeast Asia during the last few years has been limited because there has been more money in exporting the logs as such or making them into plywood than in sawing them into lumber. The reasons for this are rather complex and could be the topic of an entire symposium. I do not see any reason for saying that this situation will change through the rest of the decade--there will be some supply of lumber and lumber products, but in limited volumes. The increasing demand for lumber products locally will also cut into the availability for export.

Plywood

The plywood industry in Southeast Asia is taking a terrific beating during the current recession in the same way it is taking it all over the world. In the Philippines, only 27 of the 35 plywood mills in the country are operating. All of these are operating at a fraction of capacity, and all of the plywood companies are in financial difficulties. Indonesia is staking the future of its wood-based industry on plywood, however, and has embarked on a massive program for building plywood mills. There are at present 56 plywood mills there, and 46 more are being planned for 1985. While these figures may vary slightly according to source, there is no doubt that Indonesia is well on the way to becoming the world's largest hardwood plywood source within the next few years.

This move has caused great concern in other plywood-producing countries in the region, particularly in the Philippines--which is Indonesia's closest rival in this field in Southeast Asia. In spite of the creation of this tremendous additional capacity in Indonesia, I optimistically foresee that the Philippine mills which will remain alive after this recession will be able to hold their own in their export market by concentrating on specialties: marine plywood, thick sheets, large sizes, etc. Other factors that will come into play in plywood are:
1. The domestic demand in Southeast Asia is growing and, although at present the domestic markets are taking only a very small proportion of the region's capacity, these will be more and more important, particularly if, as predicted, the Pacific Basin becomes one of the fastest-growing areas in the world once the recession is over.

2. The demand for plywood in Japan, Korea, and Taiwan for Southeast Asian plywood is bound to grow as the supply of logs to their own plywood mills dwindles.
3. The market in the People's Republic of China is one big unknown with a tremendous potential. Increasing volumes of low-grade Southeast Asian plywood have been going into the People's Republic of China, particularly through Hong Kong, and there is no doubt that a potential demand is there. The problem lies in this country's allocation of the necessary funds to import plywood: this will depend on their economic priorities. My personal opinion is that this market will continue to grow but at a rather slow pace, and that it will not represent an important factor until the 1990s.
4. The market in the Middle East has been growing in the last few years, but I do not see it as a major factor in the near future. The population in this region is relatively small and is not growing very fast. Hence, once the present construction programs level off, the demand will drop accordingly and, while some demand will continue, it will not form a very important part of this market.

Consequently, while Southeast Asia's plywood capacity will be increasing substantially within the next few years, I do not see a serious permanent imbalance between supply and demand. Part of this new capacity will simply take over from plywood mills in Japan, Korea, and Taiwan which will have to reduce production due to difficulties in obtaining raw materials. In the Philippines, a substantial number of mills have had to close down also because of raw material shortages and financial difficulties, but the remaining mills should be able to hold their own in the export market by concentrating on specialties, increasing productivity because of their long experience, and taking advantage of growing domestic markets.

The domestic markets for lumber and plywood products in the ASEAN region are bound to grow substantially, and this will take up more and more of the region's capacity. Some plywood capacity will be built in Papua New Guinea, but I do not believe it will represent substantial volumes for export.

In plywood, we also see exciting possibilities in the use of plantations. At Paper Industries Corporation of the Philippines we have produced plywood from plantation-growth

Eucalyptus deglupta, which is practically indistinguishable from that produced from Lauan. Plywood made with *Albizzia falcataria* cores and Lauan and Deglupta faces is more promising, and is definitely something for the future. The manufacture of plywood from plantation wood will require the introduction of new technology into the area because of the small diameters of such logs, but this technology is available in developed countries.

Composition Board

I am using the term "composition board" as a generic term for wood-based panels manufactured from chips, flakes, pulp, etc. There have been no substantial exports of composition board from Southeast Asian countries to speak of and I do not foresee any remarkable growth during the rest of the decade. The reason for this is that almost any country with any type of fibers can make some kind of composition board, and Southeast Asian composition board is at a competitive disadvantage considering the high cost of glue in our part of the world as well as freight to markets, without being able to offer any unique features in these products.

This does not mean that composition board is not of interest to us in Southeast Asia. There is one company in the Philippines that has been very successful in making hardboard for the local market, and as the demand grows locally, composition board will become more and more important to satisfy this demand.

Pulp and Paper

With the exception of very small amounts of the unique abaca pulp exported from the Philippines, Southeast Asia has not been an exporter of pulp and paper products. Some projects are being looked into, particularly in Indonesia and Sabah, which, if carried through, would make this region an exporter of pulp and paper, particularly of bleached hardwood market pulp. These projects are still on the drawing board, however. Because of the beating taken by the pulp and paper industry all over the world, and particularly that in Southeast Asia because of the very large amount of dumping of pulp and paper products from developed countries into our markets during the last year, I doubt that we will see these projects implemented for some time to come.

On the other hand, the success in using plantation wood and the possibility of obtaining lower energy costs by using hydroelectric and dendrothermal energy continue to give Southeast Asia a good potential position in the pulp and paper industry. I do see an emerging trade in pulp and paper products during the second half of this decade, although limited within the ASEAN region itself.

There is also the possibility of the exportation of plantation-grown pulpwood and perhaps even certain grades of pulp made from these woods late in the second half of this decade.

Other Wood Products

The energy crisis we have lived through during the last ten years is opening new and exciting fields in the wood industry in Southeast Asia. At PICOP, we have been able to attain very substantial cost reductions by operating two of the largest wastewood-fired high pressure boilers in the world, and extensive studies are now being carried out on the establishment of industrial tree plantations for commercial dendrothermal energy. The establishment of industrial tree plantations as a raw material base for industrial chemicals is also being looked into. I do not see such products in the export market within this decade, however, except perhaps in the beginning of the export of fuelwood and its derivatives within the next few years. In the meantime, we in Southeast Asia are keeping close watch on new developments in the wood industry to see what can be done to further our participation in its international trade.

Conclusion

I have not given you quantitative projections nor tried to estimate growth rates in this paper. The main markets for our wood-based industries for the next few years will continue to be abroad. We have a saying among ourselves, "The United States catches a cold and we have to be taken to intensive care with pneumonia." I hope that what I have touched on today gives you an overview of how we see our ability to react to the markets abroad during the rest of the decade.

A MARKET FOR WASHINGTON WOOD:
THE CHALLENGE OF THE 1980s

Brian J. Boyle

As experts in the forest products industry have said, the deep recession of 1981 and 1982 will produce fundamental industry changes because it has underscored a basic marketing problem for Washington forest products companies. I want to discuss that problem, explain the role of state wood in the overall marketing scheme, and explore changes in public policy and public perceptions that can help provide more certainty and predictability in Washington's wood supply and wood markets.

Washington State is the only western state where the management of state-owned lands plays a large role in the future of our forest products industry. The Department of Natural Resources presently manages about 1.8 million acres of commercial forest land. This land supplies as much as 12 percent of the state's annual timber harvest.

In spite of this relatively small share of the annual harvest, state-owned timber is a significant part of the forest products marketing picture. Each year, the Department of Natural Resources sells about 750 million board feet of timber at public auction. Principal bidders include Fortune 500 companies, small- and medium-sized family- and cooperatively owned mills, individuals, and exporters. Approximately 60 to 70 percent of the state-owned wood now sold is old-growth western hemlock, and about 40 percent of the sales are on the Olympic Peninsula. State forest lands are close to ports and mills and generally can be logged year around.

There are no restrictions on the export of state timber to overseas markets. This is a policy which has been supported by voters and their elected state legislators for the past 15 years. Because of the high export quality of state wood, it often commands a higher price than wood sold by the federal government.

There is a very important socioeconomic link between education and our state forest lands. State land income supports education, counties, and penal and charitable institutions.[1] About 1.2 million acres of the state's commercial forest land was given to Washington when it entered the Union in 1889. These were given to the state to support schools and other state institutions. Many states sold these federal trust lands, but Washington has retained ownership of the bulk of its lands. About 60 percent of the income from Washington's commercial forest land earns money for kindergarten-through-twelfth grade school building construction and remodeling. Last year, for example, Washington's Common School Construction Fund received $93 million from timber sales. In addition, timber sales income finances between 70 and 80 percent of the building accounts at the University of Washington, Washington State University, and the three state regional universities.[2]

This relationship between state lands and schools means that schools, universities, and counties have suffered along with the industry during the current recession. In fact, revenue from state lands has dropped about 40 percent. Thus, the question of markets for state wood is important not only for forest products companies but also for the state's schools, counties, and other institutions.

Currently, about half of the timber cut each year on state-owned land is exported. We estimate about 90 percent of the exported state wood is shipped as logs, with the remainder shipped as processed wood products. Japan is the leading export customer for both logs and processed wood products, followed by China, Korea, and Canada.[3]

Our surveys show there is another trend at work in Washington's forest products industry. Since 1977, export shipments have accounted for a growing percentage of the state's lumber production. Our estimates indicate that last year lumber exports amounted to about 20 percent of the state's total lumber production, a significant increase from just a few years ago. In 1978, for example, exports accounted for only 362 million board feet out of a total state production of 4.2 billion board feet--or a little under 9 percent.[4]

---

1. Washington Constitution, article 16, section 1 (1889). Enabling Act of 1889, 25 Stat. 180, sections 10, 12, 17 (1889). Chapter 228, Washington Laws of 1927.

2. Department of Natural Resources, *Totem*, March/April 1983, p. 11.

3. Personal communication, Loren Gee, Department of Natural Resources, February 1983.

4. Personal communication, Loren Gee, Department of Natural Resources, March 1983.

This growth of Washington lumber exports is a natural result of several national and international market trends. Washington's share of the nation's softwood lumber, plywood, pulp and paper market has remained nearly constant or declined in the past 20 years. Forest products competition in the nation's domestic markets has become more fierce because of southern and Canadian production. Southern timberlands presently supply about 45 percent of the nation's wood supply and are expected to provide 55 percent of the wood supply by the century's end.[5] Southern forests are close to the area of greatest housing demand in the country—the South and sunbelt states. Labor costs are lower in the South, and Pacific Northwest mills suffer from high rail and marine transportation costs. As if the South did not represent sufficient competition in domestic markets, Canadian lumber imports now have captured 30 percent of the domestic market because of government support policies, a cheaper Canadian dollar, and lower transportation costs.

The overseas wood markets appear just as competitive. Alaska, British Columbia, the Soviet Union, New Zealand, and the Third World are expected to compete vigorously for Asian log markets. European countries are seeking woodpulp in South America and Africa. Even as foreign countries line up for Asian and Third World wood products markets, the recession has reduced the overall world demand for wood products. An added problem for this country is the high value of our dollar, a predicament underscored by the recent currency devaluations in Australia and New Zealand.

The present domestic and international marketing picture is not rosy, but I am firmly convinced now is a time to seize opportunities. It is just as much a mistake today to predict continued economic recession in the forest products industry as it was in 1979 to anticipate continuing escalation of product and stumpage prices.

The gradual rise in the percentage of Washington lumber exported overseas shows more and more mills can compete in foreign markets. State and federal policy must enhance and encourage this foreign trade. The recent U.S. Commerce Department decision regarding Canadian lumber imports shows that the national administration will be reluctant to impose tariffs which would invite retaliation on U.S. exports. This reluctance is very understandable: exports represent 19 percent of the American economy and one out of every

---

5. Bill Rose, "Southern Timber Drawing Foreign Investors," <u>Miami Herald</u>, July 4, 1982.

eight manufacturing jobs in this country. Trade barriers are just going to make it more difficult for this important sector of the economy to flourish again. A trade war with some of our major trading partners could crimp our budding forest products exports.

What is the proper thing for government to do to help this state's forest products industry? First, we must establish and maintain a consistent supply of wood for sale on public lands. Timber in public ownership accounts for about half of this state's commercial forests. It is a sole source of supply for many Washington mills. In the Department of Natural Resources, we have lowered the percentage deviation from our annual sales target. The previous policy allowed timber sales targets to vary by as much as 50 percent or 400 million board feet, from year to year. That is too wide a gap to assure mills of adequate raw materials. The first step is to make our sales policies more predictable and stable.

A second major ingredient to long-range supply is the prompt reforestation of cut-over lands. Harvested lands not only must be replanted but they must be thinned and maintained to produce commercial timber. In the 1930s, Washington State began on its lands some of the West's first forest rehabilitation projects--even before private companies began their tree farm programs. The vital ingredient in our state forest management is Washington's Resource Management Cost Account. The account receives up to 25 percent of the timber sales revenue to finance reforestation, thinning, and other forest management work. Last year, for example, this account received about $18 million from timber sales. In turn, the account financed replanting 14,600 acres, thinning more than 8,000 acres, and fertilizing more than 32,000 acres. This account assures forest management will go on year after year, whatever the political currents in the state legislature. The federal government might consider a similar dedicated fund for federal forest lands. It could do much to make federal forest management a long-term commitment rather than a passing congressional fancy.

A third element in ensuring long-term supply from public timberlands is the conservation of air, water, and soil. In the Department of Natural Resources' new Forest Land Management program, we emphasize the least road construction necessary to serve all users. Department management in this plan will consider management zones along streams, lakes, and rivers for special treatment. The plan requires our field managers to select the forest and logging management best suited to retain productivity of each site.

A fourth element of the supply issue is also related to reforestation, namely, proper enforcement of forest practices on commercial forest lands. We have sought additional money and manpower to enforce Washington's new Forest Practices Act regulations. These standards must be enforced because the reforestation and productivity of Washington's forest lands could be our long-term trump card in the marketing of our wood. I was reminded recently that the original Forest Practices Act, passed almost forty years ago, was oriented toward reforestation. I often believe the current act is oriented toward paperwork, rather than field inspection and enforcement.

Our commitment to reforestation and productivity stands in contrast to a recent USDA Forest Service report that the southern forest industry is threatened by a failure to reforest. Presently, the report said, only one-half of the acres of pine harvested in the South are being reforested. In some southern states, only one of every 10 acres is being adequately replanted.[6] There is similar concern about future timber supplies in the Third World. Our commitment to the reforestation and productivity of our forest land will mean dependability to foreign and domestic wood buyers.

I mentioned earlier that Washington State also has a very important socioeconomic link between its timber holdings and the support of its public schools and universities. Our state constitution and court decisions have directed that the state manage its lands as a trustee. That is, short-term benefits cannot be taken from these lands at the expense of future generations and beneficiaries. It means we must be careful, prudent land managers. We must ensure that our school lands retain their productivity so they produce trees both today and 60 or 120 years from now. The Department of Natural Resources' new forest management plan will carry on this commitment to long-term state forest land productivity.

Past public policy decisions have allowed the state to keep its forest base and to be a significant timber supplier today. But what about the future public policy decisions concerning forest land? I have attempted to address some of the continuity of supply issues, but other public decisions cause political factors also to influence future availability of wood.

In particular, urban growth management already has become an important policy question as population growth

---

6. Robert L. Lentz, "Forestry's Future Up to States," State Government News, 26, No. 1 (January 1983), p. 8.

extends into areas beyond Washington's existing cities and suburbs. The Washington Forest Protection Association recently predicted that the state's 17.9 million acres of forest land will shrink to 14.5 million acres by the year 2020.[7] This would result from the conversion of forest lands to urban uses. State forest land will be near much of this growth. Recent census figures[8] show that 40 percent of our state-owned timber harvested in the 1990s will occur in counties whose population growth exceeded 30 percent in the last decade. Our future timber supply will depend on how this urbanization is accommodated with forest management.

The fate of future timber supplies also will rest with the public's view of forest practices on Washington forest land. Forest management will be turned over to the courts if state government fails to enforce these regulations strongly. Such a prospect could only cast a long shadow of uncertainty over future timber supplies.

What I am outlining for state government is the role of forest manager and regulator, not designer of markets by legislation. We in state government can ensure timber supplies for state mills into the foreseeable future. We can establish the political and social policy support to maintain a timber base in Washington. State government already is equipped to do this, and has done well. It will provide considerable stability to mills that must reach out now for new markets.

A growing consensus of Washington lawmakers favors export trade assistance for small mills rather than mandatory assignment of timber supplies to them. Raw material supply without product markets is economic futility. The key is market consistency, and more legislators, citizens, and mill operators recognize this than ever before.

Moreover, and in conclusion, I think more citizens and legislators recognize the social importance of this state's forest lands. The income from these lands saves hundreds of millions of taxpayer's dollars each biennium in school construction and remodeling costs. These lands and our stewardship of them represent an inheritance to pass on to

---

7. Don Chance, "Reference Manual for the WFPA Land-Use Planning Short Course for Forest Managers," Washington Forest Protection Association, March 1982.

8. Chuck Chambers, Department of Natural Resources, Lead Biometrician, March 1983. Office of Financial Management, State of Washington, Population Trends for Washington State (Olympia: August 1981), p. 26.

future generations. For this reason, we are finding a growing awareness of the need to keep this state trust land base productive. This public awareness could curb the loss of state and private forest productivity to poorly conceived public policy. It could establish a firm public policy support for a forest base which could anchor a new Washington forest products export trade.

WE WOODED THE WORLD

W. D. Hagenstein

The most interesting biography written about the man for whom western Washington's and Oregon's principal tree is named, David Douglas, contains a fascinating forecast in its prologue. Commenting on Douglas's first trip up the Columbia River in 1825 when he observed its solid tree-lined banks including the great species which would ultimately perpetuate his name, the author says, "little does the weary voyager imagine this noble tree, destined to become the world's greatest producer of structural timber, will be named in his honor" (Harvey 1947).

Douglas himself anticipated Douglas-fir's future commercial value in his journal when he said: "The wood may be found very useful for a variety of domestic purposes: The young slender ones exceedingly well adapted for making ladders and scaffold poles,...the larger timber for more important purposes" (Douglas 1914).

While people were attracted to the Douglas-fir region primarily to develop its natural prairies and forest openings for agriculture, it was soon apparent to our pioneers that the region's greatest resource was its trees. The first commercial entrepreneurs in the Northwest were the Hudson's Bay Company employees, attracted originally to the region's fur-trapping opportunities. In a short while, their principal quarry, the beaver, was largely trapped out. Finally, when Governor Simpson decided in 1827 that the company must engage in the "timber-coasting trade," the emphasis of its activities (apart from agriculture) was changed from "fur" to "fir." In that year the company's water-powered sawmill at Vancouver, Washington, exported the first lumber from the Northwest, a small shipment of boards to the Sandwich Islands--as Hawaii was then commonly called.

When asked by the company clerk how to list the first shipment of boards on the manifest, Factor John McLaughlin responded, "It's fir, but the buyers only know pine, so call it Oregon pine." This remained the name under which Douglas-fir was shipped throughout the world for nearly three-quarters of a century. Not so bad botanically, because Douglas-fir and the other species in its genus are members of the pine family!

It was from this modest beginning 156 years ago that this region began to wood the world.

The first place I saw lumber being loaded out on a ship for export was about four miles west of here along the Lake Washington Ship Canal at the Stimson Mill Company in Ballard. I was just six. A neighbor boy and I were sauntering home after fishing for bullheads and shiners west of the Government Locks, and we stopped to watch the Japanese purchaser's representative tallying boards and timbers as they were being hoisted aboard ship. The canal was lined in those days with the industry that built this great city. Beginning at Fremont was the mill of the Bryant Lumber Company. Westward to the Locks were the sawmills of the Gould Lumber Company and the Bolcom-Canal Lumber Company on the south side, just east of the Ballard Bridge (where Champion International's plywood plant is today). Directly across on the north side was the Phoenix Shingle Company mill owned by the Ward family. I attended Salmon Bay Grammar School and Ballard High School with their boys. West of the bridge on the north side was the Motor Mill, one of the region's largest shingle producers and the employer of our neighbor, Mr. Sprague, who worked there for 50 years. Just west of the Motor Mill was the Seattle Cedar Lumber Manufacturing Company, the world's largest cedar mill. This mill manufactured both lumber and shingles, which were sold under its famous Maltese Cross label. Then came the Stimson Mill Company with its brick office building on Shilshole Avenue and, finally, the Federal Pipe and Tank Company which specialized in Douglas-fir stave pipe for water mains and aqueducts all over the world. There were also several boatyards which built wooden vessels for our north Pacific fishery fleet.

It wasn't quite a decade later when I saw the lumber shipped from the mills along the canal being off-loaded in Japan and China. I didn't know then that I would spend half a century working in the industry which produced it.

Many of the current generation of Pacific Northwest lumbermen seem to think they discovered the export market, but the fact is their great-grandfathers and even some of <u>their</u> predecessors were hard at it a century and a quarter ago.

In the next few pages I'm going to highlight the contribution our region's forests made to the upbuilding of many foreign lands and to their reconstruction after man-caused disasters of war and the natural disasters of earthquake, flood, and fire. It will not be the typical, historical presentation with scholarly footnotes; it will feature a minimum of statistics, and will make some references to a few individuals and companies. This information I have gleaned from my own half-century of association with the Douglas-fir industry and from the forest products trade journals of yesteryear when personal journalism was their hallmark.

Because I am filling the lecture slot originally slated for the eminent forest industry historian, Thomas R. Cox, of San Diego State University, I drew heavily from his book <u>Mills and Markets: A History of the Pacific Coast Lumber Industry to 1900</u> for early history of the offshore lumber trade (Cox 1974). In his preface, Cox quotes an anonymous observer who thought the industry unimpressive in its main market in San Francisco and then says himself, "but these piles of lumber, those modest little coops of offices (apparently situated amongst the lumber piles) represent thousands upon thousands of acres of land, millions upon millions of capital...[and] employment to...hands all the way from Santa Cruz on the south to Alaska on the north." Cox concludes this assessment of both the importance of the industry and the public's failure to recognize it as correct and laments that it has been widely shared since.

Cox also notes that throughout their history our cargo mills, as producers for seaborne trade were called, served not only as an important source of original employment and building materials for the American Pacific Slope, but also produced the region's leading exports and encouraged capital inflow which was greatly needed in a growing region, despite its mineral riches. One could say that timber products were our first cash crop. Now, 156 years since the first sawmill produced the first lumber for export, timber is still our number one cash crop, for it supports 42 percent of the people of Oregon and 22 percent in Washington.

Hawaii continued as the foremost foreign market for lumber for many years, but as the industry grew on the Columbia River and started on Puget Sound, its entrepreneurs sought markets in Russian Alaska, China, Australia, and Latin America. The decade prior to the Civil War saw increasing export shipments from the new, big Pope & Talbot mill at Port Gamble. That company also bought out a competitor at nearby Port Ludlow, where the mill had been in business for nearly 25 years. The Port Gamble mill has been manufacturing Douglas-fir lumber for 130 years now and is probably the oldest continuously operating business of any kind in the Pacific Northwest (Coman and Gibbs 1949).

Other mills which soon started filling foreign orders for lumber on Puget Sound began in the 1860s, notably

Captain William Renton's Port Blakely Mill Company, across the Sound from the infant city of Seattle which had grown up around Henry Yesler's mill, started on Elliott Bay in 1852. The Washington Mill Company was established at Seabeck on Hood Canal in the mid-1850s, and the Tacoma Mill Company at Tacoma in the late 1860s. While all these pioneer mills served both local and San Francisco markets, they sought to diversify their outlets by exporting their lumber across the water. Cox notes that enthusiasm for foreign trade was greatly stimulated periodically when the San Francisco market collapsed.

As early as 1855, six lumber cargoes left Puget Sound for China. Northwest lumbermen also turned to Australia for another outlet in the 1860s. Then from the late 1850s into the 1870s the west coast of South America, particularly Peru and Chile, became an active market for Northwest lumber as those nations' mineral deposits were developed by substantial railroad construction and their harbors equipped with docks and other facilities for outshipment of nitrates, guano, and copper.

After Commodore Perry's successful negotiation of a trade treaty with Japan in 1854, that market opened for Northwest lumber. Even though Japan was and is a heavily forested country, our species and their large sizes provided and are providing materials not readily obtainable from their own forests.

Another foreign market closer at hand developed as the transcontinental Canadian railroads were pushing across the Rockies into British Columbia. B.C.'s forest industry developed much more slowly than Washington's, so Washington-grown and manufactured railroad ties, bridge timbers, and car material were mainstays for the pioneer Canadian railways.

During the last 20 years of the last century, and just as the first of the transcontinental railroads crossed the Cascades, thus opening up a new market for Northwest lumber in the Midwest, foreign trade was extremely important to the growing installed capacity of cargo mills on Puget Sound and along the Columbia River and in the more recently developed timber ports of Grays Harbor and Coos Bay. The principal markets continued to be in China, Japan, Australia, and the west coast of South America. The Chinese railroad system took huge quantities of Douglas-fir ties and timbers for its extension, and the building of harbors for its shipping facilities required shiploads of long Douglas-fir piling and heavy timbers.

It was also during this period when both the Japanese and Chinese began to purchase "squares," timbers commonly from 12" x 12" to 24" x 24" in lengths up to 100 feet, destined for resawing into smaller dimensions in their own sawmills. The Chinese sawmill was just another version of the old European and American system of pit-sawing, where one man worked above the square or log and the other below as together they would ripsaw the boards by hand--a common way of making boards in pioneer days everywhere. I observed this method myself in Shanghai more than 50 years ago, and as recently as 11 years ago in Peru.

Part of the research I did for this presentation started out by asking the Multnomah County Library to dig out the bound volumes of the two principal trade journals of our industry--<u>West Coast Lumberman</u>, published at Seattle, and <u>The Timberman</u>, at Portland. They were combined some years ago into their successor, <u>Forest Industries</u>. I ambitiously asked the librarian to get out the issues of the two oldies for the years 1920 through 1932 because I believed they would be the most profitable in my quest for interesting items. Little did I realize while making such arrangements that the volumes I requested would occupy a library cart with three shelves four feet long!

But I dug in and started with <u>West Coast Lumberman</u> for October 1, 1919, as that was the beginning of a volume. After about 10 five-hour stints I had gotten through the issue of March 1, 1924, or only four and a half years of the then semimonthly periodical. I just found too much of interest and not only read hundreds of items on everything else while gleaning for items useful for this occasion, but compiled 59 pages of notes on forestry items for a history of industrial forestry in this region with which I have been struggling for two and a half years.

Let me share just a few of the choicer items with you. The issue in which they appeared is noted in my text.

October 1, 1919: An ad of Summer Iron Works of Everett claimed that "114 million, 500 thousand feet of lumber are being shipped weekly from Pacific Northwest to the far corners of the Earth to supply the lumber famine that is worldwide." Some claim! The data I have say that export lumber shipments in 1919 totaled only 386 million including 102 million from B.C. Ad writers apparently haven't changed much in 60 years!

Puget Mill Company's ad simply stated: "We make a specialty of cargo shipments to all parts of the World." Puget Mill was one of Pope & Talbot's operating companies.

A regular column labeled "The Market" said, "continued car shortage and increased movement of lumber by water particularly to foreign markets characterized the situation during the last fortnight...new ports of destination are continually appearing in the shipping records, one cargo will shortly leave for Egypt. Several are en route or will be shortly for Persia."

In a piece headed, "Northwest Cargoes Loaded Past Two Weeks" (from Puget Sound and Grays Harbor) 33 vessels had foreign destinations such as Kobe, Yokohama, Shanghai, Sydney, Genoa, Darien, Osaka, Bombay, Hong Kong, Cardiff, Balboa, Liverpool, Cape Town, Tokyo, etc.

October 15, 1919: Nebraska Congressman Jeffries introduced a bill to prohibit export of lumber from the United States for two years (apparently because of the housing shortage in the United States due to wartime population dislocation).

November 1, 1919: There were ads for the following cargo mills: St. Paul and Tacoma Lumber Company (Tacoma), Willapa Lumber Company (Raymond), Western Lumber Company (Aberdeen), Stimson Mill Company (Ballard), Grays Harbor Commercial Company (Cosmopolis), Ferry-Baker Lumber Company (Everett), Puget Mill Company (Port Gamble and Port Ludlow), Puget Sound Lumber Company, Defiance Lumber Company, and Dempsey Lumber Company (all three in Tacoma), Inman-Poulsen Lumber Company (Portland), and National Lumber and Manufacturing Company (Hoquiam).

November 15, 1919: "Four-masted schooner, 'Margaret' loaded 1½ million feet of lumber at Port Blakely Mill Company for Belgium, the first Douglas-fir shipment there since war's end."

December 1, 1919: "Morrison Mill Company (Anacortes) shipped box shook for 100,000 oil cases to Singapore. Company has another large order for same item which will be shipped monthly."

February 15, 1920: Lumber exports of 206 million feet from Washington went to India, Persia, South Seas, United Kingdom and Europe, China, Australia, west and east coasts of South America, Japan, Cuba, South Africa, and Mexico. Oregon exported 78 million to about the same places.

April 1, 1920: Headlines: "Fir Ties for Panama Railroad," "J. J. Donovan (Bloedel-Donovan Lumber Mills, Bellingham) To Discuss Export Lumber Trade," "Chile Has Large Timber Resource, Yet Needs Much Outside Lumber."

November 15, 1920: "One million, 300 thousand feet loaded on SS Eastern Belle at Stimson Mill Company in Ballard for Cuba."

February 15, 1921: "Oregon and Washington exported 64,500,000 feet of untreated railroad ties to the United Kingdom and 1,100,000 feet to China in 1920. India received 3,611,000 feet of treated ties; San Domingo, 1,473,000."

July 1, 1921: "Denmark Maru loads 3,800,000 feet of

Japanese squares at Wauna and St. Helens, Ore. and proceeds to Grays Harbor to pick up another two million."

August 15, 1921: Japan now buying "baby squares" which are 4" x 4" x 20' and cedar logs.

October 1, 1921: Article entitled "Why Japan Is Buying Northwest Lumber" explains that prices for domestic species in Japan declined, with the result that forest owners stopped cutting, and American imports filled the needs aided by the reduction of ocean freight rates to a fourth of what they were in 1920 due to increased ship availability.

November 1, 1921: "Inman-Poulsen Lumber Company (Portland) enlarging dock evincing faith in the future of the export business for its 150 million feet/annum mill...expansion will handle two large vessels simultaneously."

May 1, 1922: First issue to include more than one ad for Japanese import-export houses. They include M. Furuya Company, Terzawa and Company, Takata and Company, Suzuki and Company, and Mitsui and Company.

February 15, 1923: Item headlined, "Russians Shipping Lumber" says "up to October 1, 107 billion (sic) feet were shipped from Russian Ports on White Sea according to the U.S. Department of Commerce." Sounds like they were telling whoppers in those days too!

September 15, 1923: Carries story of the earthquake which with accompanying fires damaged a third of Tokyo and leveled Yokohama, killing 140,000 persons on September 1. Forecasts huge lumber shipments there for rebuilding in the next two years.

October 1, 1923: Item says 316,000 buildings in Tokyo or 71 percent of the total were destroyed; in Yokohama, 70,000 or 82 percent.

January 15, 1924: 2,300,000 feet of lumber for relief of Japan from nine mills on Grays Harbor were donated by the Douglas-fir Exploitation and Export Company as a gift from the independent lumbermen of Grays Harbor and were shipped to Japan on the Kohnon Maru.

These items certainly illustrate handsomely my claim in the title of these remarks that "we wooded the world."

Now let us return to our main thread.

It is difficult to measure the importance of foreign lumber shipments to our region's economy without using some numbers, but I will try to use as few as possible. The trend of volume of lumber exported offshore from Washington and Oregon for the last 85 years is indicated in the following table.

The data show that Northwest lumber exports increased steadily after the turn of the century, but decreased

Table 1. Volume of Lumber Exported from Washington and Oregon, 1895 to 1982.

Year	Volume in Millions of Board Feet
1895	131
1900	160
1905	259
1910	329
1911	412
1912	432
1913	553
1914	475
1920	502
1925	1,152
1928	1,630
1930	1,111
1935	564
1940	361
1943	114
1945	273
1950	269
1955	408
1960	404
1965	411
1970	410
1975	617
1980	985
1981	934
1982 (first half)	467

Sources: Pacific Lumber Inspection Bureau 1967 and Ruderman 1980.

substantially during World War I because of preoccupation with other matters and the severe shortage of shipping.

In the reconstruction period of the first war, export lumber shipments increased rapidly to their peak in 1928 when the declining worldwide economy led to the stock market crash in 1929 and the beginning of the devastating depression which lasted a decade.

During World War II, lumber exports were at a low ebb, with the lowest volume in 85 years in 1943. After the war, lumber exports picked up slowly until the mid-1950s, when they reached a level which was maintained into the 1970s, at which time strong marketing and trade promotion efforts were made by the industry's associations. The volume is now about twice what it was 15 years ago.

A new foreign market for Northwest forest products began about 1960 when there was an escalating demand by Japan for Washington and Oregon logs of most species. In 1961 log exports to Japan were 328 million board feet. Then disaster struck our region on October 12, 1962, when a western Pacific typhoon got confused and blew down more than 3 billion board feet of timber per hour for five hours. Following this catastrophe the increased log export demand of the Japanese from 1963 to 1965 was a heaven-sent blessing. It allowed us to salvage close to 90 percent of the concentrated blowdown, much of which would have been lost to decay and insects during that critical period. But it also started a serious controversy in the Northwest because some thought we were exporting too many logs and not enough manufactured forest products.

Late in 1967 the industry sought a 350-million board feet annual limitation on the export of logs from federal timber, but such an order was not forthcoming until after the late Senator Wayne Morse (Oregon) had held comprehensive hearings on the log export question in January 1968, and after a joint government-industry trade mission to Japan in February of that year. The trade mission was an attempt to convince the Japanese that they should buy more manufactured products and less logs. I was one of the advisers to the mission, and at a joint meeting of the American and Japanese forest industries in the Foreign Ministry in Tokyo on February 21, I said in a formal presentation:

> Japanese interest in our forests can be beneficial for our forestry because good markets for forest products under our system of free enterprise gives us the incentive to do a continuously better job of forestry. But Japanese acceleration of log purchases has created serious social and economic problems for us. If we work it out so that you take more manufactured forest products, rather than logs--it will help solve our economic and social problem and still give you the wood your growing economy needs.

Nothing much came of the mission's efforts, but later that year Congress adopted Senator Morse's proposal to prohibit all log exports from federal timber, but the Secretaries of Agriculture and Interior were permitted to exempt species or grades of logs declared surplus to domestic needs. This is still in effect.

A ballot measure in the state of Washington called for prohibition of log exports from state-owned timber in 1968, but was rejected. The Oregon Legislature had prohibited the export of logs from state-owned timber in 1961, but recently the state's attorney general has questioned the measure's constitutionality.

Log exports to Japan continued to rise significantly in the 1960s and 1970s. South Korea also began to purchase logs for its timber-short mills and, more recently, China has come into the market strongly. The log export business peaked in 1979 at 3.2 billion board feet. In the same year the Northwest exported 840 million board feet of lumber and 330 million square feet of plywood. Log exports that year constituted 22 percent of the total timber harvest in our two states.

Log exports to Japan have slowed down considerably in the last three years, but have increased significantly to China.

Another important foreign market for Northwest wood in the last 20 years has been the Japanese pulp mills, which take the chipped raw material leftovers of sawmills and veneer plants. The volume exported peaked in 1974 at over 3.5 million tons and has dropped to less than 2.5 million in 1981. The demand might have remained higher, but with domestic lumber and plywood manufacture off, the supply of chippable leftovers has dramatically decreased. But it will return.

In the last three years the Northwest press and some politicians have questioned whether the Northwest's forest industry is deserting this region by making large investments of many of its large companies elsewhere, particularly in the southeastern United States. I commented at length on this in a formal presentation to my local chapter of the Society of American Foresters in 1980 under the title, "Is Industry Moving South?" My answer to that question was an emphatic no, but I said, "it is expanding southeast and north (to Canada) and staying here with an ever-increasing intensity of forestry to grow that quarter of the forest products needs of the American people of which this region is capable because of our productive soils, species and climate."

But you must know that U.S. rail freight rates and the disappearance of American intercoastal shipping have put Pacific Northwest lumber in an unfavorable competitive position in the Midwest and East with southern pine and Canadian lumber. This has stimulated more aggressive trade promotion and merchandizing by Pacific Northwest lumber manufacturers to increase foreign markets, particularly in the Pacific Rim. Northwest plywood manufacturers are working hard to increase sales not only in the Pacific, but to Latin America and Europe too.

If the region's basic economy, which is still heavily dependent on its renewable trees, is to regain the strength it had in earlier years, it will again have to "wood the world" as it magnificently did beginning at Vancouver more than a century and a half ago.

References

Coman, E. T., Jr., and H. M. Gibbs. 1949. Time, tide, and timber: A century of Pope & Talbot. Stanford University Press, Stanford.

Cox, T. R. 1974. Mills and markets: A history of the Pacific Coast lumber industry to 1900. University of Washington Press, Seattle and London.

Douglas, D. 1914. Journal kept by David Douglas during his travels in North America, 1823-1827. Royal Horticultural Society, London.

Harvey, A. G. 1947. Douglas of the fir. Harvard University Press, Cambridge, Mass.

Pacific Lumber Inspection Bureau. 1967. Tabulation. Seattle.

Ruderman, F. K. 1982. Production, prices, employment, and trade in Northwest forest industries, second quarter 1982. Pacific Northwest Forest and Range Experiment Station, Portland.

THE PRODUCT MIX AND FLOW OF WOOD MATERIALS IN INTERNATIONAL TRADE

James S. Bethel and Gerard Schreuder

Wood is a bulky material. It costs a great deal to move it long distances. Nonetheless, it has been a very important commodity in world trade for centuries. Venice was built on pilings and timbers from North Africa. Returning from his explorations of the New World, Columbus carried mahogany from the Caribbean Islands to Europe. The ships that brought colonists to New England as often as not carried white pine and oak for ship timbers back to Old England. And wood continues to be an important commodity in world trade. In 1980, trade in forest products across international boundaries was at a level of about $34 billion.

While forests are widely distributed geographically and wood is locally available in most countries of the world, the level of consumption and production of those forest products which move in world trade is of major importance in a relatively few countries, and these are the countries that are most important to the development of world trade. Most of the wood that is harvested worldwide is used for fuel, and this wood is predominantly used within the boundaries of the country of harvest. Accordingly, wood fuel is not an important commodity for world trade. The important commodities are the so-called industrial products. These products can be separated into two broad categories--mechanically processed wood and chemically processed wood. The important commodities in the first category are lumber, plywood, particleboard, and fiberboard; and in the second category are pulp, newsprint, printing and writing paper, and packaging, personal service paper, and paperboard.

Consumption

The consuming countries constitute the principal markets or potential markets for commodities moving in international trade, and it is these countries that ought to

Table 1. Largest Consumers of Softwood (c) Lumber in 1980.

Rank	Country	Quantity $m^3 \times 10^6$	% of Total
1	U.S.S.R.	80.36	25.2
2	U.S.	75.82	23.8
3	Japan	35.59	11.2
4	China	13.26	4.2
5	West Germany	12.50	3.9
6	Canada	12.13	3.8
7	France	8.09	2.5
8	Brazil	7.15	2.2
9	U.K.	7.03	2.2
10	Italy	5.47	1.71
	Total	257.40	

Total World Consumption = $318.92 \times 10^6 m^3$
Source: FAO <u>Yearbook of Forest Products</u> 1980.

Table 2. Largest Consumers of Hardwood (nc) Lumber in 1980.

Rank	Country	Quantity $m^3 \times 10^6$	% of Total
1	U.S.	16.83	16.6
2	U.S.S.R.	12.63	12.3
3	China	7.74	7.5
4	Japan	6.89	6.7
5	Brazil	6.62	6.4
6	France	4.52	4.4
7	Italy	2.79	2.7
8	Nigeria	2.69	2.6
9	West Germany	2.61	2.5
10	India	2.43	2.4
	Total	65.76	

Total World Consumption = $102.65 \times 10^6 m^3$
Source: FAO <u>Yearbook of Forest Products</u> 1980.

be the targets for the marketing efforts of trading organizations.

Table 1 indicates the ten leading conifer lumber-consuming countries for 1980. These ten countries accounted for 79 percent of the world's consumption of softwood lumber in that year. They are clearly the prime targets of organizations undertaking to sell softwood lumber in international trade. The primary use of softwood lumber throughout the world is for building construction, and these countries have the tradition of using wood in building construction. They are primarily temperate rather than tropical countries. These softwood lumber consumers represent a relatively stable market in terms of geographical location, even though the actual level of consumption may vary from year to year. Eight of the ten leading consumers of conifer lumber in 1980 were also the leading consumers in 1970. The exceptions were Brazil and Italy, which replaced Poland and Sweden in the 1970 listing. In 1970 the consumption of conifer lumber was even more concentrated than in 1980. A decade earlier the top ten consumers accounted for 87 percent of world demand. Consumption of softwood lumber increased a modest 2.3 percent between 1970 and 1980. Although this rate of growth in consumption is relatively small, it reflects a significant volume of material since conifer lumber represents the greatest volume consumption among the major forest products.

Worldwide hardwood lumber consumption represents less than one-third of softwood lumber consumption. Like softwood lumber, hardwood lumber is primarily used by the building construction industry, though most commonly in nonstressed components, furniture, and fixtures. Table 2 shows the top ten consumers of hardwood lumber among the countries of the world. Consumption of hardwood lumber is more dispersed than is the case of softwood lumber. For 1980, the fraction of hardwood lumber demand attributable to the top ten consuming countries was 64 percent. In 1970 it was 69 percent. As was the case for softwood lumber, the makeup of the list of top ten consumer countries was stable. Nine of the ten countries were the same ones in 1970 and 1980. The exception was that Australia, which was ranked seventh in 1970, was replaced by Nigeria, which was ranked in the eighth position in 1980. Increase in consumption of hardwood lumber from 1970 to 1980 was in excess of 11 percent.

Plywood was originally as nicely segregated as to conifer and nonconifer as was lumber, and they represented what were essentially two separate and distinct industries serving separate and distinct markets. Since World War II,

Table 3. Largest Consumers of Plywood in 1980.

Rank	Country	Quantity $m^3 \times 10^6$	% of Total
1	U.S.	16.75	42.8
2	Japan	8.40	21.5
3	Canada	1.87	4.8
4	U.S.S.R.	1.72	4.4
5	West Germany	0.87	2.2
6	U.K.	0.72	1.8
7	France	0.71	1.8
8	Brazil	0.64	1.6
9	South Korea	0.62	1.6
10	Netherlands	0.49	1.2
	Total	32.80	

Total World Consumption = $39.16 \times 10^6 m^3$
Source: FAO <u>Yearbook of Forest Products</u> 1980.

Table 4. Largest Consumers of Particleboard in 1980.

Rank	Country	Quantity $m^3 \times 10^6$	% of Total
1	West Germany	6.57	16.2
2	U.S.	6.41	15.8
3	U.S.S.R.	4.43	11.0
4	France	2.18	5.4
5	Italy	2.16	5.3
6	U.K.	1.85	4.6
7	Canada	1.31	3.3
8	Poland	1.24	3.1
9	Spain	1.05	2.6
10	Japan	1.02	2.5
	Total	27.21	

Total World Consumption = $40.46 \times 10^6 m^3$
Source: FAO <u>Yearbook of Forest Products</u> 1980.

these distinctions have become blurred. Increasingly, hardwood and softwood plywood is manufactured in the same plant, and hardwoods and softwoods are mixed in the same panels. Some plywood is all veneer, and some represents a mixture of veneer with lumber core or particleboard core. Some plywood is sold as stock panels to standard sizes and thicknesses and some as custom panels designed to meet customer size and thickness requirements.

Table 3 shows the top ten plywood consuming nations in 1980. These ten countries accounted for 84 percent of world demand in 1980. In 1970 the top ten consuming countries represented 89 percent of world demand. Eight of the ten leading consumer countries in 1980 were also on the list in 1970. From the 1970 list, the Philippines and Italy were replaced in 1980 by the Republic of Korea and Brazil. Consumption of plywood increased 17 percent worldwide between 1970 and 1980. The U.S. was the dominant consumer of plywood, representing 47 percent of demand in 1970 and 43 percent of demand in 1980.

Particleboard is the newest of the mechanically processed wood commodities, and its use is growing faster than that of any of the others. Table 4 is a list of the top ten particleboard consuming countries in 1980. These countries accounted for 67 percent of world demand in 1980. Consumption of particleboard grew a spectacular 111 percent from 1970 to 1980. Unlike softwood lumber, hardwood lumber, and plywood, particleboard showed much more volatility in the makeup of the roster of leading actors. Seven of the countries among the top ten consumers in 1970 were still in that group in 1980. These included West Germany, the United States, the U.S.S.R., France, Italy, the United Kingdom, and Spain. East Germany, Belgium-Luxembourg, and the Netherlands among the leaders in 1970 were replaced by Canada, Poland, and Japan by 1980. Europe, where the particleboard industry had its origin, is still the dominant region in particleboard consumption. In 1970, Europe accounted for 65 percent of the world consumption of particleboard. While this share was reduced by 1980, it was still a substantial 59 percent.

Fiberboard as a product has a longer history than particleboard, but its most spectacular growth in acceptance has been during the past thirty years. Fiberboard varies in properties, and accordingly has a broad spectrum of uses. The property of principal interest is density. Fiberboard density varies from about 1.2 grams per cubic centimeter at the high end of the range to 0.15 grams per cubic centimeter at the low end of the range. Low-density fiberboard has a density range of 0.15 to 0.4 grams per cubic centimeter;

Table 5. Largest Consumers of Fiberboard in 1980.

Rank	Country	Quantity $m^3 \times 10^6$	% of Total
1	U.S.	5.60	34.4
2	U.S.S.R.	2.70	16.6
3	Canada	0.73	4.5
4	Poland	0.68	4.2
5	Brazil	0.61	3.7
6	Japan	0.56	3.4
7	West Germany	0.52	3.2
8	China	0.46	2.8
9	U.K.	0.38	2.3
10	East Germany	0.36	2.2
	Total	12.60	

Total World Consumption = $16.27 \times 10^6 m^3$
Source: FAO <u>Yearbook of Forest Products</u> 1980.

Table 6. Largest Consumers of Market Pulp in 1980.

Rank	Country	Quantity $mt \times 10^6$	% of Total
1	U.S.	46.09	36.6
2	Canada	12.64	10.0
3	Japan	12.08	9.6
4	U.S.S.R.	8.22	6.5
5	Sweden	5.69	4.5
6	Finland	5.58	4.4
7	West Germany	4.31	3.4
8	France	3.48	2.8
9	China	2.62	2.1
10	Brazil	2.51	2.0
	Total	103.22	

Total World Consumption = $125.90 \times 10^6 mt$
Source: FAO <u>Yearbook of Forest Products</u> 1980.

medium-density fiberboard is in the range of 0.4 to 0.8 grams; and high-density fiberboard in the range of 0.8 to 1.2 grams. The low-density boards are primarily insulating boards and are used principally in North America, whereas the medium- and high-density boards are used extensively in Europe, Asia, and the U.S.S.R. The ten countries that were the largest consumers of fiberboard in 1980 are shown in table 5. These countries accounted for 77 percent of world consumption. Their counterparts in 1970 accounted for 79 percent of world consumption. Fiberboard's list of leading consumer countries was quite stable. Nine of the top ten consumers were identical in 1970 to 1980. World consumption of fiberboard was 14 percent greater in 1980 than in 1970.

Among the principal consuming countries, the U.S., U.S.S.R., and West Germany were the only ones that were among the top ten consumers for all of the mechanically processed products in 1980.

Market pulp of various types is sold worldwide to paper mills for conversion to secondary products. Table 6 indicates the ten most important consumers of market pulp in 1980. These ten countries represented 83 percent of world consumption for that year. In fact, the U.S., which was the world's leading consumer, used 37 percent of total world consumption. Consumption of market pulp worldwide was 23 percent greater in 1980 than in 1970. Eight of the ten leading consumers of market pulp in 1980 were also on the list of ten largest consumers in 1970.

Table 7 shows the top ten consumers of newsprint among the countries of the world in 1980. They accounted for 78 percent of total world consumption. The same ten countries were the leading consumers in 1970.

Table 8 illustrates the major consumers of printing and writing papers in 1980. For this product, the top ten consuming countries represented 76 percent of world demand in 1980. Eight of the top ten consumers in 1980 were also in the list of top ten for 1970.

The major consumers of packaging, personal service, and miscellaneous paper and paperboard in 1980 are listed in table 9. These ten countries represented 77 percent of world consumption for that year. Eight of the ten largest consumers of this commodity in 1980 were also among the ten major consumers of the same product in 1970.

The principal consuming countries clearly represent the best market opportunities for those who produce and sell wood products whether the producer is in a major consuming country and therefore selling domestically or is outside of the consuming country and selling in export. There were

Table 7. Largest Consumers of Newsprint in 1980.

Rank	Country	Quantity mt x 10^6	% of Total
1	U.S.	10.54	39.8
2	Japan	2.59	9.8
3	China	1.67	6.3
4	U.K.	1.38	5.2
5	West Germany	1.22	4.6
6	U.S.S.R.	1.07	4.0
7	Canada	0.91	3.4
8	Australia	0.49	1.9
9	France	0.43	1.6
10	Netherlands	0.37	1.4
	Total	20.67	

Total World Consumption = 26.45 x 10^6 mt
Source: FAO Yearbook of Forest Products 1980.

Table 8. Largest Consumers of Printing and Writing Paper in 1980.

Rank	Country	Quantity mt x 10^6	% of Total
1	U.S.	14.25	34.4
2	Japan	3.63	8.8
3	West Germany	3.19	7.7
4	China	2.62	6.3
5	France	2.10	5.1
6	Italy	1.63	3.9
7	U.S.S.R.	1.28	3.1
8	India	1.16	2.8
9	Canada	1.01	2.4
10	Brazil	0.80	1.9
	Total	31.67	

Total World Consumption = 41.46 x 10^6 mt
Source: FAO Yearbook of Forest Products 1980.

Table 9. Largest Consumers of Packaging, Personal Service, and Miscellaneous Paper and Paperboard in 1980.

Rank	Country	Quantity mt x 10^6	% of Total
1	U.S.	38.08	36.9
2	Japan	11.27	10.9
3	China	5.11	5.0
4	U.S.S.R.	5.05	4.9
5	West Germany	4.94	4.8
6	U.K.	3.81	3.7
7	France	3.51	3.4
8	Italy	3.44	3.3
9	Canada	2.19	2.1
10	Brazil	2.09	2.0
	Total	79.49	

Total World Consumption = 103.23 x 10^6 mt
Source: FAO Yearbook of Forest Products 1980.

Table 10. Largest Producers of Softwood (c) Lumber in 1980.

Rank	Country	Quantity m^3 x 10^6	% of Total
1	U.S.S.R.	87.20	27.0
2	U.S.	58.30	18.1
3	Canada	40.50	12.6
4	Japan	30.70	9.5
5	China	13.29	4.1
6	Sweden	11.08	3.4
7	Finland	9.15	2.8
8	West Germany	8.44	2.6
9	Brazil	7.33	2.3
10	Austria	6.30	1.9
	Total	272.29	

World Production = 322.44 x $10^6 m^3$
Source: FAO Yearbook of Forest Products 1980.

just twenty-one countries listed on the nine lists of ten major consuming countries, indicating that there was a lot of duplication from product to product. Four countries were among the leading consumers in every product category. These were the U.S., the U.S.S.R., Japan, and West Germany. These four countries collectively consumed well over half of all structural and fiber wood products in 1980.

Production

The demands of the consuming nations for forest products must be met by the production units of the producer nations. If a country is both a major consumer and a major producer of a particular product, then its product supply budget may be essentially balanced internally. If its requirements exceed its production capacity, it will have to import. If its requirements are consistently less than its production capacity, it will have to export its surplus.

Table 10 indicates the ten largest producers of softwood lumber among the nations of the world for 1980. These ten countries accounted for 84 percent of world production of softwood lumber. Softwood lumber leads all industrial wood products in quantity of material produced. Of the ten leading producers of softwood lumber in 1980, eight were also among the top ten producers in 1970.

The major producers of hardwood lumber are indicated in table 11. Two-thirds of the world's production of hardwood lumber is manufactured by the ten leading producer countries. Given that hardwood forests are more widely dispersed throughout the world than are softwood forests, it is not surprising that production of hardwood lumber is less geographically concentrated than that of softwood lumber. As was the case for softwood lumber, eight of the ten leading producers in 1980 were also on the list of top ten in 1970. While production of hardwood lumber in 1980 was less than one-third the volume of softwood lumber produced in the same year, its production was growing faster than the production of softwood lumber.

The ten largest producers of plywood are shown in table 12. The FAO data upon which this table is based do not distinguish between hardwood and softwood plywood. The softwood plywood is produced almost exclusively in the U.S. and Canada. All the other producer countries are basically hardwood plywood producers. The North American countries produce hardwood plywood as well as softwood plywood, but the fraction is relatively small. In the United States the ratio of softwood to hardwood plywood produced is typically

Table 11. Largest Producers of Hardwood (nc) Lumber in 1980.

Rank	Country	Quantity $m^3 \times 10^6$	% of Total
1	U.S.	17.04	16.6
2	U.S.S.R.	12.40	12.1
3	China	7.81	7.6
4	Brazil	6.74	6.6
5	Japan	6.41	6.2
6	Malaysia	5.15	5.0
7	France	4.24	4.1
8	Indonesia	3.40	3.3
9	Nigeria	2.69	2.6
10	Romania	2.46	2.4
	Total	68.34	

World Production = $102.80 \times 10^6 m^3$
Source: FAO Yearbook of Forest Products 1980.

Table 12. Largest Producers of Plywood in 1980.

Rank	Country	Quantity $m^3 \times 10^6$	% of Total
1	U.S.	16.00	39.7
2	Japan	8.40	20.9
3	Canada	2.34	5.8
4	U.S.S.R.	1.99	4.9
5	South Korea	1.57	3.9
6	China	1.56	3.9
7	Brazil	0.76	1.9
8	Finland	0.64	1.6
9	Philippines	0.55	1.4
10	France	0.55	1.4
	Total	34.36	

World Production = $40.27 \times 10^6 m^3$
Source: FAO Yearbook of Forest Products 1980.

on the order of eight to one. In 1980 the United States was the dominant producer of hardwood plywood. It should be noted that whereas the volume of softwood plywood produced in the United States was many times that of hardwood plywood, the volume of hardwood plywood produced was sufficiently great to place the United States second or third among hardwood plywood producing countries.

Among the countries producing plywood, the top ten producers accounted for 85 percent of total world production. Nine of the ten leading producers in 1980 were also on the list of ten leading producers in 1970.

The principal producers of particleboard in 1980 in table 13 account for 69 percent of total world production. All of these producers are in either Europe or North America. This pattern of production leadership has been maintained over the past several years. Nine of the ten countries that were on the list of leading producers in 1980 were also on the same list in 1970.

The United States is the principal producer of fiberboard and it has been in that role for some time. Table 14 illustrates the ten major producers of this commodity in 1980. While the United States accounted for more than a third of the world production of fiberboard, its output has not been growing. In contrast, world production of fiberboard in 1980 was 14 percent greater than in 1970. In contrast to the relative stability of the production base for most commodities, the production base for fiberboard is relatively volatile. Only six of the countries among the top ten producers in 1980 were on that list in 1970. The most spectacular change was that demonstrated by Brazil. Not listed among the top ten producers in 1970, it was in third place behind the United States and U.S.S.R. in 1980, as shown in table 14. Seventy-nine percent of the world's production of fiberboard was represented by the top ten producers listed in table 14.

The chemically processed wood products--market pulp, newsprint, printing and writing paper, and packaging, personal services, and miscellaneous paper and paperboard products--represent very large capital investments. Accordingly the pattern of manufacture from country to country tends to be fairly stable over time. In the case of the packaging, personal services, and miscellaneous paper and paperboard category, the same ten countries were the leading producers in 1970 and 1980. In all the other categories of fiber products, nine of the top ten producers were the same in 1970 and 1980. Tables 15, 16, 17, and 18 show the ten leading producers of the fiber products for 1980. In the case of market pulp and newsprint, the ten

Table 13. Largest Producers of Particleboard in 1980.

Rank	Country	Quantity m^3 x 10^6	% of Total
1	West Germany	6.26	15.5
2	U.S.	6.10	15.1
3	U.S.S.R.	4.69	11.6
4	France	2.17	5.4
5	Italy	1.77	4.4
6	Belgium-Lux	1.66	4.1
7	Spain	1.33	3.3
8	Austria	1.31	3.2
9	Canada	1.27	3.1
10	Sweden	1.19	2.9
	Total	27.75	

World Production = 40.33 x 10^6 m^3
Source: FAO <u>Yearbook of Forest Products</u> 1980.

Table 14. Largest Producers of Fiberboard in 1980.

Rank	Country	Quantity m^3 x 10^6	% of Total
1	U.S.	5.62	34.0
2	U.S.S.R.	3.01	18.2
3	Brazil	0.75	4.5
4	Canada	0.72	4.4
5	Poland	0.67	4.1
6	Sweden	0.61	3.7
7	Japan	0.57	3.5
8	China	0.46	2.8
9	Romania	0.31	1.9
10	Spain	0.31	1.9
	Total	13.03	

World Production = 16.51 x 10^6 m^3
Source: FAO <u>Yearbook of Forest Products</u> 1980.

Table 15. Largest Producers of Market Pulp in 1980.

Rank	Country	Quantity mt x 10^6	% of Total
1	U.S.	45.83	35.2
2	Canada	19.74	15.6
3	Japan	9.97	7.9
4	U.S.S.R.	8.73	6.9
5	Sweden	8.7	6.9
6	Finland	7.44	5.9
7	Brazil	3.34	2.6
8	China	2.35	1.9
9	West Germany	2.00	1.6
10	France	1.91	1.5
	Total	110.01	

World Production = 126.75 x 10^6 mt
Source: FAO Yearbook of Forest Products 1980.

Table 16. Largest Producers of Newsprint in 1980.

Rank	Country	Quantity mt x 10^6	% of Total
1	Canada	8.62	32.9
2	U.S.	4.10	15.6
3	Japan	2.57	9.8
4	Finland	1.57	6.0
5	Sweden	1.53	5.8
6	China	1.50	5.7
7	U.S.S.R.	1.35	5.2
8	West Germany	0.61	2.3
9	Norway	0.56	2.1
10	U.K.	0.36	1.4
	Total	22.77	

World Production = 26.21 x 10^6 mt
Source: FAO Yearbook of Forest Products 1980.

Table 17. Largest Producers of Printing and Writing Paper in 1980.

Rank	Country	Quantity mt x 10^6	% of Total
1	U.S.	13.81	32.8
2	Japan	3.77	9.0
3	West Germany	2.80	6.7
4	China	2.63	6.3
5	France	2.03	4.8
6	Finland	2.03	4.8
7	Italy	1.80	4.3
8	Canada	1.54	3.7
9	India	1.15	2.7
10	U.S.S.R.	1.14	2.7
	Total	32.70	

World Production = 42.06 x 10^6 mt
Source: FAO *Yearbook of Forest Products* 1980.

Table 18. Largest Producers of Packaging, Personal Service, and Miscellaneous Paper and Paperboard in 1980.

Rank	Country	Quantity mt x 10^6	% of Total
1	U.S.	41.22	38.9
2	Japan	11.52	10.9
3	U.S.S.R.	5.24	4.9
4	China	4.71	4.4
5	West Germany	4.09	3.9
6	Sweden	3.65	3.4
7	Canada	3.25	3.1
8	Italy	2.99	2.8
9	France	2.90	2.7
10	U.K.	2.49	2.4
	Total	82.06	

World Production = 105.92 x 10^6 mt
Source: FAO *Yearbook of Forest Products* 1980.

principal producing countries account for 87 percent of world production in 1980. For printing and writing paper, the percentage is 78 percent; and for packaging, personal services, and miscellaneous paper and paperboard, the share is 77 percent. There were twenty-four countries included on the nine lists of ten leading producer countries. Two countries, the United States and the Soviet Union, were listed among the leading producers in every product category. In fact, these two countries account for more than 40 percent of world production of all nine commodities.

World Trade

The data on consumption and production of wood commodities clearly indicate that as a general rule major producing countries tend also to be major consumers of forest products. This is not surprising given the pervasive characteristic of forest geography and the fact that wood is a very bulky and relatively low-value material that is expensive to transport great distances. Among the twenty-one countries represented on the lists of principal consumers of the wood commodities reviewed in this report, seventeen were also on the list of major producers of the same commodities. Furthermore, the producers and consumers were well matched within commodity categories, as indicated in table 19. the role of world trade in forest products is to bring about a balance between supply and demand.

Tables 20 through 28 show the ten leading importers and exporters of the nine commodities studied for the year 1980.

Since importing countries are the principal targets of the marketing efforts of firms that engage in world trade or that aspire to engage in world trade, it is perhaps useful to examine some of the reasons that a country imports significant quantities of wood structural and fiber commodities. This sort of analysis is perhaps useful when decisions are being made to target major marketing efforts.

Forest Production Base

An obvious reason that a country will import wood structural and fiber commodities is that it simply does not have the productive forest base to support an industry of the size required to meet the demands of its population. When a nation's land base is small relative to its population, land may be needed for the production of food or for cities or towns, roads, railroads, and airports and

Table 19. Top Ten Consuming Countries that Were Also Top Ten Producing Countries in 1980.

Commodity	Number of Common Countries
Softwood lumber (c)	9 out of 10
Hardwood lumber (nc)	7 " " "
Plywood	7 " " "
Particleboard	7 " " "
Fiberboard	8 " " "
Market pulp	8 " " "
Newsprint	7 " " "
Printing and writing paper	8 " " "
Packaging, personal service, and miscellaneous paper and paperboard	9 " " "

Table 20. Largest Importers and Exporters of Softwood Lumber in 1980.

	Imports		Exports	
Rank	Country	Quantity $m^3 \times 10^6$	Country	Quantity $m^3 \times 10^6$
1	U.S.	22.17	Canada	28.97
2	U.K.	5.98	U.S.S.R.	6.95
3	Japan	4.90	Finland	6.90
4	West Germany	4.50	Sweden	5.89
5	Italy	4.36	U.S.	4.65
6	France	2.63	Austria	4.25
7	Denmark	1.23	Chile	1.26
8	East Germany	1.21	Czechoslovakia	1.04
9	Belgium-Lux	1.03	Portugal	1.00
10	Hungary	0.82	Poland	0.69

Table 21. Largest Importers and Exporters of Hardwood Lumber in 1980.

	Imports		Exports	
Rank	Country	Quantity $m^3 \times 10^6$	Country	Quantity $m^3 \times 10^6$
1	Italy	1.35	Malaysia	3.14
2	Singapore	1.10	Indonesia	1.19
3	West Germany	0.99	Singapore	1.13
4	France	0.75	U.S.	0.88
5	Canada	0.75	Yugoslavia	0.86
6	U.S.	0.68	Philippines	0.74
7	Netherlands	0.65	Brazil	0.57
8	U.K.	0.63	France	0.47
9	Thailand	0.56	Romania	0.42
10	Japan	0.53	West Germany	0.37

Table 22. Largest Imports and Exports of Plywood in 1980.

	Imports		Exports	
Rank	Country	Quantity $m^3 \times 10^6$	Country	Quantity $m^3 \times 10^6$
1	U.S.	1.00	China	1.24
2	U.K.	0.77	South Korea	0.95
3	Netherlands	0.49	Singapore	0.62
4	West Germany	0.49	Canada	0.55
5	Saudi Arabia	0.37	Finland	0.53
6	Finland	0.33	Malaysia	0.47
7	Singapore	0.25	Philippines	0.36
8	Hong Kong	0.24	U.S.S.R.	0.31
9	Belgium-Lux	0.21	Indonesia	0.25
10	Denmark	0.16	U.S.	0.25

Table 23. Largest Imports and Exports of Particleboard in 1980.

Rank	Imports Country	Quantity $m^3 \times 10^6$	Exports Country	Quantity $m^3 \times 10^6$
1	U.K.	1.28	Belgium-Lux	1.16
2	West Germany	0.97	Austria	0.68
3	U.S.	0.54	West Germany	0.66
4	Netherlands	0.53	Sweden	0.47
5	Italy	0.48	Finland	0.37
6	France	0.39	France	0.37
7	East Germany	0.18	Spain	0.28
8	Poland	0.17	U.S.S.R.	0.27
9	Denmark	0.14	Switzerland	0.25
10	Belgium-Lux	0.11	Romania	0.23

Table 24. Largest Imports and Exports of Fiberboard in 1980.

Rank	Imports Country	Quantity $m^3 \times 10^6$	Exports Country	Quantity $m^3 \times 10^6$
1	U.K.	0.32	U.S.S.R.	0.30
2	West Germany	0.30	Sweden	0.28
3	U.S.	0.18	U.S.	0.20
4	Netherlands	0.13	Brazil	0.15
5	Italy	0.13	France	0.14
6	France	0.12	Finland	0.12
7	Poland	0.10	Spain	0.12
8	East Germany	0.08	Poland	0.09
9	Belgium-Lux	0.07	Austria	0.07
10	Denmark	0.06	Canada	0.07

Table 25. Largest Imports and Exports of Market Pulp in 1980.

	Imports		Exports	
Rank	Country	Quantity mt x 10^6	Country	Quantity mt x 10^6
1	U.S.	3.65	Canada	7.24
2	West Germany	2.43	U.S.	3.39
3	Japan	2.21	Sweden	3.05
4	U.K.	1.86	Finland	1.90
5	Italy	1.76	Brazil	0.89
6	France	1.75	U.S.S.R.	0.68
7	Netherlands	0.65	South Africa	0.62
8	South Korea	0.46	Norway	0.53
9	Belgium-Lux	0.42	New Zealand	0.47
10	Norway	0.35	Portugal	0.44

Table 26. Largest Imports and Exports of Newsprint in 1980.

	Imports		Exports	
Rank	Country	Quantity mt x 10^6	Country	Quantity mt x 10^6
1	U.S.	6.59	Canada	7.71
2	U.K.	1.08	Finland	1.43
3	West Germany	0.69	Sweden	1.24
4	Australia	0.27	Norway	0.52
5	Netherlands	0.25	U.S.S.R.	0.30
6	France	0.18	New Zealand	0.24
7	China	0.17	U.S.	0.16
8	Denmark	0.15	Brazil	0.13
9	Brazil	0.15	Japan	0.10
10	India	0.14	West Germany	0.08

Table 27. Largest Imports and Exports of Printing and Writing Paper in 1980.

	Imports		Exports	
Rank	Country	Quantity mt x 10^6	Country	Quantity mt x 10^6
1	West Germany	1.08	Finland	1.75
2	U.K.	0.82	West Germany	0.68
3	U.S.	0.65	Canada	0.65
4	France	0.60	France	0.53
5	Netherlands	0.39	Sweden	0.51
6	Belgium-Lux	0.34	Italy	0.34
7	Australia	0.18	Belgium-Lux	0.29
8	U.S.S.R.	0.17	Netherlands	0.28
9	Italy	0.16	Norway	0.23
10	Sweden	0.16	U.S.	0.21

Table 28. Largest Imports and Exports of Packaging, Personal Service, and Miscellaneous Paper and Paperboard in 1980.

	Imports		Exports	
Rank	Country	Quantity mt x 10^6	Country	Quantity mt x 10^6
1	West Germany	1.71	U.S.	3.75
2	U.K.	1.62	Sweden	2.88
3	France	1.14	Finland	1.64
4	Netherlands	0.71	Canada	1.20
5	Italy	0.69	West Germany	0.86
6	U.S.	0.61	Japan	0.70
7	Belgium-Lux	0.51	U.S.S.R.	0.64
8	China	0.50	France	0.53
9	Japan	0.45	East Germany	0.53
10	U.S.S.R.	0.45	U.K.	0.30

other nonforestry uses. Sometimes a nation's forest land is relatively unproductive and its output of forest products modest. Thailand is an example of the first case. Once a major exporter of hardwood lumber, it is now a net importer of this commodity. Much of its most productive forest land has been converted to nonforestry uses. In 1980, Thailand imported 31 percent of its requirements for hardwood lumber and exported essentially none. Saudi Arabia is a different case. In 1980 it was the fifth largest importer of plywood in the world. Essentially a desert country, it has very little potential for production of forest; but as a country with a small population and high per capita income, its demand for forest products, particularly structural products, is great.

But the status of a country as a producer of wood for manufacture of forest products is not just a matter of land allocation or land productivity. It may be a function of the social priorities within a country. In some countries where forests are often grown on modestly productive sites but where the production of wood-based structural and fiber commodities has a high social priority, the output of these commodities is high. The Scandinavian countries are examples of this phenomenon. Some countries that are major importers of wood commodities have significant areas of forest land but have landownership and land tenure systems that discourage the type of forest management required to support significant forest products industries. Mexico is a case in point. Finally, some countries have temporary shortages of wood commodities that are supplied through imports until increased productivity can meet local needs. Japan drew down its growing stock during World War II and has been rebuilding its forestry production base for the past thirty years. In the interim it has been a very large importer of many forest-based commodities. While it may never supply all of its needs for wood commodities from its own land base, it will soon be able to supply a larger fraction. Many countries enter periods where they must go into world markets for wood when they go through the transition from exploitation of natural forests to continuous production from intensively managed forests. Some countries go through this transition without significant shortfall. Some have a temporary need to increase imports while managed forest growing stock is built up. Some countries never get through the transition.

Consumer Preference

Another major reason for countries to import wood commodities from other countries is consumer preference for the imported product. This preference may be based upon species not available in the importing country or it may be based upon product quality. The United States, which was the world's largest producer of hardwood lumber in 1980, also grows far more hardwood timber that it uses. It was the fourth largest exporter of hardwood lumber in 1980. Nonetheless, the U.S. was also the sixth largest importer of hardwood lumber in the same year. These hardwood imports often represented preferences for species and product quality not available in adequate quantities from domestic sources. Over the years, the U.S. has been the beneficiary of a worldwide preference for Douglas-fir as a premier structural timber. For the production of pulp and paper commodities where strength is an important characteristic, there has always been a preference for long-fibered softwoods largely produced in the temperate zone countries. These commodities have often been sold in wood-rich tropical countries. Clearly the firm that sells a product that enjoys a high consumer preference rating does not need to avoid wood-rich countries in seeking potential markets. But consumer preference can be transitory. If supply of a preferred item is not reliable, consumers may seek alternative supplies during periods of shortage and not return as customers when the shortage disappears.

Price Advantages

Sometimes a country can produce a preferred commodity from its own wood supply but can purchase it at lower cost abroad. This is often the case where the low-cost producer is exploiting rich natural forests or where labor costs are very low relative to those in the consumer country. In 1948-49, the U.S. Hardwood Plywood Institute and the Southern Plywood Manufacturers Association undertook a large trade promotion campaign targeted at home builders. They promoted the idea of using hardwood plywood-faced flush doors and at least one hardwood plywood paneled wall in each new home. The promotion scheme was successful, but about that time dipterocarp plywood began to be imported from Asia for sale at significantly lower prices. It captured this newly developed domestic market for hardwood stock panels and door skins. This sort of price advantage can be transitory. Increases in labor rates, shifts in currency exchange rates, and depletion of stocks of wood from natural stands can all erode price advantages.

Financial and Technological Advantages

A country may import wood structural and fiber products even when it has the raw material supply base for domestic manufacture to meet its own requirements, if its financial and technological base is inadequate. This is particularly true of industries such as pulp and paper production, fiberboard production, and particleboard production where the cost of building a factory to manufacture the commodity is great, the energy needed to produce the commodity either exceeds local supply or is prohibitively expensive, or the manufacturing and marketing operation is so sophisticated as to require scientific, technological, and managerial skills not available in the country. In the less developed countries, these financial and technological inputs are sometimes obtained by inviting multinational firms to operate in the country or by engaging in joint ventures with multinational firms. These sorts of partnerships between multinational firms and less developed countries or their nationals can be mutually beneficial arrangements or disasters, and there are many examples of each. Where they are good, they can provide a feasible way to develop a resource in a less developed country that will meet some of its needs for structural or fiber commodities and perhaps provide a surplus for export. Each of the partners to such a joint venture brings certain assets to the partnership. The country of origin typically provides the raw material supply and most of the labor force. The multinational often provides the basis for investment capital, managerial skills, technological skills, and knowledge of the market and its behavior. Understandably, each member of the partnership tends to exaggerate the importance of its contribution vis-à-vis that of the other.

Trade Restrictions

In a completely free trade environment, the countries with competitive advantages in the manufacture and marketing of forest products would produce wood commodities to meet their own needs and would export their surpluses to countries less well endowed in the forestry sector. But a completely free trade environment exists only in the textbooks. Countries impose biases on free trade through the mechanisms of tariffs, import and export restrictions or quotas, currency controls, and restrictions on the disposition of corporate profits. The reasons for imposing restrictions on free trade are often not as sound as they might appear at first blush. Also a reason which may be sound for a developed country may not hold for an undeveloped country, and vice versa. The following sections discuss some of the reasons usually given as arguments to interfere with free trade.

Faulty Signals by the Market Price Mechanism

It has been argued that there generally are systematic discrepancies between existing market prices of factors of production and their true opportunity or scarcity costs, or between private and social costs and benefits. This would lead one to reject the market signals which, for example, in the case of the state of Washington or Indonesia, have induced a flourishing export of unprocessed logs. One could do better than the market seems to indicate by prohibiting free exports of logs or partly processed products, and by forcing processing at home. In place of expected net discount profit as the main economic force, a number of alternative criteria have been suggested by economists to judge investments.

Such discrepancies obviously exist in the centrally planned communist or socialist countries. It has been argued that they also prevail in the developing countries. They should be largely absent in the capitalistic countries. The rejection of the market mechanism in favor of a set of alternative criteria would thus be applicable only to those countries. The question then becomes: is investment in wood product processing so attractive as judged by these criteria that it warrants interference with the free import and export of logs or other partly processed products?

In an exhaustive review of these criteria as applied to forestry activities, including downstream wood processing, the answer appears to be mixed at best (G. F. Schreuder, Commonwealth Forestry Review, December 1970). Using these criteria means selecting projects with high labor intensity, low capital/output ratios, high export value plus import substitution values/capital ratios, positive balance-of-payments effects, high social marginal productivities, high marginal per capita reinvestment quotients, high marginal growth contributions, high forward and backward linkages, and so forth. It is shown that only lumber manufacturing appears to be attractive enough to warrant consideration of interference with free exports. Veneer, plywood, fiberboard, particleboard, and pulp and paper manufacturing tend to be relatively unattractive. One is forced to conclude that even if the market mechanism is indeed imperfect, one is hard pressed to use this as an argument against the free export of logs or partly processed forest products in favor of forced processing at home.

Value Added

Another argument for biasing the direction and quantity of world trade in wood--an argument which does not rely on market imperfections--centers on the effort to increase in-country value added by restricting the export of partly processed products. This is a regulatory device that is being used with increasing frequency by developing countries, but it is by no means confined to these countries. The U.S. government has engaged in this practice with respect to federal lands. The government of Canada has also imposed such restrictions, as have the governments of most of the Northwest states with the exception of Washington State.

Figure 1 is a simple schematic diagram illustrating the principle of value added in the manufacture of plywood as it is generally understood. Value added by manufacture represents that fraction of the income from sale of the product that is represented by wages and salaries, interest on capital, depreciation, taxes, marketing expenses, energy costs, and profits. Figure 1 illustrates the relation of these processing costs to input to and output from the production systems.

Figure 1. Schematic Illustrating Value Added in the Manufacture of Plywood.

Input	Processing	Output

Cost of:
  Logs
  Adhesives
  Tape
  Sandpaper
  Miscellaneous
    supplies
	Cost of:
	  Salaries and wages
	  Interest on capital
	  Depreciation
	  Taxes
	  Marketing
	  Energy
	  Profits
		Sale of:
		  Finished product

Sometimes the country that imposes restrictions on trade in order to capture the value added due to processing does in fact achieve this purpose, but often the gains turn out to be more apparent than real. It is important to analyze the potential for capturing value added very carefully before an assumption is made that in-country processing is necessarily in the interest of the producer country. It is not always true that expenditures for value added in manufacture accrue to the people and governments where manufacture occurs. There are several ways that apparent gains to the producing country, as a result of value added through manufacture, may in fact not be gains at all. Among these sources of leakage of value added are:
1. When the timber-producing country does not have personnel with the technological and management competence required and where expatriates must therefore be employed, some of the wages and salaries may not accrue to the citizens of the country and in fact may, in part, be exported.
2. If manufacturing equipment must be imported, depreciation on the equipment is exported.
3. If foreign equity is required to build and operate the manufacturing operation, then some of the corporate profits may be exported.
4. If borrowed foreign capital is needed to finance the manufacturing operation, the interest on these investments will be exported.
5. It it not at all uncommon for log export prices to include a number of taxes and fees paid to the government of the exporting country that are not assessed against logs that are domestically processed. These incomes are lost.

As noted, there are many ways that the value added in domestic manufacture may in fact be exported, thus substantially reducing the apparent gain from local manufacture. If the requirement for local processing results in a substantial reduction in the level of timber utilization, accompanying losses in revenue may more than offset real gains due to value added in manufacture.

Employment

Politically, one of the most attractive arguments in favor of imposing restrictions on the free trade of logs or partly processed products is that by forcing downstream processing at home one keeps jobs at home, or in fact increases jobs at home. Upon careful analysis, the validity

of this argument rests on at least two assumptions:
1. The products thus produced (say lumber or plywood) can just as easily be sold as the original products (say logs). It assumes that the purchaser of the logs has no other option nor any incentive to go elsewhere to get what he wants. The evidence to sustain this argument is generally not explored, nor is it likely to be available.
2. The employment gained by downstream processing more than offsets losses in employment associated with handling the partly processed products (i.e., increase in sawmilling jobs outweighs the loss in jobs for longshoremen). Even in the absence of retaliatory trade actions by the purchasing countries, the argument does not necessarily hold true (whether employment multipliers are employed or not).

Risk

In forcing the downstream processing of wood products at home, several elements of risk are not generally considered:
1. Investments in wood processing are inherently risky, just like investments in other economic sectors. Does the country have the necessary manpower to run the plants? In developed countries there often is a lack of unskilled labor, in developing countries a lack of skilled labor.
2. Are the home-processing plants willing and able to produce for the export market? In the United States we have witnessed a lack of interest until recently in producing to metric or other export specifications. In many developing countries, the quality control is so poor that many of the products are not salable on the demanding export market; quality control, of course, can also be a problem in developed countries.
3. Both developed and undeveloped countries may not possess the marketing contacts and skills to sell the processed products.
4. Forced downstream processing often goes at the expense of competition for logs and then tends to depress the price of logs. This discourages tree growing and in turn increases the long-term risk for product manufacturing.

Considering these factors, again the case against interference with free trade appears to be stronger than the case in favor of it. Temporary tariffs or other interferences with free trade to protect infant industries can be beneficial and justifiable. The problem historically has been to keep these measures temporary, to do away with them after some time. Where efforts take the form primarily of protective tariffs that shelter inefficient domestic manufacturing operations or inferior raw material supplies, the domestic consumer is the victim and the domestic economy is worse off.

In summary, the flow of wood products in world trade is likely to be driven by demand for commodities by the major consumer countries. Most of these countries are also major producers, so that movement of wood products across national boundaries tends to represent an effort to fill in shortages. Some of the major consumers are leaders in both export and import trade in the same commodity. The United States is a classic example of such a country. This trade in the presence of plenty may reflect accommodation to sophisticated customer preference or it may reflect maximizing trade efficiency in terms of geography.

The most stable elements in the world wood trade picture are the countries whose export of commodities is based upon a reasonably constant flow of materials from intensively managed forests through well-established manufacturing facilities designed to process those materials efficiently. The least stable elements in the world wood trade picture are the countries whose production of commodities is based exclusively on the exploitation of natural forests. In many of these countries it is not clear what the long-term role of forestry is likely to be. If, as in the case of Thailand, the original forests are largely replaced by farms, then the country may drop out of the list of important exporters, and indeed it may become a major importer. If the Chilean pattern is followed, then a country will emerge as an important and probably stable exporter based upon an intensively managed timber resource perhaps based upon plantations of exotic species.

One thing is clear. The pattern of flow of products in world trade is relatively stable. Changes occur slowly and most frequently reflect changes in the countries whose forestry manufacturing and marketing economy is based upon natural stands.

PRODUCTION PROBLEMS IN LUMBER AND PANEL MANUFACTURING

W. Ramsay Smith and David G. Briggs

Introduction

In evaluating prospective international trade involving a firm operating in a foreign market, it is useful to distinguish two situations which differ in some fundamental ways in terms of technological impact on the manufacturer and in terms of product quality.

The first case occurs when the manufacturer modifies his product line and production facilities in order to manufacture commodities presently accepted by the foreign marketplace. In this situation, the foreign market is already accustomed to the sizes and grades of products, although there may be unfamiliarity with the species. If these products are significantly different in manufacturing requirements, the manager of the production facility may encounter an array of technical problems in the plant. These could include unexpected changes in product recovery and grade yield or difficulties with process flow and in-process inventory. This case is illustrated in trade with Japan where U.S. firms are now producing lumber for traditional Japanese housing construction.

The second case occurs when the manufacturer attempts to market his conventional product line in a foreign country. Technological problems internal to the production facility are minimized, but considerable problems may arise in gaining acceptability of the product in a marketplace that is unfamiliar with the producer's sizes, grades, and species. An example of this approach is the attempt to establish a U.S. style frame housing market in Japan.

With either of these situations, the manufacturer faces two changes from his accustomed dealings in conventional markets. One of these changes is the process of packaging, shipping, and delivering of the product to the consumer. This is likely to be a more elaborate and risky procedure than that in domestic markets. The second change from accustomed domestic procedures is the process of product quality assurance and inspection.

Technical Problems Within Mill

Problems which can occur in production of a new wood product for a foreign market will depend on the specific product being produced, the foreign country involved, and specifics of the existing mill layout. It is the purpose of this section, therefore, to discuss areas which should be considered during the decision-making period of whether or not to enter foreign markets. It will not cover specific product grades and specifications, since these are covered in a following paper.

It is assumed that the market potential for a product has been shown to be economically promising. Given this, there are two steps the interested producer must take. Step one is identification of the specific product and all of its size variations for the particular foreign market targeted. Step two is determining the technical feasibility to produce for that foreign market, using the information obtained in step one. Included in the step two analysis are five major areas of concern. These include: (1) differences in widths, thicknesses, and lengths from those presently being produced; (2) training of mill personnel for changes in product quality; (3) feasibility of producing a product within desired specifications; (4) actual product mix best suited for plant flow; and (5) correct pricing structure required for the new product.

Products produced for foreign markets will almost always use metric sizes. When these are converted into English units, odd dimensions result which can at first be confusing. The new machine settings may not be too difficult for thicknesses and widths, but sawing accuracy may be. Since the basic unit is the millimeter (mm) (0.039 inches), tolerances may need to be this tight, depending on product and country. The specified tolerance may become more critical as markets become saturated, as in the United States; especially when a long-term contract is held. Be sure, therefore, that tolerances are well understood between both parties and that the mill can hold to such tolerances.

Another consideration for differing product dimensions is material flow. Specified sizes may not permit the ability to produce a desired dimension from a miscut piece in a single pass. A merry-go-round may therefore be required, which may disrupt material flow.

Differing lengths due to metric sizes may be more of a problem than the varying thicknesses and widths. This is due to the necessity of a correct decision being made at the bucking saw. Larger quantities of an odd length--for example 13 feet--may be required, and this should be considered as far back as the woods operation unless tree length material is being brought to the mill site. Log merchandizing is also affected. It has been found that a computer-controlled bucking station greatly helps relieve this situation.

Additional space for sorting may be required, depending on the number of dimensions and lengths decided on. Available space must therefore be assured, and proper handling equipment considered before production is started and the flow becomes overly congested.

The same consideration must be made when additional grades are required. These, combined with an increase in the number of size classes, multiply rapidly. Familiarity with the product grades is of course mandatory. This is not only necessary to comply with the product standards, but also to produce the highest value product. This could very well entail the training of various mill personnel who have decision-making jobs throughout the process. This in itself could dictate the profit margin. This area is more critical in the lumber market than the others.

The total plant product mix also must be determined. Some of the considerations include whether or not to run domestic market products and foreign market products at the same time, separately, or to produce for foreign markets only. Combining both markets at the same time may increase efficiency of log allocation; however, this would have to be done at the expense of increased equipment and space for two production lines, which, when allocated for a particular market, would have to be kept separate.

When planning to produce for the different markets at different times, the quantity optimization for a given run has to be considered. A run too short or too long could either deplete or overburden the inventory, resulting in lost revenue or extra costs. The optimization will also vary with the market; therefore, an updating facility is desirable.

Last but by no means least is choosing the correct pricing structure to use. The exact structure will depend on product, country, and existing market conditions as well as the following factors.

- <u>Market entry</u>. It has been estimated that at least two to five years are required for a firm to enter successfully in a foreign market. The time and cost of mistakes during this period must be taken into account.
- <u>Learning curve cost</u>. A certain amount of time will be required to train personnel, as described previously, and minimize mistakes. This also includes marketing personnel.

- Total recovery. Changes in dimensions may increase or decrease total recovery or yield. This should be well understood. Included in this category would be losses due to mismanufacture or variation from the specified target size.
- Increased downtime. Downtime may be increased if mill is switching product lines from domestic to foreign and vice versa. This is simply due to increased set-up time required.
- Increased handling. Increased handling and lower productivity may be encountered through increased number of dimensions desired or remaining odd sizes.
- Effect on domestic market. Entering into a foreign market may exclude producing for the domestic market. This could cause long-term effects such as losing a share of the domestic market.
- Sales expenses. Sales and marketing expenses required to obtain a place in the foreign market may be high. This may include product or species acceptance costs as well as travel expenses and advertising. These cannot be taken too lightly.

Marketing

When a firm attempts to enter a foreign market, it may encounter significant problems associated with product acceptance. Any deviation from customary grades, sizes, species, or uses in the foreign market may lead to a barrier that must be overcome.

As an example, a firm may believe that one of its domestic species, commonly used for structural purposes and having ample technical and design data available, would provide an excellent opportunity to export to a foreign country where there is a shortage of structural timber. The firm therefore produces for the foreign country structural products that, in all respects except species, are identical to those used in the foreign country. The firm's products may not be accepted, however. Designers and workers in the foreign country may be unfamiliar with the species, and the technical data available from the firm may be in an unfamiliar form and in unfamiliar units of measurement. These problems should be addressed and overcome by an early promotional effort to gain understanding and acceptance of the species prior to any production and shipment. Unfortunately, this promotional homework may not be undertaken until some of the product is improperly employed, leading to ill-fated experiences that may be difficult to overcome.

These problems magnify with each increment of difference between domestic and foreign market. In the extreme, the manufacturer may attempt to export his traditional domestic product line, which may differ in species, grade, and dimensions from products used in the foreign country. Although this benefits the firm, since technical problems of producing for export are minimal, it creates major product acceptance problems. An exception could be when the foreign country finds that the species is acceptable and that it can remanufacture certain sizes and grades received from the firm into its own products. Otherwise, a substantial effort in gaining product acceptance is required. As an example, there has been substantial effort during the 1970s to promote and gain acceptance of conventional U.S.-style frame housing in Japan as a means for developing an export market for finished products from U.S. firms. This has included transferral of technical design data on U.S. species, agreement on lumber grades and tolerances, and workshops and demonstrations of U.S. frame housing production methods.

Product Shipment and Delivery

As mentioned earlier, and covered elsewhere in this symposium, the procedures involved in packaging, shipping, and delivery of a product entering an export market may be more elaborate, complex, and risky to the firm than the accustomed procedures in the domestic market. The product may need extra packaging, either by requirement of a shipper or to ensure minimal damage and loss during transit. The paperwork associated with the logistics of two or three modes of transportation (truck, rail, ship), loading and unloading, and crossing national boundaries is more complex than domestic shipping.

Quality Assurance and Acceptance

Introduction

An important consideration in transferring product from producer to consumer is the necessity of making a decision at some point as to the acceptance or rejection of the product. It is important to recognize that some portion of the product will be defective. In the case of lumber, defectives could be in the form of pieces that are off-grade, miscut, or perhaps even of the wrong species.

Much of acceptance inspection is by sampling. Often 100 percent inspection is impractical or uneconomical and may also be less effective than we might believe, due to inspector fatigue. Sampling inspection exerts more effective pressure for quality improvement than is possible with full inspection, where an attitude may develop that even if items are not all manufactured properly, the 100 percent screen will remove them. Sampling inspection leading to rejection of whole lots based on a sample provides a strong incentive to manufacture products properly in the first place. On the other extreme is no inspection, where it is assumed that quality exceeds some minimum accepted standard. Historical records may reveal a relatively constant, stable process average fraction defective, and an economic evaluation incorporating cost of inspection versus the cost of accepting a defective may reveal that no inspection is presently the best plan. In reality, however, some kind of sampling inspection would still be called for in order to give assurance as to the stability of the process average and to detect changes in its level.

Sampling plans may be based on two general types of measurement. They may be performed on manufactured products with a decision made only as to whether each sampled item is acceptable or not. This approach is called "attributes sampling." "Variables sampling" is an alternative approach, where the degree to which each item conforms to a standard is determined, as by measuring lumber thickness with a micrometer. Although variables inspection sampling provides more information than attributes inspection sampling, it is also generally more expensive.

Lot-by-Lot Acceptance: Single Sampling by Attributes

Here each submitted lot of product is considered separately, and the accept/reject decision is based on one or more sample sets chosen randomly from the lot. When the decision is always made on the basis of one sample, the acceptance plan is described as a single sampling plan. Such plans require specification of three numbers: the number of items in the lot (N), the number of randomly selected items in the sample (n), and the acceptance number (c), which specifies the maximum number of defective items allowed in the sample. More than c defectives causes rejection of the lot. C may be zero or some higher integer.

For example, the plan N = 1000, n = 20, c = 4 is interpreted as meaning: take a random sample of 20 from the 1000 items in the lot and reject the lot if the sample contains more than 4 defective items. In addition to single sampling, plans can be developed that involve double sampling, sequential sampling, and others.

Rectifying and Nonrectifying Plans

If the lot is rejected, two possible courses of action may be pursued. First, the rejected lot can be returned to the producer with no further action. This is called nonrectifying inspection, and could involve considerable expense and hardship to both the producer and consumer. The second course of action is to retain the rejected lot but subject it to 100 percent inspection, removing all defectives in the process. In this situation, defectives may or may not be replaced. This is called rectifying inspection. Rectifying inspection with replacement of defectives may be more feasible to conduct at the producer's location than at the consumer's location.

The parties involved in sampling inspection must be aware that some defective items are likely to be passed on by the sampling scheme. Since the whole lot is not inspected during sampling, two types of errors may occur. First, lots that in reality are acceptable may be rejected, and second, lots which are in fact bad may be accepted. Even in rectifying plans, lots that are accepted on the basis of the initial sampling may contain defectives among the unsampled items. If defectives are removed or replaced, however, the overall average quality of accepted lots is improved.

The statistical approach to acceptance sampling faces these facts and attempts to evaluate the risk assumed under various sampling procedures and to make a decision as to the degree of protection needed in a particular situation. A sampling scheme can then be chosen that gives the desired degree of protection with due consideration to the costs involved.

OC Curve--Single Sampling Plans

In order to measure a plan's ability to discriminate between lots of acceptable and unacceptable quality, a graph may be constructed that plots the probability of acceptance versus different levels of fraction defective (figure 1).

This curve is called an operating characteristic (OC) curve and can be used to compare the performance of different plans over a range of possible quality levels of submitted product. Naturally, the producer wants lots with a small fraction defective to have a high chance of acceptance. This is called the producer's risk point and might be defined by a producer indicating that he wants a plan that 95 percent of the time will accept lots containing 2 percent defective. Similarly, the consumer wants lots with a high fraction defective to have a low chance of acceptance. This is called the consumer risk point, and might be defined by a consumer indicating that he wants a plan that only 5 percent of the time will accept lots with 3 percent defective. When the consumer and producer make these quantitative statements, a sampling inspection plan can be devised that meets or exceeds their specifications. It should be pointed out that the fraction defective levels being discussed will vary greatly with the type of product. A fraction defective of 10 percent may be acceptable in the case of lumber but not in terms of aircraft engines.

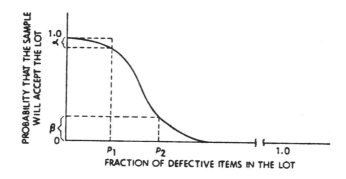

FIGURE 1. Operating characteristic (OC) curve.

Figure 2 presents OC curves for three sampling plans for lots containing 1000 items that meet a consumer's requirement of desiring a 10 percent probability of acceptance for lots that are 2.2 percent defective. If, for example, the producer desires that 90 percent of lots with 0.5 percent defective be accepted, then the plan n = 240, c = 2 would exceed his desires and thus be acceptable to both parties. The plan n = 170, c = 1 would accept about 80 percent of the producer's product and the plan n = 100, c = 0 would accept about 60 percent of the producer's product.

In the process of devising sampling plans, a common misconception is the idea that plans with equal percent

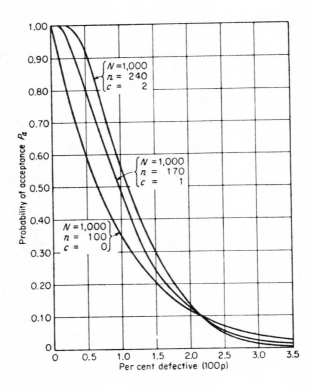

FIGURE 2. OC curves for plans with a 0.1 probability of acceptance of a 2.2 percent defective lot.

sampling provide the same protection. Figure 3 illustrates four plans with 10 percent samples and shows that the degree of protection offered by a 10 percent sampling of a lot of size 50 is very different from a 10 percent sampling of a lot of size 1000. At 2 percent fraction defective, the probability of acceptance of the former is 90 percent and the probability of acceptance of the latter is only 12 percent. Figure 4 shows four plans for different lot sizes having a constant sample size of n = 20 and an acceptance number of zero. Here the fixed sample size tends toward more constant quality protection. That is, the absolute size of the random sample is more important than its relative size compared to the lot size.

The practical difficulty in devising a sampling plan is that its selection depends on the risks that the parties are willing to face, which in turn are economic decisions that depend on a number of considerations.

The owner of a mill may ask how he can obtain information on his fraction defective so he knows where his current position is. This can be done by inspecting

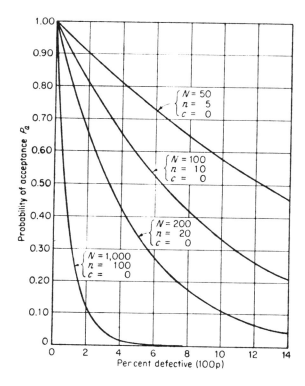

FIGURE 3. OC curves for four sampling plans with 10 percent samples.

packages of his current finished product and keeping records of fraction defective by another quality control device called a P-chart.

Implications for Trade

The implications for trade of this discussion are twofold: that some proportion of the commodity is going to be defective, and the producer and consumer had better agree in advance on a plan for verification of quality and acceptance. Table 1 presents a list of 10 points that may be covered during contract negotiations. The parties need to define their risk points, decide whether or not there will be rectifying or nonrectifying inspection, devise the plan, determine who will conduct the inspection and where it will be conducted, indicate appropriate grading rules, species nomenclature, dimensions, and tolerances, indicate any special restrictions, and agree upon details of grade

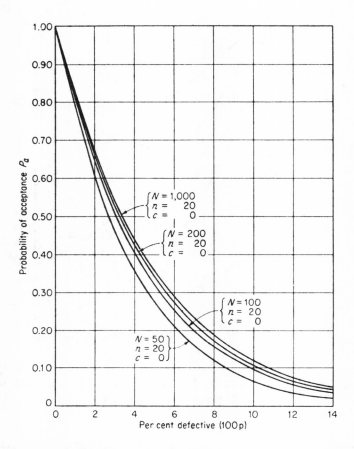

FIGURE 4. OC curves for four sampling plans with samples of 20 and acceptance number of 0.

marking. Any special national or international standards or documents should be referenced as appropriate, and finally, an agency or procedure should be agreed upon for settling disputes.

Once the risk points are defined, consultation with any standard book on quality control will reveal alternative plans that can meet the objectives. Factors of cost may help in deciding on a particular plan. The question of where to conduct the sampling is more difficult to address, although there are strong reasons with wood products for conducting it at the producer's location, since rectifying inspection, if chosen, may be facilitated. Also costs of handling and shipping an unacceptable product are avoided.

A major difficulty in conducting acceptance sampling at the producer's location may be the consumer's mistrust. One

Table 1. Ten Factors To Be Agreed Upon in a Quality Acceptance Procedure.

---

1. Definition of risk points
2. Rectifying or nonrectifying inspection
3. Sampling plan type and nomenclature including appropriate reference
4. Location where acceptance sampling will be conducted
5. Grading rules applied and grading rule authority (references)
6. Commercial or standard names of species
7. Nominal and actual dimensions and tolerances, etc.
8. Special restrictions of rules with respect to species, sizes, and grades or restrictions on use application
9. Details of marking related to grade, species, etc.
10. Other
    a. Relation of rules to national and international standards if unclear from 1-9
    b. Reference to documents upon which national approval is based
    c. Agreed upon agency or procedure for settling disputes

---

solution could be to have the consumer's agent inspect at the producer's location or for the two parties to agree on a third party inspection bureau. There may be a need for some type of international inspection and grading agency to ensure fairness to buyer and seller on the international market and to arbitrate disputes.

FACTORS INFLUENCING THE INCREASED WORLDWIDE USE OF WOOD

Stephen B. Preston

Introduction

There are very few manufactured products for which there are no technically suitable alternative raw materials. Thus, with few exceptions, the choice of an industrial material for any commodity is likely to be governed substantially by its influence on the cost and acceptability of the product to the consumer. Most industrial uses for which wood has found acceptance certainly are not exceptions.

Wood has established functional and economic advantages in many parts of the world for such products--among others--as residential houses and their furnishings; commercial buildings; bridges; dock installations and other engineered structures; packaging materials; and printing, writing, and hygienic papers. Its potential as a chemical feed stock has been technically demonstrated for fuels and other products which are not primarily based on petroleum. In these and other products for which wood is, or can become, a socially acceptable material, its increased use throughout the world will depend on the level of demand for this group of commodities and the degree to which wood can be economically competitive with alternatives. Trends indicate that the demand for many commodities for which wood has established technical suitability can be expected to increase accompanying increased personal income, particularly in developing countries. This paper concentrates on factors which have a high probability of improving wood's economically competitive position and of increasing the volume of its use in established and expanding markets.

Renewability--A Factor in Maintaining Supply

The comparative economic advantages of renewable or nonrenewable raw materials which are technically substitutable in the manufacture of a product are exceedingly complex and beyond the scope of this paper. Social and institutional factors, prior capital investments, organizational structures, and the mix of human skills, are all involved. The fact remains, however, that most nonrenewable materials are decreasing in accessible quantity, and this is accompanied by increasing cost of extraction and transportation. Moreover, the recent oil embargo has clearly demonstrated the economic impact of a curtailment in supply of critical nonrenewable resources. A clear advantage of wood--a renewable resource--is control of quantity, quality, and often location of the supply. Additionally, and of increasing importance, the natural environment is enhanced by managed forests rather than degraded, as is often a result of mineral extraction.

As the natural forests of the world which have grown with minimum or no financial investment are replaced with silviculturally managed ones, the cost of wood obviously must reflect the cost of producing it. Recent and impressive advances in forest and wood sciences, however, keep the cost of wood from the renewed forest resource competitive with alternatives. Foresters are succeeding in steadily increasing forest productivity per unit of land and investment, and wood scientists and technologists are finding ways of harvesting and using short rotation material and an ever-increasing percentage of the biomass produced. Thus, unlike nonrenewable materials, the availability of wood at a known, modest cost is under human control.

Understandably, because of the availability of an adequate supply of wood largely from extensively managed forests and the long life span of trees, the state of the art of silviculture is far less advanced than that of agriculture. Successes in genetic manipulation and crop management for agricultural products hint at what the future may hold for increased forest productivity through silvicultural practices. Even now, the potential for increased wood material supply through the application of available knowledge is very significant. For example, it has been estimated that in the United States the production of commercial forests could be nearly doubled within 50 years with the widespread application of proven silvicultural practices. It could be approximately tripled with intensive silviculture and complete tree use of both hardwoods and softwoods (National Research Council 1976). This, together with productivity now being realized in plantations and the potential that is projected by industrial leaders through the use of intensive culture and modest genetic improvement, leads to the conclusion that potential supply is unlikely to limit increases in wood use.

Energy Considerations

In spite of the current downward trend in the price of imported crude oil, the cost of energy can be expected to be an increasingly important factor in the cost of manufactured products. Wood enjoys a striking energy advantage over alternative raw materials for essentially all products for which it has established acceptance. Relatively low net energy requirements in processing contribute to this advantage in the manufacture of structural products from solid-wood and fiber-based packaging supplies which are functionally interchangeable with those from alternative materials. In the manufacture of both solid-wood and fiber products, a substantial percentage of process energy can be derived from by-products that otherwise would represent waste. Additionally, the in-place systems of processing and burning forest biomass can be an advantage to the wood products industry in supplementing fuel from manufacturing residue with forest residue. The considerable energy advantage of wood-based products over alternatives was studied and reported in some detail in 1976 by the National Academy of Sciences/National Research Council through the Committee on Renewable Resources for Industrial Materials. Although this study is now some eight years old--a period in which the industry has responded to the increased cost of fossil fuel with substantially increased energy self-sufficiency and improved energy efficiency in processing--the findings remain generally valid.

Solid-wood Products

Energy requirements for solid-wood products with particular attention to their use in residential construction were studied by the CORRIM Panel on Wood for Structural and Architectural Purposes (Boyd et al. 1976). From data supplied by industrial sources in western and southern regions of the U.S., total and net energy requirements for harvesting, transport of the raw material to the mill, and processing were determined for eight principal primary building products (hardwood and softwood lumber and plywood, wetform and dryform hardboard, insulation board, and particleboard) and four forms of composites (three types of flakeboard and lumber laminated from veneer) not then in production but now being manufactured in forms not unlike those projected by the panel. Net energy requirements were calculated on the assumption that all plant residue unsuitable for other products was used for electrical power and steam but not for transport fuel. Similarly, energy requirements for extraction through primary processing were ascertained for nonwood building materials commonly used as alternatives to, or in conjunction with, wood-based materials. This permitted comparison of energy requirements on a weight basis for primary wood products and common alternatives.

Differences in energy requirements on a weight basis for producing wood-based and alternative materials are indeed striking. Softwood lumber, for example, requires about one-fiftieth the energy of steel. On the basis of one ton this is the equivalent of approximately 10 barrels of oil. The difference in energy required to produce plywood and the more energy-intensive aluminum is the equivalent of nearly 40 barrels of oil.

The difference in energy requirements for a functional product made from different materials is not always readily apparent from the difference in requirements for the manufacture of primary materials. For many if not most products, the mass of the product may vary considerably, depending upon the design and the material used. Moreover, processing plants for alternative materials may be located at considerably different distances from the end-use location. Also, in such uses as building construction, combinations of materials differ considerably, and realistic comparisons can best be made on the basis of conventional systems of construction. For this reason, the CORRIM panel selected and analyzed alternative designs of roofs, interior and exterior walls, and floors. These designs included the most important systems for each now in common use and, in addition, some feasible but uncommonly used ones. Sections of each design with an area of 100 square feet were used for comparison. Included in the calculations were the energy required for processing all components of each design and that required to transport it to the building site.

Alternative components serving major functions were found to differ strikingly in energy requirements. For example, steel floor joists require approximately fifty times as much energy from extraction to the building site as wood counterparts. Aluminum framing for exterior walls is about twenty times as energy intensive as is wood. Wood trusses require approximately one-seventh the energy of steel rafters. Aluminum and brick siding exceed plywood and fiberboard counterparts five and twenty-five times, respectively.

When sections of commonly used designs incorporating conventional combinations of materials were considered, differences in energy requirements, although not as striking as for components, were still impressive. Brick-sided exterior walls required seven to eight times the energy of all-wood counterparts. Exterior and interior walls using metal framing require about twice the energy of wood-framed construction. Concrete slab construction is nearly ten times as energy intensive as floors from wood materials.

The energy advantages of wood-based materials over alternatives on a bulk material basis or in residential housing cannot be expected to be the same in all parts of the world as it is in the U.S. Moreover, the CORRIM data were based on efficient, integrated manufacturing facilities in which all possible input raw material is used for a product and that unsuited for a product is used for fuel. Broad averages were used for transport distances. Nevertheless, the implications of the study are clear. Wood-based structural materials and construction utilizing them can be substantially less energy intensive than nonwood alternatives--a factor which should increasingly contribute to wood's comparative advantage.

Fiber and Fiber Products

Pulp and paper manufacture is inherently energy intensive. But on a per-ton basis the industry requires less energy than such materials as plastics, rolled aluminum, and plate glass (Jahn and Preston 1976). Unlike other energy-expensive industries, the paper industry can supply a substantial part of its requirements from manufacturing by-products which otherwise would be process waste.

From the standpoint of substitutability, there are currently no serious competitors to wood for printing and writing papers, although other materials are used in relatively small quantity. For packaging materials, the other major segment of the paper market, however, an array of plastics are commonly used alternatives. Such substitutes for paper-based products as plastic bottles for paper milk cartons, polyethylene bags for paper ones, polystyrene food trays for molded woodpulp, foam plastic padding materials for paper-based cushioning, and many others are commonplace throughout the world today. The CORRIM study examined the relative energy costs of three widely used products made interchangeably from petrochemicals and wood fiber. Specifically selected for comparison were one-half gallon paper and polyethylene milk containers, size 6, woodpulp and polystyrene meat trays, and polyethylene and kraft paper bags. Energy requirements for the input raw material, the manufacture of the product, and the energy content of the raw material were calculated and compared.

On a weight basis, the energy required to manufacture the plastic products exceeds that for the wood products by ratios of approximately 1.6:1 for milk containers, 3.1:1 for meat trays, and 3.3:1 for bags. Because of the weight differences between the heavier wood fiber and lighter plastic products, the energy differences for the products are considerably less than those for the materials on a weight basis. Nevertheless, they are still impressive for milk containers and plastic bags, for which the ratios for plastic to wood-fiber products are 1.5:1 and 1.65:1 respectively. Inasmuch as woodpulp meat trays are approximately four times heavier than polystyrene ones, it is not surprising that they are nearly equal in energy content and requirements for manufacture.

Technology in Relation to Consumption Trends

Consumption Trends

Trends in the apparent consumption of products made from wood throughout the world, on both total volume and per capita bases, provide interesting insights to factors which are influencing the increasing use of wood and the potentials for further increases, assuming the wood-based products remain competitive with potential substitutes. Several factors become apparent from tables 1 and 2. It may be seen from table 1 that three regions of the world, Africa, Asia, and Latin America, account for nearly three-fourths of the world's population. Yet they consume less than 20 percent of the world's industrial wood. Moreover, the per capita consumption of these three regions combined is but 0.08 cubic meters, approximately one-fourth of the world average (0.31 cubic meters) and only one-twelfth of that of the rest of the world combined (0.99 cubic meters). It is not surprising that the affluent regions of the world are the largest per capita consumers of industrial wood, implying a positive correlation between affluence and industrial wood use. What is suggested is the tremendous increase in industrial wood use which potentially

Table 1. Apparent Worldwide Wood Consumption (1980) (volume).

Region	Populations (millions)	Roundwood			Products						
		Total volume 000 m³	Per capita volume		Sawnwood and sleepers		Wood-based panels		Woodpulp		Paper and paperboard
			Fuel-wood m³	Industrial wood m³	Total 000 m³	Per capita m³	Total 000 m³	Per capita m³	Total 000 mt	Per capita kg	Total 000 mt	Per capita kg

Region	Pop.	Total vol.	Fuel m³	Ind. m³	Sawn Total	Sawn pc	Panels Total	Panels pc	Pulp Total	Pulp pc	Paper Total	Paper pc
Africa	469	427,583	0.82	0.09	9,051	0.019	1,364	0.003	807	2	2,214	5
Asia (excluding Japan)	2,441	966,345	0.34	0.06	44,057	0.018	5,826	0.002	4,787	2	16,828	7
Latin America	368	361,282	0.78	0.20	23,153	0.063	3,874	0.011	4,529	12	8,951	24
Developing Countries	3,278	1,755,210	0.46	0.08	76,261	0.023	11,064	0.003	10,123	3	27,993	9
Europe	484	351,868	0.09	0.64	99,721	0.206	35,939	0.074	35,190	73	48,159	100
Japan	117	84,739	0.02	0.71	42,660	0.365	10,322	0.088	12,081	103	18,226	156
North America	246	463,261	0.08	1.81	106,565	0.433	33,125	0.135	58,738	239	66,988	272
Oceania	22.8	24,027	0.26	0.89	5,753	0.252	1,142	0.050	1,550	68	2,394	105
U.S.S.R.	267	340,990	0.29	0.99	92,897	0.348	9,384	0.035	8,218	31	8,398	31
Developed Countries	1,137	1,264,885	0.13	0.98	347,596	0.306	89,912	0.079	115,777	102	1,144,165	127
World	4,415	3,021,101	0.37	0.31	423,907	0.096	100,975	0.023	125,897	29	172,157	39

Sources: Volume data calculated from FAO Yearbook of Forest Products 1980 (Rome); Population data from U.N. Demographic Yearbook 1980. Columns may not add due to rounding.

Table 2. Annual Change in Apparent Worldwide Wood Consumption 1970-80 (%).

Region	Industrial Roundwood		Sawnwood and Sleepers		Wood-based Panels		Woodpulp		Paper and Paperboard	
	Total	Per capita	Total	Per capita	Total	Per capita	Total	Per capita	Total	Per capita
Africa	2.9	---	4.6	1.7	6.6	4.1	3.7	---	2.8	---
Asia (excluding Japan)	4.4	1.8	3.4	1.2	9.7	6.9	8.4	6.9	7.8	2.2
Latin America	4.5	1.6	3.9	1.4	8.8	6.1	7.0	4.1	4.8	1.8
Europe	0.8	0.3	0.8	0.3	4.5	3.9	1.5	1.0	2.2	1.7
Japan	-1.0	-0.2	-0.7	-0.2	2.3	1.2	2.2	1.1	3.7	2.7
North America	0.7	-0.1	0.4	-0.5	1.6	0.8	1.8	0.9	2.3	1.4
Oceania	---	-0.4	-0.8	-2.5	3.1	1.5	1.7	---	2.6	0.9
U.S.S.R.	-0.7	-1.6	-1.9	-2.8	5.6	4.6	2.3	1.4	2.6	1.8
World	0.9	-0.3	0.3	-0.6	3.6	3.0	2.1	1.5	3.0	2.0

Source: Calculated from FAO Yearbook of Forest Products 1980, U.N. Demographic Yearbook 1980.

can accompany the increasing affluence of the developing world.

The annual percentage change in apparent per capita consumption of industrial wood for major products in various regions of the world, as shown for the 1970 to 1980 decade in table 2, is revealing. Of interest are the different patterns of consumption between sawnwood and sleepers, panel, and fiber products. These data indicate a worldwide decrease in per capita consumption of sawnwood, although modest consumption increases from a very small base have exceeded population increases for Africa, Asia, and Latin America. Among the developed regions, only Europe, with but a 0.3 percent annual per capita increase, shows an upward trend in sawnwood consumption. During the same period, and contrasting with sawnwood, per capita increases for products based on modern technology--panels, pulp, and paper and paperboard--show impressive worldwide annual increases resulting from increases for all of these products in every region. Also of interest is the fact that the less affluent and most populous regions of the world show the greatest annual increases during the decade in both total volume and per capita consumption for all of these products except paper and paperboard. Although the base is still small, the increase in Asia (excluding Japan, with its 2.4 billion people) is most impressive. Consumption statistics for certain selected countries for which the trade balance has radically changed may illustrate potential market expansion. For example, during the 1970 to 1980 decade, wood-based panels imported and consumed in Saudi Arabia increased from 13,000 cubic meters to 417,000 cubic meters, representing increases of 32-fold in total volume and 20-fold (0.002 cubic meters to 0.04 cubic meters) on a per capita basis.

Of particular interest from the standpoint of increasing wood use is the pattern of consumption of pulp and paper products. The contrast in current per capita consumption of these products in developing and developed regions of the world is indeed striking. As seen from table 1, the heavily populated developing regions of the world use on an annual per capita basis an average of only 9 kilograms (19.8 pounds), as opposed to 127 kilograms (279.4 pounds) for the developed countries, a 14-fold difference.

Trends in fiber consumption in the industrialized regions indicate an increasing reliance on paper products as the economy expands. For example, the per capita consumption in the U.S. has approximately doubled in the last 40 years. The per capita use in Japan has nearly tripled in only 20 years. The possibility of comparable consumption increases in the less affluent parts of the world as the economy of these regions develops suggests a very substantial increase in the future world market for fiber. This is supported by the pattern developing in the People's Republic of China in which the per capita consumption of paper and paperboard has increased from 10 pounds in 1970 to 20 pounds in 1980, representing an annual rate of growth of 7 percent.

Technological Trends

It is unquestionably more than coincidence that the products which are derived by technology to meet human needs are those for which consumption is increasing. The driving force many years ago which led to the technical concepts on which these products were developed was unquestionably the need for these special materials in economically developing societies. Since their inception, impressive advances in technology have succeeded in making these products substantially more useful, economically available to the consumer, and competitive in performance and cost with alternatives. This has been achieved in spite of increasing costs of input wood of decreasing size and quality. Indications are that technology will continue to improve wood's availability and competitive position with other materials to combine with the other previously discussed driving forces in increasing the world's use of wood.

Sawnwood, in spite of its very modest increase in consumption on a total volume basis, remains a product to which over half of the world's industrial wood is converted. Moreover, its consumption continues to increase substantially in developing countries. Thus, technological advances which lead particularly to more efficient raw material use, including small log conversion and reduced per-unit production costs, can be important factors in increased use of sawnwood. Essentially, all recent technological innovations have these objectives. Electronic scanners, computer control of operations, mechanized grading, and other applications of high technology increase yield from material of decreasing size and quality, and provide for increased production per units of capital and labor.

The most impressive technological advances in products based on solid wood are in the manufacture of composites, particularly panels for which the worldwide per capita consumption is increasing at a more rapid rate than for any other form of product. The current state of the art in producing economical and fully acceptable panels for interior and exterior use from a variety of veneers, flakes, strands, particles, and fibers in various combinations makes this type of product essentially independent of log size and tree species. A major technological obstacle to be overcome for exterior panels remains the development of an inexpensive adhesive as an alternative to phenol formaldehyde, suitable for exterior use and not based on petrochemicals. Much research attention is now being directed toward this need.

It appears that there is available an improving technology to make inexpensive, reconstituted wood panels designed to meet the needs of society. Because the technology for this product is indiscriminate in raw material requirements, an increased use of wood to serve this expanding market is assured.

Technological advances in the fiber industry have certainly played a dominant role in the increasing use of wood for this purpose. Not only has the industry been successful in developing and improving products to meet changing societal needs, but additionally it is making impressive progress in processing efficiency and in flexibility in raw material requirements. These advances are exemplified by technical adjustments in the industry in the U.S. to bring about such changes as decreases in average wood requirements per short ton of fiber from 1.6 cords in 1960 to 1.51 in 1973, substantial increases in hardwood in the raw material mix, increasing capability of using whole tree chips, and others. Advances in pulping mixed chips of widely different wood properties as has been demonstrated with tropical hardwoods substantially broaden the raw material base. The potential of new technologies to increase yields of improved pulp--as, for example, the emerging thermomechanical process--suggest the possibility of further major advances in pulping methods. Industrial integration which facilitates an increase in use as raw material of by-products from the manufacture of lumber and other products and of forest residue is a substantial force in reduced product cost. It appears that the fiber industry is technically prepared to meet the certain increase in demand for fiber products.

Summary

Data for worldwide apparent consumption of wood-based products from 1970 to 1980 indicate a clear trend of increases in total wood use and significant per capita increases in panels and fiber products. Patterns of increasing use are strongest in the most populous and least affluent but economically developing parts of the world. There appears to be the potential for very substantial increases in wood use worldwide in products for which a growing market has been established, if wood can remain economically and functionally competitive with alternative raw materials and the products can be afforded--particularly by the people of the developing world. Renewability of wood as a raw material, in contrast to mineral and petrochemically derived alternatives, inherent energy advantages in wood processing, and advancing technology--particularly in panel and fiber products--are reasons for optimism for continued increases in worldwide wood use.

References

Boyd, C. W., P. Koch, N. B. McKean, C. R. Morschauser, S. B. Preson, and F. F. Wangaard. 1976. Wood for structural and architectural purposes. Wood and Fiber 8(1):1-72.

FAO. 1982. Yearbook of forest products 1980. Food and Agriculture Organization of the United Nations, Rome.

Jahn, E. C., and S. B. Preston. 1976. Timber: More effective utilization. Science 191:757-761.

National Research Council. 1976. Renewable resources for industrial materials. National Academy of Sciences, Washington, D.C.

United Nations, Department of Economic and Social Affairs. 1982. Demographic yearbook 1980.

FACTORS INFLUENCING INTERNATIONAL TRADE IN FOREST PRODUCTS,
AS SEEN FROM A EUROPEAN VIEWPOINT

T. J. Peck

Introduction

The purpose of this paper is to review the position of Europe in the international market of forest products, to discuss trends, and--to the extent possible--prospects in Europe and the regions with which it trades in these products. Europe's importance in international trade is not confined to its role of importer. It has also been expanding its exports to other regions. What its future position in the international market will be depends not only on trends in the supply/demand balance within Europe but equally on developments in its main trading partners--North America, the U.S.S.R., and the tropical regions as its main suppliers, as well as North Africa and the Middle East as its customers.

Europe's Place in the International Trade of Forest Products

The volume of world exports of forest products in 1980 was approximately $55.6 billion (U.S.) (f.o.b.) and of imports $61.6 billion (c.i.f.). European countries accounted for 44 percent of the exports--$24.6 billion--and 52 percent of the imports--$31.9 billion (table 1). The region coming nearest to Europe in share of world trade was North America, with 31 percent of total exports and 13 percent of total imports. Table 2 shows that Europe accounted for over half the value of world exports of paper and paperboard and of wood-based panels; and for imports, well over half of all paper and paperboard, woodpulp, sawnwood, and wood-based panels. The only product group for which it has less than a leading place in world trade is unprocessed wood (roundwood, chips, and residues).
  Percentages of world totals are somewhat misleading in the sense that for both Europe and North America they include intraregional trade: in Europe trade between 24

Table 1. Value of International Trade in Forest Products* in 1980, and Comparison with 1970.

	Exports (f.o.b.)			Imports (c.i.f.)		
	Value in 1980	Share of world total		Value in 1980	Share of world total	
Region/country		1970	1980		1970	1980
	(million U.S.$)	----(%)----		(million U.S.$)	----(%)----	
World	55,572	100.0	100.0	61,575	100.0	100.0
Europe	24,616	42.9	44.3	31,880	56.8	51.8
North America	17,454	34.7	31.4	8,279	17.3	13.4
Japan	7,160	10.1	12.9	9,527	13.2	15.5
Other Asia				6,120	5.9	9.9
U.S.S.R.	2,673	6.7	4.8	5,769	6.8	9.4
Other	3,669	5.6	6.6			

*Roundwood (including chips and residues), sawnwood, woodpulp, wood-based panels, paper, and paperboard.
Source: FAO Yearbook of Forest Products 1981.

Table 2. Value of European Exports and Imports of Forest Products in 1980, by Product Group and Comparisons with 1970.

	Value in 1980 (million U.S.$)	Share of European Total		Share of World Total	
		1970	1980	1970	1980
		(%)		(%)	
------Exports (f.o.b.)------					
Total, of which:	24,616	100.0	100.0	42.9	44.3
Roundwood*	1,427	5.4	5.8	15.3	16.5
Sawnwood	5,062	20.7	20.6	41.5	41.1
Wood-based panels	2,593	10.2	10.5	48.1	51.3
Woodpulp	3,325	21.0	13.5	45.3	34.9
Paper and paperboard	12,209	42.7	49.6	53.1	61.0
------Imports (c.i.f.)------					
Total, of which:	31,880	100.0	100.0	56.8	51.8
Roundwood*	3,386	10.9	10.6	32.3	27.3
Sawnwood	8,254	23.4	25.9	62.4	59.4
Wood-based panels	3,435	9.0	10.8	61.4	66.7
Woodpulp	5,618	27.6	17.6	83.9	58.0
Paper and paperboard	11,188	29.1	35.1	50.8	54.5

*Including chips and residues.
Source: FAO Yearbook of Forest Products 1981.

countries; in North America between Canada and the U.S. Excluding intraregional trade, Europe's exports to the rest of the world, in terms of cubic meters, equivalent volume of wood in the rough (m^3 eq.),[1] amounted to 27 million m^3 eq. in 1980 and its imports from other regions 81 million m^3 eq. This puts Europe in first place for imports, ahead of Japan with 74 million m^3 eq., and North America with less than 8 million m^3 eq. For exports, it trails behind North America, which shipped 104 million m^3 eq. to other regions in 1980, Southeast Asia with 45 million m^3 eq., and the U.S.S.R. with 36 million m^3 eq.

Europe's exports to other regions are perhaps of less interest in the present context than its imports from other regions. As tables 3 and 4 show, however, there was a quite substantial increase in Europe's exports of forest products to non-European destinations during the 1970s: these exports in total rose by 8.4 million m^3 eq., or 45 percent over the decade to 1980, and more than held their share of total European exports. This is perhaps surprising, given Europe's deficit position in forest products. It reflects, however, the strong growth in import demand of the oil-producing countries of the Middle East and North Africa. Partly to meet their heavy construction programs, European exports of sawn softwood rose markedly--by 56 percent to all non-European destinations, of which the Middle East/North Africa took the largest share. There were also sizable increases in Europe's exports of pulp, paper, and paperboard to non-European markets. It may be mentioned that this is one sector where the forecasts in "European Timber Trends and Prospects, 1950 to 2000" (abbreviated to ETTS III) appear to have been too conservative. It was assumed in that study that Europe's exports to other regions would, at best, show modest growth between the mid-1970s and 2000, reaching between 20 and 25 million m^3 eq. compared with 21 million m^3 eq. in 1974 (FAO/ECE 1976). As noted above, the volume had already reached 27 million m^3 eq. in 1980.

For imports, not only does Europe hold the largest share of total interregional trade, but compared with the other major importing areas, it is notable on account of the wide distribution of its imports, in terms both of its suppliers and of the products. Japan is still essentially an importer of raw materials, despite a noticeable trend in

---

1. Differs from roundwood equivalent by including the transfer of industrial wood residues and the recycling of waste paper in the factor for converting product volume.

Table 3. European Exports of Forest Products, 1970, 1974, and 1980, by Product and Main Destination (1000 units).

Product	Unit	Total			To: Europe			Non-European Destinations		
		1970	1974	1980	1970	1974	1980	1970	1974	1980
Pulpwood (round)	m³	9,415	9,935	11,044	9,410	9,933	11,021	5	2	23
Chips and residues	"	1,330	2,399	3,952	1,330	2,399	3,952	--	--	--
Softwood logs	"	1,694	3,175	3,671	1,674	2,978	3,481	20	197	190
Hardwood logs	"	1,561	2,224	2,579	1,476	2,106	2,456	85	118	123
Pitprops	"	694	660	329	628	594	273	66	66	56
Total roundwood	"	14,694	18,393	21,575	14,518	18,010	21,183	176	383	392
Sawn softwood	"	19,256	19,374	22,283	16,928	17,760	18,643	2,328	1,614	3,640
Sawn hardwood	"	2,458	2,635	3,012	1,978	2,030	2,589	480	605	423
Sleepers	"	343	310	323	330	289	281	13	21	42
Total sawnwood	"	22,057	22,319	25,618	19,236	20,079	21,513	2,821	2,240	4,105
Veneer sheets	"	207	246	373	169	218	323	38	28	50
Plywood	"	1,161	1,111	1,477	940	886	1,252	221	225	225
Particleboard	"	1,840	3,537	4,847	1,789	3,447	4,617	51	90	230
Fiberboard	"	1,530	1,501	1,392	1,256	1,242	1,175	274	259	217
Total wood-based panels	"	4,738	6,395	8,089	4,154	5,793	7,367	584	602	722
Mechanical woodpulp	mt	1,045	1,120	662	988	993	575	57	127	87
Chemical woodpulp (paper)	"	6,878	7,069	6,654	6,279	6,239	5,734	599	830	920
Total woodpulp	"	7,923	8,189	7,316	7,267	7,232	6,309	656	957	1,007
Newsprint	"	2,643	2,797	3,594	2,103	2,293	2,734	540	504	860
Printing and writing	"	2,797	4,664	5,694	1,692	3,365	4,139	1,105	1,299	1,555
Other paper and paperboard	"	5,644	7,837	8,885	4,429	6,283	7,190	1,215	1,554	1,695
Total paper and paperboard	"	11,084	15,298	18,173	8,224	11,941	14,063	2,860	3,357	4,110

Sources: FAO Yearbook of Forest Products.
FAO/ECE Timber Bulletin for Europe.
OECD The Pulp and Paper Industry.

Table 4. Changes in Product Share and Distribution of European Forest Products Exports, 1970 to 1980.

Product	Total 1970	Total 1980	To: Europe 1970	To: Europe 1980	Non-European Destinations 1970	Non-European Destinations 1980
	\multicolumn{6}{c}{(1000 m³ eq.*)}					
Roundwood	14,694	21,575	14,518	21,183	176	392
Sawnwood	36,285	42,139	31,642	35,390	4,643	6,749
Wood-based panels	8,461	13,536	7,313	12,205	1,149	1,332
Woodpulp	35,626	33,594	32,609	28,960	3,018	4,634
Paper and paperboard	37,575	61,606	27,879	47,674	9,695	13,933
Total	132,642	172,451	113,962	145,412	18,680	27,039
	\multicolumn{6}{c}{(Percent change 1970 to 1980)}					
Roundwood		+ 47		+ 45		+123
Sawnwood		+ 16		+ 12		+ 45
Wood-based panels		+ 60		+ 67		+ 16
Woodpulp		− 6		− 11		+ 54
Paper and paperboard		+ 64		+ 71		+ 44
Total		+ 30		+ 28		+ 45
	\multicolumn{6}{c}{(Percent share of total exports, by main destination)}					
Roundwood	100.0	100.0	98.8	98.2	1.2	1.8
Sawnwood	100.0	100.0	87.2	84.0	12.8	16.0
Wood-based panels	100.0	100.0	86.4	90.2	13.6	9.8
Woodpulp	100.0	100.0	91.5	86.2	8.5	13.8
Paper and paperboard	100.0	100.0	74.2	77.4	25.8	22.6
Total	100.0	100.0	85.9	84.3	14.1	15.7
	\multicolumn{6}{c}{(Percent share of total exports, by product)}					
Roundwood	11.1	12.5	12.7	14.6	0.9	1.5
Sawnwood	27.3	24.4	27.8	24.3	24.9	25.0
Wood-based panels	6.4	7.9	6.4	8.4	6.1	4.9
Woodpulp	26.9	19.5	28.6	19.9	16.2	17.1
Paper and paperboard	28.3	37.7	24.5	32.8	51.9	51.5
Total	100.0	100.0	100.0	100.0	100.0	100.0

*Cubic meters, equivalent volume of wood in the rough. Converted from product volume taking as a basis factors used in ETTS III.

recent years toward more imports of semiprocessed products, sawnwood, and woodpulp, while the bulk of its imports come from three sources: Southeast Asia, North America, and the U.S.S.R. North America's imports from other regions are mostly tropical sawnwood and plywood from Southeast Asia and paper from Europe. In contrast, Europe is a substantial importer of all types of forest products and is a major customer for exports from virtually all regions. Even in the case of Latin America, which has been slow to realize its potential as a supplier to world markets, there has been a noticeable rise in exports to Europe in recent years of such products as woodpulp, fiberboard, plywood, and sawn hardwood.

The main pattern of Europe's imports of forest products and changes between 1970 and 1980 are set out in tables 5 and 6. Total imports from other regions rose from 63 million m^3 eq. in 1970 to 81 million m^3 eq. in 1980, an increase of 18 million m^3 eq. or 29 percent. It is noteworthy that, although the volume growth was larger than that of Europe's exports to other regions, the rate of growth was apparently less. The frequently given image of Europe as becoming increasingly import-dependent should thus be treated with reservation. Among the more noteworthy features of Europe's imports from other regions in the 1970s may be mentioned:

1. Virtual stagnation in imports of raw materials (roundwood and chips) and of sawn softwood
2. Marked growth in imports of sawn hardwood and plywood, with Southeast Asia and North America figuring prominently
3. A substantial rise in woodpulp imports, notably from Canada, but with other countries, including the U.S.S.R. and Brazil, also making their mark
4. A not very impressive growth in paper and paperboard imports

The last point may be further elaborated. The paper industry in Europe, especially in the net importing countries such as those of the European Economic Community (EEC), has been undergoing a structural change in the face of increasing competition from producers in countries with the potential to establish large-scale, integrated industries. To take an example: paper and paperboard in the United Kingdom, which in 1970 was the second largest producer in Europe, fell by 1.1 million tons during the 1970s, or by 23 percent, while its imports rose by 1 million tons, or 40 percent. While Europe's total imports rose by 5.6 million tons between 1970 and 1980, those from North America actually fell slightly. Almost all of the increase

Table 5. European Imports of Forest Products, 1970, 1974, and 1980, by Product and Main Origin (1000 units).

Product	Unit	Total 1970	Total 1974	Total 1980	From: Europe 1970	Europe 1974	Europe 1980	Non-European Sources 1970	Non-European Sources 1974	Non-European Sources 1980
Pulpwood (round)	m³	15,922	15,971	17,828	9,625	9,478	11,121	6,297	6,493	6,707
Chips and residues	"	2,395	3,702	4,747	1,676	2,107	3,949	719	1,595	798
Softwood logs	"	3,563	5,936	6,102	2,020	3,463	4,426	1,543	2,473	1,676
Hardwood logs	"	8,192	9,299	9,711	1,593	2,564	3,529	6,599	6,735	6,182
Pitprops	"	1,221	1,094	719	455	378	143	766	716	576
Total roundwood	"	31,293	36,002	39,107	15,369	17,990	23,168	15,924	18,012	15,939
Sawn softwood	"	27,408	27,251	28,109	17,539	17,921	17,847	9,869	9,330	10,262
Sawn hardwood	"	3,704	4,234	6,085	1,960	2,215	2,154	1,744	2,019	3,931
Sleepers	"	357	416	464	262	282	336	95	134	128
Total sawnwood	"	31,469	31,901	34,658	19,761	20,418	20,397	11,708	11,483	14,321
Veneer sheets	"	364	557	704	249	353	404	115	204	300
Plywood	"	2,076	2,172	3,047	1,030	944	1,173	1,046	1,228	1,874
Particleboard	"	1,906	3,668	4,742	1,746	3,238	4,407	160	340	335
Fiberboard	"	1,569	1,563	1,560	1,268	1,179	1,160	301	384	400
Total wood-based panels	"	5,915	7,960	10,053	4,293	5,804	7,144	1,622	2,156	2,909
Mechanical woodpulp	mt	1,100	1,040	635	1,026	1,000	549	74	40	86
Chemical woodpulp (paper)	"	9,467	10,047	10,918	6,150	6,269	5,668	3,317	3,778	5,250
Total woodpulp	"	10,567	11,087	11,553	7,176	7,269	6,217	3,391	3,818	5,336
Newsprint	"	2,592	3,343	3,557	1,957	2,580	2,697	635	763	860
Printing and writing	"	1,848	2,926	4,168	1,792	2,752	4,098	56	174	70
Other paper and paperboard	"	6,403	8,196	8,748	4,713	6,321	6,838	1,690	1,875	1,910
Total paper and paperboard	"	10,843	14,465	16,473	8,462	11,653	13,633	2,381	2,812	2,840

Sources: FAO Yearbook 1981.
FAO/ECE Timber Bulletin.
OECD The Pulp and Paper Industry.

Table 5. (Continued)

Product	Unit	From: U.S.S.R.			North America			Other non-European Sources		
		1970	1974	1980	1970	1974	1980	1970	1974	1980
Pulpwood (round)	m^3	5,233	6,311	5,807	1,054	167	540	10	15	360
Chips and residues	"	630	1,595	233	89	---	565	--	--	---
Softwood logs	"	1,523	2,400	1,614	14	19	29	6	54	33
Hardwood logs	"	73	135	109	115	149	266	6,411	6,451	5,807
Pitprops	"	764	716	558	1	---	---	1	--	18
Total roundwood	"	8,223	11,157	8,321	1,273	335	1,400	6,428	6,520	6,218
Sawn softwood	"	6,698	6,350	5,820	2,726	2,705	3,994	445	275	448
Sawn hardwood	"	1	10	---	121	183	586	1,622	1,826	3,345
Sleepers	"	28	36	16	9	2	3	58	96	109
Total sawnwood	"	6,727	6,396	5,836	2,856	2,890	4,583	2,125	2,197	3,902
Veneer sheets	"	---	---	---	33	32	83	82	172	217
Plywood	"	229	241	222	468	529	804	349	458	848
Particleboard	"	145	218	305	1	94	16	14	28	14
Fiberboard	"	206	208*	254	36	79*	62	59	97	84
Total wood-based panels	"	580	667	781	538	734	965	504	755	1,163
Mechanical woodpulp	mt	6	---	---	68	40	70	---	---	16
Chemical woodpulp (paper)	"	395	435	710	2,671	2,957	3,660	251	386	880
Total woodpulp	"	401	435	710	2,739	2,997	3,730	251	386	896
Newsprint	"	119	176	212	516	585	628	---	2	20
Printing and writing	"	---	40	---*	50	106	36	6	28	34
Other paper and paperboard	"	250*	487	535	1,391	1,335	1,229	49	53	146
Total paper and paperboard	"	369*	703	747	1,957	2,026	1,893	55	83	200

Table 6. Changes in Product Share and Distribution of European Forest Products Imports, 1970 to 1980.

Product	Total 1970	Total 1980	From: Europe 1970	Europe 1980	U.S.S.R. 1970	U.S.S.R. 1980	America 1970	America 1980	Other 1970	Other 1980
				(1000 m^3 eq.*)						
Roundwood	31,293	39,107	15,369	23,168	8,223	8,321	1,273	1,400	6,428	6,218
Sawnwood	51,760	57,073	32,496	33,454	11,036	9,573	4,689	7,534	3,540	6,512
Wood-based panels	11,003	17,885	7,626	11,846	1,106	1,407	1,195	2,123	1,076	2,508
Woodpulp	48,192	53,994	32,085	28,579	1,911	3,408	12,991	17,743	1,205	4,264
Paper and paperboard	36,758	55,843	28,686	46,216	1,251	2,532	6,634	6,417	186	678
Total	179,006	223,902	116,262	143,263	23,527	25,241	26,782	35,217	12,435	20,180
				(Percent change 1970 to 1980)						
Roundwood		+ 25		+ 51		+ 1		+ 10		− 2
Sawnwood		+ 10		+ 3		− 13		+ 61		+ 84
Wood-based panels		+ 63		+ 55		+ 27		+ 78		+133
Woodpulp		+ 12		− 11		+ 78		+ 37		+254
Paper and paperboard		+ 52		+ 61		+102		− 3		+265
Total		+ 25		+ 23		+ 7		+ 31		+ 62
				(Percent share of total imports, by origin)						
Roundwood	100.0	100.0	49.1	59.2	26.3	21.3	4.1	3.6	20.5	15.9
Sawnwood	100.0	100.0	62.8	58.6	21.3	16.8	9.1	13.2	6.8	11.4
Wood-based panels	100.0	100.0	69.3	66.2	10.0	7.9	10.9	11.9	9.8	14.0
Woodpulp	100.0	100.0	66.6	52.9	4.0	6.3	26.9	32.9	2.5	7.9
Paper and paperboard	100.0	100.0	78.0	82.8	3.4	4.5	18.1	11.5	0.5	1.2
Total	100.0	100.0	64.9	64.0	13.1	11.3	15.0	15.7	7.0	9.0
				(Percent share of total imports by product)						
Roundwood	17.5	17.5	13.2	16.2	35.0	33.0	4.7	4.0	51.7	30.8
Sawnwood	28.9	25.5	28.0	23.3	46.9	37.9	17.5	21.4	28.5	32.3
Wood-based panels	6.1	8.0	6.6	8.3	4.7	5.6	4.5	6.0	8.6	12.4
Woodpulp	26.9	24.1	27.6	19.9	8.1	13.5	48.5	50.4	9.7	21.1
Paper and paperboard	20.6	24.9	24.7	32.3	5.3	10.0	24.8	18.2	1.5	3.4
Total	100.0	100.0	100.0	100.0	100.0	100.0	100.0	100.0	100.0	100.0

*Cubic meters, equivalent volume of wood in the rough. Converted from product volume taking as a basis factors used in ETTS III.

was accounted for by the growth in intra-European trade, the two main components being intra-EEC movements and EEC imports from the Nordic countries.

A closer look at North American exports of paper and paperboard to Europe shows the preponderance of two grades--kraft liner and newsprint--which between them made up over 80 percent of the total in 1980 (table 7).

While North American producers have established themselves very strongly in the European market for two mass-production grades, they did not share in the growth of the European paper and paperboard market in total during the 1970s and still hold relatively minor shares of the market for other grades.

Consumption of Forest Products in Europe

Since the Second World War, the evolution of consumption of forest products in Europe can be divided into two periods: up to and including 1973; and 1974 to the present. The quarter-century up to 1973 is now identified as a period of economic growth almost without precedent. To a considerable extent this growth was linked with reconstruction and recovery in the aftermath of war. Dwelling construction, for example, rose to an all-time peak of over 4 million units in 1973. By and large, consumption of forest products was associated with this general expansion.

The second period, which began with the oil-price shock of 1973 and 1974 and the subsequent worldwide recession, has been marked by faltering growth in most European economies, with the rate of growth of GDP^2 between 1973 and the present averaging around 3 percent annually, compared with over 5 percent up to 1973. This slower growth was combined with the seemingly intractable problems of high levels of inflation and unemployment. Its impact on consumption of forest products, which reached record levels in 1973, fell back sharply in 1974-75, recovered to a new peak in 1979, and then subsided again into the second major recession in six years from which only today are there signs that recovery may be occurring.

These trends are apparent from table 8. Two aspects of consumption trends in Europe since 1973 are of possible

---

2. GDP = gross domestic product. It is the equivalent of GNP (gross national product) minus the product originating outside the country of interest--Ed.

Table 7. North American Exports of Paper and Paperboard to Europe, by Main Grades, 1970 and 1980.

	Canada		U.S.		North America		Percent of total	
	1970	1980	1970	1980	1970	1980	1970	1980
	------------------(1000 mt)------------------							
Total	922.3	940.1	1,094.5	1,070.4	2,016.8	2,010.5	100.0	100.0
Newsprint	569.3	605.6	18.5	14.4	587.8	620.0	29.1	30.8
Kraft liner	181.7	185.4	922.7	833.4	1,104.4	1,018.8	54.8	50.7
Other wrapping and packaging	138.5	118.2	49.4	155.0	187.9	273.2	9.3	13.6
Other paper and paperboard	32.8	30.9	103.9	67.6	136.7	98.5	6.8	4.9

Source: OECD.

Table 8. Europe: Apparent Consumption of Forest Products, 1965 to 1981.

Year	Sawnwood	Wood-based Panels	Paper and Paperboard	Total*	Of which: Products of Sawlogs and Veneer Logs†	Products of Pulpwood‡	Wood Used in the Rough§
	(million m³)		(million mt)		(million m³ eq.)		
1965	85.7	15.3	29.1	395.5	153.0	124.8	117.7
1966	84.7	15.7	30.9	393.5	151.1	130.9	111.5
1967	85.7	17.3	32.1	396.9	153.3	137.0	106.6
1968	87.9	19.3	34.0	409.7	157.9	145.9	105.9
1969	90.3	21.2	37.4	423.9	162.2	160.4	101.3
1970	93.4	23.2	38.8	434.5	168.1	167.5	98.9
1971	95.2	25.1	38.9	436.3	171.4	169.8	95.1
1972	96.6	28.7	41.0	446.2	175.1	181.6	89.5
1973	102.9	32.7	44.2	473.9	187.3	197.2	89.4
1974	99.3	31.8	45.8	466.5	179.1	202.8	84.6
1975	87.8	30.2	38.8	417.9	159.0	177.4	81.5
1976	95.0	33.7	44.0	452.7	172.8	198.5	81.4
1977	97.2	33.9	44.3	454.1	175.7	200.3	78.1
1978	98.7	34.9	45.7	458.1	178.4	205.5	74.2
1979	105.7	36.5	49.4	485.2	191.2	219.4	74.6
1980	104.6	36.2	48.9	483.3	188.1	218.4	76.8
1981	97.0	35.0	48.1	468.7	175.4	213.5	79.8

*Including wood used in the rough, dissolving pulp.
†Sawnwood, sleepers, plywood, veneer sheets.
‡Paper, paperboard, particleboard, fiberboard, dissolving pulp.
§Fuelwood, pitprops, poles, piling, posts, etc.

significance. One is the performance of the "structural" forest products—sawnwood and wood-based panels—relative to that of their single most important market, dwelling construction. From a peak in 1973, the latter has declined to a level in the early 1980s some 27 percent lower. Sawnwood and panels, however, recovered to reach new record levels at the end of the 1970s. Unfortunately, inadequate end-use information makes it difficult to analyze this in the necessary depth, but it could be surmised that a diversification of utilization has been taking place into such sectors as repairs and maintenance and pallets.

A second phenomenon was the quite sudden change in the rhythm of expansion of particleboard. Up to 1973 growth had been more than 10 percent a year. Thereafter, it has followed much more closely the trend of the other forest products. This suggests that the possibilities for substitution have been largely exhausted. But the possibility should not be excluded that other products may enter the field in the future and keep the product pattern evolving. MDF is having some impact in Europe at the present time; perhaps there will be others in the future.

With regard to the present situation, the recession which began in 1980 in Europe may not have been quite as deep as the 1974-75 one, but has had an even more debilitating effect on the forest products sector because of its longevity. Business confidence has been sapped by the repeated dashing of hopes of recovery, and this may explain why, even in the spring of 1983 when there are clear signs that recovery is under way in North America, Europe remains more cautious with regard to the short-term outlook. Among the reasons for this is that the unemployment problem is a long-term structural one, and would require a long period of active, sustained growth to deal with it. Such growth would, it is feared, quickly renew inflationary pressures which have eased during the recession but remain high in most West European countries by historical standards.

In addition to unemployment, there is also in Europe overcapacity in many industries in relation to foreseeable demand, and this has a restraining influence on investment. Also in the investment field, it is felt in many countries that dwelling construction has finally caught up with the backlog and that in the future the rate of building new dwellings will remain well below the levels of a decade ago. Given that there are very many substandard buildings, however, it is also argued in several countries that there is need for a more active program of renovation and maintenance, and in some cases replacement, of older dwelling stock. There is more general recognition that changing demographic patterns--smaller families, higher proportion of elderly and single people--demand different types of accommodation, including smaller units.

The probable impact of these developments--and above all of the prospect for a lower rate of growth of the overall economies in most European countries than had been true up to 1973--on consumption of forest products in the medium to long term is the subject of intensive study at the present time. ETTS III was prepared during the 1974-75 recession, at which time there was great uncertainty whether the recession was just a big "hiccough" after which growth would resume as before or whether it marked a turning point in the world and the European economy. Today, the latter view is generally accepted, and even though the lower of ETTS III's GDP growth assumptions for 1970 to 2000 was, at 3.8 percent a year for Europe as a whole, well below the 5.2 percent recorded during the 1950s and 1960s, it appears now that 3.8 percent is probably the most optimistic assumption that could be made.

FAO and ECE, more specifically the Timber Section of the FAO/ECE Agriculture and Timber Division in Geneva, are examining these questions at the present time in preparation for the next of the long-term studies for Europe, due for publication in 1986. Assumptions regarding the future prospects for the fixed variables, such as GDP, population, housing, furniture making, packaging, etc., which would be used in the projection models for consumption of forest products, have not yet been established. Nor is it likely to be easy to do so under today's unsettled and uncertain conditions. Nevertheless, it is probable that for Europe as a whole, the assumption of GDP average growth could be for a range of 2.5 to 3.5 percent a year in the 1980s and 1990s--that is, similar to the performance since 1973. Even the higher growth rate would, however, do little to resolve the grave unemployment problem, so that a question mark must remain about possible sociopolitical developments and their impact on European countries' economies.

Without prejudicing the results of modeling work currently in hand, it would be the author's guess that a 3 percent average annual growth rate of GDP could be achieved in Europe and that this might be reflected in 1 percent growth for sawnwood consumption, 2 percent for wood-based panels, and 2 to 3 percent for paper and paperboard. In the case of the latter products, the net impact of the electronic revolution is difficult to assess, but it could be that what paper loses "on the swings"--e.g., national newspapers--it could largely make good "on the roundabouts"--e.g., computer printout paper. On the packaging side, oil price trends will have an important bearing on the competition between paper and plastics: in today's chaotic conditions in the oil market, nothing more can usefully be said on this subject for the time being.

To summarize the foregoing, one scenario of consumption of forest products in Europe could be for the relationship between its average annual growth and that of GDP, the income elasticity, to be similar to that foreseen in ETTS III, around 0.5, but for GDP growth itself to be lower than foreseen in ETTS III with corresponding effects on the volume of forest products used. For paper and paperboard and for wood-based panels in particular, the previous projections are likely to be reviewed markedly downward in the new study, which will be known as ETTS IV. Nonetheless, in volume, there could still be appreciable growth in European consumption of forest products in the next two decades.

Europe's Sources of Supply of Forest Products

The principal source of the forest products consumed in Europe is the European forest itself. This is supplemented by supplies from overseas, the second most important source. To an increasing extent, two other sources have contributed to supply: the transfer of industrial wood residues as raw material for other wood-processing industries, and the recycling of waste paper. To take these sources into account, the unit used for expressing both consumption and supply is the cubic meter, equivalent volume of wood in the rough (m^3 eq.). There is thus an element of double counting on both sides of the supply:consumption balance, so that m^3 eq. is not strictly comparable with cubic meters, roundwood equivalent.

Bearing this in mind, Europe's consumption of forest products in total of around 480 million m^3 eq. in 1980 was supplied in the following way:

	Volume (million m^3 eq.)	Percent of total supply (%)
Removals from European forests	343	71
Transfer of industrial residues	47	10
Recycling of waste paper	37	8
Total domestic supply, of which:	427	89
Exported outside Europe	27	
Net domestic supply		400	83
Imports from outside Europe		81	17
Total supply		481	100

Europe in aggregate was thus 83 percent self-sufficient in forest products in 1980. The self-sufficiency of many individual European countries, however, was on average much lower, because of the importance of intraregional trade--143 million m^3 eq. This volume, together with the 81 million m^3 eq. imported from other regions, means that European countries' total imports accounted for nearly half of their consumption. In a few countries, such as the United Kingdom, Belgium, and the Netherlands, imports made up 90 percent or more of consumption. In others, notably the Nordic countries and Austria, supply exceeds domestic consumption by a wide margin. Intra-European trade remains, therefore, a very important element in keeping the region from being heavily dependent on supplies from other regions.

The 81 million m^3 eq. in European imports of forest products from other regions in 1980 constituted an important but by no means predominant share of total supply to the European market. In the case of certain species and grades of tropical hardwoods, these imports fill niches in the market which cannot be supplied from other sources. Generally, however, imports from other regions have achieved their market share in the face of direct competition from European products.

Factors in Europe Which Shape the Markets for Forest Products

Despite its dense population and relatively high per capita consumption levels, Europe has remained reasonably self-sufficient in forest products. This has been made possible by exploiting the forest resource at a harvesting intensity that is high by world standards, whether expressed as the ratio of removals to growing stock, removals to increment, or removals to area of forest. At the same time, in the majority of European countries, the forest could sustain a higher level of harvesting, especially of hardwoods and of the smaller dimensions (thinnings), without reducing the volume or quality of the forest resource. Growing stock has, in fact, been steadily growing, as increment has remained above fellings and natural losses.

It would even benefit the long-term health of many of Europe's forests for the rate of cutting and regeneration to be stepped up. Up to the present time, however, economic conditions have not made this possible, even if the forest owners had wished to increase their harvests, which has not always been the case. The economic conditions have been set by the international market, and domestic producers have had to compete against stiff competition from imports from regions which for various reasons have enjoyed lower production costs.

Europe's wood production costs are relatively high, the reasons including the fragmented structure of forest holdings in many countries, labor costs, climatic impact on growth rates, environmental constraints, and--not least--the widespread application of the principle of sustained yield. This means that regeneration and management costs have to be covered in the price of the wood sold from the forest.

There have been other factors accounting for a level of productivity, both in harvesting and in converting wood in Europe, below that in some other parts of the world. In the case of industry, these factors include inadequate economies of scale and investment for modernization, environmental constraints, and sometimes a rather conservative management and labor force.

Are conditions changing in Europe in ways which would alter the competitive position of its forestry and forest industries in the future? First, its forests are for the most part already established on a sustained-yield basis, accessible, and within reasonable reach of industry and markets. Second, unlike some other regions, there has been little evidence of a long-term rise in the real price of wood in Europe, suggesting that improvements in productivity have kept pace with cost increases. Third, a continuous restructuring of industry has been taking place, which has tended to accelerate during the current recession, leaving larger, leaner units better able to compete in the international market.

To a greater or lesser extent these developments, with the possible exception of price stability, are occurring everywhere. Which regions can in the long run take the lead in this process and still have sufficient resources to export to the timber-deficit regions?

External Factors Shaping the European Market for Forest Products

It was suggested above that European wood production had in the past been kept below its biological possibilities by the competition from overseas suppliers. These are, in order of importance, North America, the U.S.S.R., and the tropical regions.

The Tropical Regions

European imports from the tropics, in which are included small quantities from temperate-zone regions such as New Zealand, Australia, Japan, and South Africa, amounted to 20 million m^3 eq. in 1980 and accounted for one quarter of Europe's imports from outside Europe. The increase of 62 percent compared with 1970 was appreciably faster than that of total imports (25 percent), but it is necessary to distinguish two quite different trade flows: that emanating essentially from the natural tropical hardwood resource, and that from plantations. Europe's imports of tropical hardwoods reached an all-time high of over 15 million m^3 eq. in 1973. Since then, the trend of its imports of tropical hardwoods in total has been flat (Peck 1983). Southeast Asia has gained market share through the rapid rise in trade with Europe of sawn hardwood and plywood, mainly at the expense of West Africa, the chief supplier of logs to Europe.

With regard to the future, European industry will no doubt continue to adapt to the increasingly difficult tropical log supply situation, as tropical producers take measures, including curtailment of log exports, to divert them to domestic industries and to export part of the resulting production. European imports of tropical hardwood logs will therefore gradually decline. Those of processed and semiprocessed tropical hardwood products will be converted from logs brought from less and less accessible areas, and it seems inevitable that relative costs of supplying them to overseas markets must rise. The moist tropical forest is, unlike many other types of forest, a nonrenewable resource and one which is increasingly under threat from many pressures, of which commercial exploitation is only one and perhaps not the most serious one. World opinion is beginning to recognize the necessity, for fundamental ecological reasons, of conserving at least a part of the remaining natural tropical forest, and this is likely to have an impact on the long-term availability and cost of wood from this source.

The European import of pulp and paper from tropical plantations was small in 1970 but rose rapidly during the 1970s. The principal supplier is Brazil. Very high increment rates in these plantations appear attractive from the investment point of view. On two counts, however, there may be reasons for uncertainty whether such sources should be relied on in the long term as major suppliers to the European market. The first is the potentially huge market for forest products in the tropical countries themselves, which could probably absorb all available domestic or regional production. The second is the long-term ecological stability of plantations, which are normally monocultures, under tropical and subtropical conditions. They may prove sound ecologically, but only time will tell. Given the fragility of many soils and the climatic extremes in the tropics, it cannot be assumed automatically that early successes with tropical plantations can be sustained in the long run.

The U.S.S.R.

In 1980, Europe imported 25 million m³ eq. of forest products from the Soviet Union. This was only 7 percent more than in 1970, and its share of overseas supplies to the European market declined from 37 percent in 1970 to 31 percent to 1980. Exports account for a relatively minor part of total removals in the Soviet Union--about 7 percent in 1980, which compares with 83 percent in Canada, 50 percent in Europe, and even 21 percent in the U.S.[3] Its exports to the market economy countries of Western Europe and Japan are an important source of hard currency, however, while within the Council for Mutual Economic Assistance (CMEA) it plays a significant role as supplier of raw material to the timber-deficit countries of Eastern Europe.

The commodity composition of Europe's imports from the U.S.S.R. differs markedly from that of intra-European trade and imports from North America, in that the former are still heavily weighted toward roundwood and sawnwood. These accounted for 71 percent by volume of Europe's imports from the Soviet Union in 1980 (down from 82 percent in 1970) which compared with 40 percent for intra-European trade and only 25 percent for imports from North America. The U.S.S.R. is therefore in a position somewhat similar to the tropical countries as being still a supplier to Europe predominantly of unprocessed and semiprocessed wood products, even though the pattern is gradually changing.

Removals in the U.S.S.R. have been declining in recent years. Overcutting has taken place over a long period to

---

3. These figures are exports in m³ eq. as percentages of removals in m³. For reasons given earlier, the percentages exaggerate the share of exports, but what is interesting here is the relative position of the countries and regions mentioned.

the west of the Urals, where most of the forest industries are located. Policy has been directed, therefore, toward opening up Siberia and establishing new logging enterprises and industries there. Immense difficulties are involved of a climatic, infrastructural, and demographic nature, and it does not seem to have been possible yet to raise removals enough to the east of the Urals to compensate for declining product to the west.

No doubt the problems of shifting the center of gravity of the forest industries sector eastward will be overcome in time. The question for European importers, however, is what will be the economics of producing and transporting forest products to the European market over the long distances from Siberia. One option for the Soviet Union would be to give increased emphasis to higher unit value products, such as plywood, paper, and paperboard, provided the quality was right. Alternatively, it could give more attention to other markets, not only Japan where it is already well established as a roundwood supplier, but other Pacific Basin markets as well as southern and western Asia. ETTS III concluded that European imports from the U.S.S.R. would rise gradually between the mid-1970s and 2000, but this trade would account for a declining share of total U.S.S.R. exports--and of total European imports. Nothing has happened since the study was prepared to alter the view.

North America

Europe imported about 35 million m^3 eq. of forest products from North America in 1980 or 44 percent of the total from other regions. Canada and the U.S. thus consolidated their position during the 1970s as, between them, the most important overseas suppliers to Europe (43 percent in 1970). As noted above, exports represent a major share of Canada's total output and a not insignificant share of the U.S.'s. Their exports to Europe accounted for 16 percent of Canada's total forest products exports in 1980 and 23 percent of the U.S.'s. If intra-North American trade is excluded, these shares rise to 51 percent and 26 percent, respectively. Europe is thus Canada's largest offshore market; the second largest for the U.S. after Japan.

In terms of m^3 eq., woodpulp remained by far the most important of the forest products traded between North America and Europe, with 50 percent of the total in 1980. This was followed by sawnwood (21 percent) and paper and paperboard (18 percent). At the individual product level, the growth of chemical (sulphate) woodpulp and sawn softwood exports from Canada to Europe during the 1970s was noteworthy. Also marked, but on a smaller scale, was the expansion of U.S. exports of hardwood logs and sawnwood as well as of softwood plywood and chips. As discussed earlier, neither country increased its exports appreciably of paper and paperboard in total to Europe during the 1970s.

Trends up to 1980 have confirmed thus far the expectations of ETTS III that North American exports to Europe would rise over the period to 2000 and that the share of this trade in total North American exports would be stable or possibly rise slightly.

Since ETTS III was published, however, a number of studies have appeared which suggest that softwood supply in North America will become increasingly tight toward the end of the century and that there will be increases in relative prices for softwoods. (The supply/demand situation for hardwoods, it is felt, is easier, at least during the present century.) The USDA Forest Service's "An Analysis of the Timber Situation in the United States, 1952-2030" (1982) refers to a "growing economic scarcity of timber, and the associated increases in relative stumpage and timber product prices." The Canadian Forestry Service, in "Policy Statement--A Framework for Forest Renewal" (1982), states that under present management conditions the forest resource (in Canada) will not be sufficient to meet expected future needs. A study prepared by an Industry Working Party appointed by the FAO Advisory Committee on Pulp and Paper (1982) drew certain conclusions for global trends in the forest products sector from the expectations for a tightening softwood supply situation in North America.

The author has some slight reservations whether these assessments take adequately into account certain positive aspects of supply, such as substitution possibilities by hardwoods for softwoods, reduction of harvesting losses, improved yields, and so on. Nevertheless, as highly authoritative forecasts, their implications should be given serious attention. For Europe, it could mean that forest products from North America will become more expensive. Whether, as a result, they would be less competitive depends on cost trends in Europe and on other sources of supply. The discussion thus returns once again to prospects for the relative trend of costs and prices for wood and wood products in different regions--a matter as much for personal judgment as for rational analysis.

Discussion and Conclusions

Barriers to Trade

The question of measures--of whatever kind and imposed for whatever reason--that result in a distortion of trade and the international location of industry has not been raised earlier in this paper, not entirely out of caution, but because it raises issues which go far beyond the forest products sector. It could be argued, in fact, that this sector has been relatively lightly touched by protectionism. Protectionism is in the long term counterproductive, not least for the country that applies it. Most excuses for indulging in it--for example, that it is necessary to support a fledgling industry--would carry more weight if there was not a tendency to perpetuate the protective system long after the reason for it has disappeared.

Some progress has been made through the Kennedy and Tokyo rounds of negotiations of GATT to lower tariff structures, including those for forest products. Imports of unprocessed and semiprocessed tropical products, for example, enter the industrialized countries duty-free and there has been a general easing of duties on pulp and paper. Generally speaking, however, tariff structures tend to discriminate against the more highly processed products, such as furniture and building components, where the industries in the developed countries feel most threatened by competition from countries with markedly lower product costs, especially of labor.

In addition to tariffs, which are a visible barrier, there are many others which are to a greater or lesser extent invisible. That is to say, the fact that they may or do inhibit trade is not the stated or principal purpose of their existence. Among these, the following may be mentioned:

- building codes and regulations
- performance specifications or product standards
- units of measurement
- discriminatory purchasing practices by government departments or other buyers
- product reclassification
- import quota systems
- customs clearance procedures
- reinspection and regrading of goods in importing country
- phyto-sanitary regulations
- foreign exchange controls
- freight rate discrimination

- dumping allegations
- preferential trading agreements (free trade areas, most favored nation agreements, etc.)

It would be wrong to suggest that governments are deliberately using these as mechanisms to protect their own industries and workers. Nevertheless, it is usually difficult to generate momentum for dismantling or changing them, when it can be shown that they are inhibiting trade, as the Ministerial Meeting of GATT last December demonstrated.

Countries with export interests also have a range of measures at their disposal intended to improve the competitiveness of their industries in overseas markets. These may include:
- subsidies, "development" grants, or loans at preferential interest rates
- devaluation
- export licensing or quota systems
- export bans, usually of raw materials with the object of diverting them to local industries
- export guarantees and insurance schemes
- government involvement in overseas promotion activities

It is virtually impossible to determine in reliable terms the impact of discriminatory measures on the level and pattern of trade and consumption of forest products. The assisted sectors may gain, at least in the short run, but possibly at the expense of other sectors against which retaliatory measures are taken. In any event, the losers are the taxpayers and the consumers; the one paying to keep an uncompetitive industry alive, the other paying higher prices for the goods he needs.

Generally speaking, European countries appear to be no better and no worse than others with regard to the existence of visible and invisible barriers to trade in forest products. Undoubtedly, the diversity of building codes and of dimensions and qualities of sawnwood and wood-based panels create difficulties for exporters wishing to sell to several European markets. Even in the United States, though, there are considerable variations from state to state in building regulations. Tariffs in European countries tend to be increasingly discriminatory with the extent of further manufacturing, but that seems to be a phenomenon almost everywhere. It is no argument for justifying trade barriers to say that other countries also have them. The start of a serious debate leading to negotiations for easing them, however, must be to recognize that their existence is universal.

While practices which distort international trade may be condemned on economic grounds, it must be recognized that other grounds may exist on which the governments concerned, even supported by international opinion, feel justified in exercising controls. Trade in endangered species and their products (furs, tusks, etc.) come into this category. It is even being suggested that the tropical forest resource, as an immensely complex ecosystem and potentially vital gene bank, also deserves such protection.

Environmental arguments may also be used to explain the expansion of trade. It has been suggested, for example, that the industrialized countries, rather than add to their already serious pollution problems, prefer to "export" pollution along with manufacturing capacity to those countries where economic and social development may have higher immediate priority than the quality of the environment. The degree of pollution control exercised in a given country may well be the determining factor as to whether the industries concerned are competitive or not on the international market.

The problem for the policy maker is where to draw the line--how to determine what is in the true interests of his country and of society as a whole.

Relative Cost and Price Trends in the Forest Products Sector

Measures taken by governments which distort international trade and the location of industry, either deliberately or indirectly, are one factor to be taken into account in assessing the possible future developments in the forest products sector. Government policies with regard to the forest resource itself are another. Earlier, it was pointed out that there have been past policies in many exporting countries outside Europe to allow timber exploitation to proceed without any organized attempt at regeneration: what has been called "timber-mining." As a result, the forests in the more accessible regions have either been destroyed and the land transferred to other uses (or to no use at all), degraded through inadequate natural regeneration, or "creamed" of commercially attractive species and qualities. Exploitation has had to move into less economically accessible areas. Good natural regrowth has occurred on many sites, but little of this timber has yet reached commercial size, and it may not compare favorably in quality with that from the previous stands.

Recognition of the need to put forestry onto a sustained-yield basis has been gaining ground in many of these countries for some decades now. The inclusion of regeneration and management costs would raise the price of wood, however, as would the harvesting costs of wood from less accessible areas.

It follows, therefore, that the trends in relative prices for wood could be expected to rise in such countries as Canada, the U.S., and the U.S.S.R., as well as the tropics, whether the reason is the application of the principle of sustained yield or costlier exploitation. Indeed, forecasts to that effect have been made in several of these countries. For Europe, as a major importer from these regions, the question is not so much whether the supply will be there, but at what prices, and how those prices will compare with those of domestically produced forest products. As noted earlier, there has not been a distinct trend, upward or downward, in relative prices for wood in Europe. If that situation persisted, it could put European forestry and forest industries gradually onto a more competitive footing than in the past.

For the European forest products sector to take advantage of such a situation would not be easy, given its record of poor profitability and the problems it might experience in raising the necessary investment both for the improvement of forest productivity and for the further rationalization and modernization of the forest industries. Nevertheless, there does appear to be a strong incentive for policy to be directed toward maintaining and, where economically viable, expanding the European forest resource and toward improving cost structures.

Final Remarks

However dynamic European countries' forest policies may become, the region will remain a large net importer of forest products. It is quite possible, however, that the trend of increasing import dependency, which was marked up to the "turning point" of the mid-1970s, could be replaced by a certain equilibrium between domestic and overseas supplies, especially if, as has been postulated, forest management in the overseas supplying countries generally adopts the principle of sustained yield, which has for long been practiced in Europe.

The views expressed in this paper are perhaps somewhat controversial in the sense that they suggest the possibility of a change from past trends in Europe: from a tendency toward increasing import-dependency to a state of equilibrium, where Europe can maintain its present level of self-sufficiency even if consumption should continue upward. This hypothesis is based on the belief that, whereas in Europe forest management practices will not have to change drastically, even if there is still considerable scope for improvement, other regions, which have traditionally been suppliers to Europe, are likely to face increasing wood production costs, either by replacing "timber-mining" by forest management or by having to exploit less economically accessible tracts. The paper will have served its purpose if it stimulates reactions to this hypothesis and generates further thought on the implications of the universal adoption of a sustained-yield policy.

References

Canadian Forestry Service. 1982. Policy statement--A framework for forest renewal. Environment Canada, Ottawa.

FAO. 1982. World forest products demand and supply 1990 and 2000. Food and Agriculture Organization, Rome. Forestry Paper 29.

FAO/ECE. 1976. European timber trends and prospects, 1950 to 2000. Supplement 3 to volume XXIX of the Timber Bulletin for Europe, United Nations. Geneva.

Peck, T. J. 1983. Overview of trends and challenges for tropical wood products in European markets. Presented at the Conference on Asian Timber, Singapore.

USDA Forest Service. 1982. An analysis of the timber situation in the United States, 1952-2030. FRR-23. U.S. Government Printing Office, Washington, D.C.

TRANSPORTATION FACTORS IN THE MOVEMENT OF VARIOUS FOREST PRODUCTS IN INTERNATIONAL TRADE

Charles E. Doan

Introduction

The transportation of forest products to foreign markets has changed remarkably in the past 10 years. Since trade began from the forest-rich Pacific Northwest in the late 1800s, it has revolved primarily around lumber. But as shipping and handling technology evolved from sail to steam to motor, the variety of products shipped also evolved to include everything from a round raw log to finished products, which in addition to lumber include poles, wood chips, pulp, paper, and board.

Today we are experiencing major changes in the handling of these goods, including bulk, breakbulk, unitization, and containerization. New ships have been designed to accommodate these developments and to improve the quality of handling, thus the economics of getting the goods to market has improved considerably. The transport cost factors are still a significant element in the marketability of these products. These factors include inland transport, port facilities, ocean shipping, and quality control.

Inland Transport

At the turn of the century, most mills were on tidewater where they had the advantage of being adjacent to their supply and transport. As the supply moved inland, so did most of the sawmills. Only the capital intensive pulp/paper mills stayed at tidewater locations.

Moving a product from forest (logs) or mill to port for loading primarily involves either rail or truck, although some volume of product does move via river barge, such as paper and milk carton stock from Lewiston, Idaho, to Portland, Oregon, where all three methods of transportation are available. The choice is usually economic. Obviously the lack of a river transport system (regionally we have only the Columbia River) or the lack of rail infrastructure will limit the choices.

Generally speaking, rail and barge are viable only if the distance to be traveled is over 300 miles, or less if the volume can generate unit trains of ± 50 rail cars (logs).

Typical Transport Loads

	Rail Car	River Barge	Ocean Barge	Truck
Logs	9,000fbm	400Mfbm	--	4500fbm
Lumber	49,000fbm	1,000Mfbm	1,000Mfbm	30,000fbm
Pulp	75st	1,500st	1,500st	23st
Paper	75st	1,500st	1,500st	23st
Linerboard	65st	1,500st	1,500st	23st

Truck hauling is generally faster and certainly more flexible, since logs and chips can go from the cutting area directly to the mill or shipside; whereas rail hauling requires loading from and delivering to a marshaling area most likely served by truck.

Comparable Transport Costs

	Rail Car	River Barge	Ocean Barge	Truck
Logs	$1,000Mfbm	$43Mfbm	--	$56Mfbm
Lumber	76st	11st		53st
Paper	44st	15st	11st	48st
Pulp	40st	15st	10st	48st
Linerboard	44st	15st	--	48st

Port Terminal Handling

In the Pacific Northwest (excluding Alaska) most shipping of forest products to foreign markets is through public port facilities, be they leased from or operated directly by municipal port authorities. This was not the case a few years ago. Economics of scale in shipping has given us larger vessels with high operating costs. Many tidewater mill docks in Washington, Oregon, and British Columbia cannot aggregate enough volume economically to induce a vessel for loading direct, so they transship (via truck, rail, or barge) to a regional port for consolidation with other mill suppliers.

Thus the port must furnish the receiving, staging, and manipulation services as well as the dispatch areas for a wide variety of products: logs, green lumber, dry lumber, plywood, pulp, paper, chips, etc. Properly equipped and well-serviced with ocean carriers regularly serving world trade routes or time charters, the port can expect to generate sufficient cargoes to pay for it all. This infrastructure costs:

900 foot berth	$ 5,400,000
Transit shed (150,000 square feet)	3,750,000
CFS (100,000 square feet)	2,600,000
100 acres backup land	4,750,000
Container crane	3,000,000
Straddle carrier	500,000
Log handler	400,000
Forklift	75,000
Total	$ 20,475,000

Hardware suffers wear and tear, and forest products can brutalize a port's equipment and facilities. Generally speaking, maintenance of these items is running 10 percent of revenues.

The biggest element of cost in handling is, of course, labor. At most Pacific Northwest ports, including mill docks, the ILWU represents labor. The prevailing wages for the USWC segment are noted:

ILWU Labor Rates (7/82-6/83)
(per hour cost--straight time)

	Basic	Supervisor	Foreman
Base rate	$13.52	$16.73	$18.23
Taxes and insurance	1.78	2.09	2.24
Fringes	8.75	8.55	8.00
Total	$24.05	$27.37	$28.47

Port charges for handling forest products are fairly uniform and consist of the following items:
- Dockage: the charge assessed against an ocean vessel for berthing at a wharf, pier, bulkhead structure, or bank, or for mooring to a vessel so berthed.

- **Wharfage**: a charge assessed all cargo passing, or conveyed over, onto, or under wharves, or between vessels (to or from barge, lighter, or water), when berthed at a wharf, piling structure, pier, bulkhead structure, or bank, or when moored in slip adjacent to wharf. Wharfage is solely the charge for use of wharf, and does not include charges for any other service.
- **Service and Facilities**: that charge assessed against ocean vessels, their owners, agents, or operators which load or discharge cargo at the terminals for the use of terminal working areas in the receipt and delivery of cargo to and from vessel, and for services in connection with the receipt, delivery, checking, care, custody, and control of cargo required in the transfer of cargo between vessels and shippers, consignees, their agents, or connecting carriers.
- **Handling**: the charge made against vessels, their owners, agents, or operators, for physically moving cargo from the end of ship's tackle to a point of rest, or from point of rest to within reach of end of ship's tackle. It includes ordinary sorting, breaking down, and stacking.
- **Loading/Unloading**: the services performed in loading cargo from wharf premises on or into rail cars and unloading cargo from railroad cars onto wharf premises. The service includes ordinary breaking down, sorting, and stacking.

Car loading and unloading charges are assessed against the cargo when not absorbed by carriers.

For typical charges see the chart on the page opposite.

Typical Charges (Export)

Commodity	Wharfage	S & F	Load Unload	Handling	Stevedoring
Logs (per 1,000 foot board measure, in Mfbm)	$4.65	$1.59 to 3.62	@ man-hour rates	@ man-hour rates	$25.00
Lumber (per Mfbm)	2.75	2.78	11.62	8.17	11.45
Paper (per metric ton (mt)	2.25	2.57	9.87	7.19	3.48
Pulp (per mt)	1.80	2.57	9.10	7.68	2.91

Stevedoring is not a port tariff item, but is privately contracted.

Loading containers: paper in rolls $7.84/1000 kg; paper in bales or bundles $14.08/1000 kg.

Handling containers from inland conveyance to or from container yard (CY) $45/container. Handling from CY to or from ship's tackle varies by amount of productivity; i.e., 0-14/hour=$38.00/container, 25 or over/hour=$22.30.

Dockage is based on length of vessel, 236 m to 244 m is $3,157 per 24 hour day.

Equipment Rental (without operator):
　　Container crane　　　　　　　　　　$254.00/hour
　　Lift trucks (@ 2.7 to 11.3 mt)　　$ 24.50/hour
　　Container straddle carrier　　　　$ 55.00/hour

Ocean Shipping

History

As regards exports, the earliest information goes back to 1786, when John Mears, a retired lieutenant of the British Royal Navy launched a ship at Nootka to export to China a deckload of spars and masts. Until 1832, the shipments of forest products remained infrequent and the quantities small. Then Hudson's Bay Company shipped a cargo of sawmill material from Port Vancouver. In 1862-64, Oregon and Washington territories exported 20 million board feet. The modern history of lumber exports from the Northwest began about the middle of the last decade of the 19th century. In 1895, Washington and Oregon exported 131 million board feet of lumber and timber products to foreign markets:

Australia	39%
South America	27%
Other Pacific	10%
Non-Pacific	24%

At the turn of the century, the ships carrying that cargo were in the 1,200-gross-ton category capable of carrying about 600,000 cubic feet. Twenty-five years later there was a small growth to 1,800 gross tons with capacity of 1.2 million cubic feet. Fifty years later (1975), we had experienced quantum jumps to 20,000 dead weight ton (dwt) ships, and the forecast to year 2000 is for 100,000 dwt specific forest product carriers.

Ships

In most trade routes, the multipurpose general cargo ship is serving a good portion of the world's market. These run from breakbulk (B/B) to combination B/B container, to fully containerized carriers. There is, however, a large segment of ship tonnage specifically designed for forest products:

Wood chip carriers	Woodpulp carriers
Timber (log) carriers	Newsprint carriers

Most of these special carriers operate under time, space, or voyage charter contracts. Economies of scale are realized with such a vessel's volume, and these ships can usually unitize their loads to facilitate and further economize in their loading operations--for example, bundling the logs, strapping multiple bales/rolls of pulp and paper. The primary charterers of such vessels are large-volume

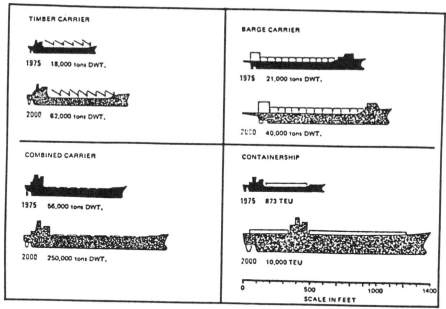

FIGURE 1. Comparison of typical contemporary ships with the expected largest ships serving the Pacific Northwest in the year 2000.

shippers who can aggregate loads from their operations to ship for limited or multiple receivers.

On the other hand, smaller volume shippers can utilize B/B and container ships in these same trade routes, although with a noticeable penalty in cost of shipping.

Costs/Logs

A typical Japanese log ship charter from the Pacific Northwest will be a ship of 18,000 dwt (2.5 million fbm) to 30,000 dwt (3.5 million fbm), costing about $80 U.S. per 1000 Scribner. The round-trip time will be approximately 36 days (18 days each direction):
   3 days loading (1 port call, 2 shifts/day)
   12 days sailing
   3 days unloading

If more than one port or berth is used to load or unload, an additional charge of $5,000 each is standard, and stevedoring will probably increase by 25 percent. Some of you may be interested to know that there is currently about a 30-day congestion lay at PRC ports as they unload to skin of the dock or direct to rail cars (they are loading about 4.5 million fbm per vessel); such lays can cost $5 to $10,000 per day in ship demurrage.

Noted earlier were stateside port charges for loading logs; the comparable charges for unloading at Nagoya are $41/Mfbm to water or $68/Mfbm to the dock.

Costs/Lumber

Lumber shipments in space charters were being effected two or three years ago in conventional vessels of 16,000 to 18,000 dwt. The size is now up to 24,000 to 27,000 dwt, which can hold 9.5 to 11.2 million fbm. The smaller vessel will load seven to nine million fbm, and at today's prevailing rate of $70/fbm, requires a minimum of 6 million fbm to break even.

Such a vessel (16,000 to 18,000 dwt) in this lumber trade will cost about $2,500 per day to operate (including crew)--with bunkers an additional $4,000 per day. These ships are usually self-geared, using three to four gangs to load ± 500,000 fbm/shift at a stevedoring cost of $16 and port terminals of $13 per 1000 fbm.

Ironically, shipping lumber in smaller volume in containers is currently cheaper, at about $68/fbm. This is because there is a need for repositioning empty containers, and thus the container carriers offer bargain rates. The availability of a steady and sufficient volume of container slots is not, however, reliable.

Current rates and load factors for containerized lumber shipments to Japan are:

Container Rates

	20 ft	40 ft
Load (fbm)	750-850	1,400
Rate	$500	$900

The rate difference is even more dramatic when you consider that these rates are generally door-to-door, whereas the B/B bookings are dock-to-dock. Container loading productivity for a cellular container vessel at Tacoma is about 20 per hour; thus an aggregate shipment of 40-foot containers (working two hatches) would load at only about two-thirds the rate of loading a B/B lumber ship. Another type of ship that has been effectively used for lumber shipping is the roll-on/roll-off vessel. Ro/Ro-type ships can achieve high levels of productivity: about 750,000 fbm/shift. Although more expensive to operate (at ± $5,000/day, plus bunkers of about $475/day) than the conventional B/B ship, their productivity cuts down berth time. Stevedoring and terminal charges are about 25 percent less expensive.

Costs/Pulp and Paper

Pulp and paper products are handled similarly, with the rates only slightly different--primarily the reflection of value and vulnerability to claims. These products lend themselves to unitizing into various-size lifting units and load effectively into containers. The prevailing load factors and rates for pulp and paper products from the Pacific Northwest to Japan (add $10 to $15 for Korea) are noted below, with the charter rates applying from mill docks (add terminal receiving charges of about $14/mt if at public port dock).

<u>Comparable Transport Costs per Metric Ton</u>

	Unitized* B/B Charter	20' cont.	40' cont.
Pulp (baled)	$47	$70.25	$70.25
Paper (kraft)	$68	$71.25	$71.25

*Usually minimum mill dock or port call 1500-3000 mt.
 Productivity on unitized baled pulp ± 200 mt per hour.

Loading to container is a cost element which must be considered. If space and facilities can be allocated, it is best to load at the mill; in addition to the obvious advantages of quality/quantity control, there can be cost savings.

As in the case of lumber, the container loads make a convenient delivery vehicle package for the receiver.

Typical Port Container Loading Rates

	Rate per metric ton	20' cont.	40' cont.
Paper, rolls	$7.84	11.4 mt/cont. $89.00	24 mt/cont. $188.00
Pulp, bales	$6.27	17.7 mt/cont. $111.00	20 mt/cont. $125.00

Summary

I hope this paper has helped to illustrate the means and costs of transporting various forest products to the foreign marketplace. The transportation modes are flexible to fill the needs of the high- or small-volume shipper. The economies of scale in ship design which have evolved in this later part of the 20th century have helped to keep the cost of transport low and retain the industries' ability to compete in these important markets.

Proportional Elements of Transport Costs

Land Transport	37%
Port Terminals	11%
Ocean Transport	52%

In the foreign marketplace, forest products transportation represents about 20 percent of the total cost of that commodity (at landed port of entry).

STRUCTURING AN EXPORT TRADING COMPANY

Richard V.L. Cooper

Introduction

The recent enactment of the Export Trading Company (ETC) Act presents a new challenge to the American business community. Part of this challenge is to find ways of structuring a trading venture. The structuring issue is especially timely now. Most discussions about ETCs to date have focused more on the reasons why American trading companies should and will be formed, rather than on how they will be structured. Yet, as the concept of the American ETC begins to take hold, it becomes correspondingly more important to consider how ETCs will be structured and operated.

The purpose of this paper is thus to suggest some ways for thinking about how to structure an ETC. This discussion is necessarily broad, since each specific ETC will be driven by its own particular circumstances. At the same time, there are certain factors that all investors must take into account as they consider how to establish a trading company. This paper is intended to provide a common analytic framework for incorporating these factors into the process by which a trading company is established.

There are three sets of issues that need to be addressed in the course of structuring an ETC: strategic factors, structural factors, and development and organizational factors. First, the prospective ETC investor needs to understand how an ETC might fit into his broader strategic objectives. Second, he needs to consider how the ETC might be structured to accomplish its objectives. Finally, he needs to understand what must be done to develop and organize this ETC. Each of these issues will be discussed in turn.

Before proceeding, though, let me note that much of this paper is devoted to the application of some basic strategic and business planning techniques to the development of a trading company. Although many of the issues and techniques that will be addressed here are thus a standard part of traditional strategic and business planning, I believe their consideration is important for two reasons. First, they provide a very useful way for approaching the ETC structuring problem. Second, to my knowledge, they have not before been applied to ETCs.

Strategic Factors

As one considers whether and how to establish an ETC, the most appropriate place to begin is with some important strategic considerations. An ETC investor must first define his objectives, constraints, and the trade-offs among them; assess the strengths and weaknesses of his existing organization; and then analyze the potential synergies and advantages created by alternative ETC investments.

Objectives and Constraints

Perhaps the most important strategic issues for any ETC investor to consider are objectives and constraints. Specifically, what does the investor hope to gain from forming an ETC and what constraints, both real and perceived, will be faced in the process?

There can be many reasons for forming a trading company. Although profit will be the driving motive for most, if not all, private investors, profit can come in a variety of different ways. Thus, it is important to consider how a trading company can contribute to the investor's overall profit objective. Moreover, some potential ETC investors may be not-for-profit concerns, such as port authorities, which will have objectives other than profit.

The most obvious reason for an investor to consider an ETC is to earn a profit from the trading operation itself. To be sure, the ETC's own profit will be an important consideration for any potential investor, but for some investors other factors are likely to be more important. For example, manufacturing and service companies may wish to use an ETC as a vehicle for increasing the market penetration and sales of their existing products or services. In this instance, profit may flow back to the investor, not from the trading operation itself, but rather from the increased volume of products or services sold through the trading company.

Similarly, some investors may see the ETC as a vehicle for entering, sustaining, or expanding international trade.

Others may see the ETC as helping to make use of excess capacity. Alternatively, corporations with an existing international presence may find that they can use an ETC to defray some of the costs of their present international infrastructure. Those with many divisions or subsidiaries already engaged in international trade my find the ETC useful for rationalizing and restructuring their international operations.

Banks and other institutions may decide to pursue the ETC opportunity for offensive or defensive reasons, or for both. Offensively, they may find that the ETC provides an attractive mechanism for expanding into new markets or for providing new services to existing clients, while building on their present strengths. Defensively, they may find that an ETC can help fend off encroachment from competitors who will offer the service if they do not.

These are just a few of the many possible reasons why some will choose to invest in an ETC.[1] The more important point, though, is that the investor needs to define his objectives carefully, since the optimal ETC structure will depend greatly on the investor's reason for investing.

Just as an investor needs to consider his objectives, he also needs to consider the constraints that will be encountered. Some constraints represent genuine obstacles. These will most likely be legal. For example, although the ETC legislation permits certain banks to invest in ETCs, there are limitations on such investment. Similarly, the Shipping Act constrains certain activities that some trading companies might want to undertake. For instance, an ETC cannot take a forwarder's commission on products in which he has a beneficial interest.

The more important constraints, however, are likely to be self-imposed. For instance, some investors may decide not to consider joint ventures. Others may be willing to join in a joint venture, but may decide against doing so with competitors. Some investors may be willing to consider only certain foreign markets, or may not want to engage in any importing. And most investors will impose financial constraints on their investment, such as requiring some specified rate of return.

Together, these objectives and constraints make up the backdrop against which any potential ETC investment must be

---

1. For a more complete discussion of these reasons, see Richard V.L. Cooper, <u>Potential Investors in Export Trading Companies</u>, Coopers & Lybrand, May 1982.

considered. Again, they are important because they help bound the problem.

Trade-offs

Although it is important to consider the objectives and constraints of any investment, these deserve special consideration in the case of ETCs. Because there is no set ETC format or structure, there is virtually an unlimited number of alternatives open to the investor. The investor's objectives and constraints thus become an important part of the process by which the investor can narrow the available options to a manageable number.

Concurrently with the consideration of these objectives and constraints, the investor needs to consider what <u>trade-offs</u> he is willing to make among them. This, too, is an important part of the winnowing process. Specifically, because it is unlikely that the ETC will be able to meet all the investor's objectives, or at least to an equal degree, it becomes important to understand which objectives and constraints are most important, and which can be sacrificed.

For example, and particularly important in the case of a trading company, the investor must consider his willingness to trade short-term profitability for longer-term profitability. The Japanese have been particularly adept at this over the past two decades, trading short-term profits for market share, thereby realizing longer-term profitability.

This short-term/long-term trade-off may be the most important that the investor has to consider. Trading companies are traditionally high-volume operations. If an investor wants his ETC to become a large-volume operation, he may have to sacrifice quick returns on the investment. Alternatively, if short-term profitability is of primary importance, the investor may have to settle for a lower-volume, more specialized ETC.

Consideration of the trade-offs among one's objectives and constraints is thus an important part of the objective setting process.

Existing Operations

The prospective ETC investor next needs to analyze the composition of his own organization. By understanding his own strengths and weaknesses, the investor can structure the

ETC to take maximum advantage of his strengths, while at the same time minimizing the impact of his weaknesses.

This self-analysis is especially important because most ETCs will be built from the bottom up. Thus, the task of structuring a new ETC can be simplified substantially to the extent that parts of the ETC's infrastructure can be adapted or borrowed from the investor's present organization. To that extent, the prospective investor should thus look to build on his existing organization.

To illustrate: if the prospective investor's organization already conducts trade with some countries, then the ETC's initial trading activities probably ought to focus on these countries. In this case, the investor is likely to have already established important trade channels in these countries, such as agents, distributors, and knowledge of these countries' markets. The ETC could almost surely benefit from these existing relationships.

Alternatively, because banks are expert in the various aspects of international finance, such as letters of credit and foreign currency exchange, a bank-sponsored ETC should clearly include financing arrangements as one of its major services.

Synergies

Finally, when considering participation in an ETC, the prospective investor should look for possible synergies with both his existing operations and his longer-term objectives. An ETC should not be viewed in total isolation, but rather in terms of how it could contribute to the sponsor's other activities, either present or planned.

For example, I suggested earlier that a multinational company could use an ETC to unify the present international trade operations of its divisions and subsidiaries. As another example, a multinational corporation could structure an ETC to take advantage of its present international infrastructure. This would not only make the ETC more efficient, but would enable the corporation to defray some of the costs of its international infrastructure.

These synergies are not limited to multinationals, however. For some prospective investors, the ETC might provide countercyclical or counterseasonal markets—that is, foreign markets to offset domestic recessions or off-peak seasons. Other companies may see the ETC as a vehicle for entering new overseas markets. By offering products complementary to their own, they can achieve sufficient trade volume to justify market entry. In this case, the ETC would be synergistic with the company's goal of entering more overseas markets.

Banks and other financial institutions eager to broaden their customer and service base are especially likely to find that an ETC can provide important synergies. By broadening the range of services provided by a bank, the ETC could provide the bank with a valuable tool for retaining its present customers and attracting new customers.

An ETC can thus complement an investor's other corporate objectives in many seemingly unrelated ways that may not show up in traditional financial analysis.

The Context for Developing an ETC

The development of a trading company is an open-ended problem. There is no prescribed format, structure, or approach. The prospective ETC investor, therefore, has great latitude in developing a specific trading company. Given the latitude available, the investor should thus seek to develop the ETC that most closely conforms to his other objectives, constraints, present operations, and potential synergies.

In summary, to maximize the value of the ETC, the investor would do well to focus on five key issues:

- <u>Objectives</u>: What are the investor's objectives, and how might an ETC fit into these?
- <u>Constraints</u>: What constraints, real or self-imposed, does the investor face?
- <u>Trade-offs</u>: What trade-offs among these objectives and constraints is the investor willing to make?
- <u>Existing operations</u>: What are the investor's present strengths, and how can he build on these to form an ETC?
- <u>Synergies</u>: What synergies can an ETC create for the investor?

Structural Factors

Having thought about the strategic factors that should underlie the development of an ETC, we can now consider how the ETC might be structured. Again, it is important to recognize that there is no prescribed ETC structure. It is therefore most appropriate to view this structuring problem in terms of the elements or factors that will collectively compose the ETC.

Although there are a myriad of specific factors that should be considered in the process of structuring a trading company, eight categories stand out:
- Products
- Markets
- Domestic Scope
- Activities
- Services
- Distribution Network
- Investors
- Financial Resources

Indeed, I will argue that once one has settled on the specifics of these eight factors, the ETC has been largely structured. The key to structuring the ETC, at least in a broad sense, thus depends on choosing the right mix of these eight factors.

Products

Choosing which products to export is one of the key decisions that any ETC must make. In making this decision, the prospective ETC investor must keep two things in mind: (1) the marketability of the candidate products for export, and (2) the ETC's access to those products, including its comparative advantage in marketing these products versus others. The first of these is obvious—there must be a foreign market for the products in order for the ETC to succeed in selling abroad.

The second is less obvious, but is also important. Specifically, the ETC needs adequate access to products. Other things equal, it will have more success exporting those products for which it has better access. For example, we would expect an ETC located in the Northwest to have more success in exporting wood products than in exporting electronic components. This is not because there is greater world demand for wood products, but because the ETC in the Northwest is likely to have better access to wood products, given the industry makeup of the Northwest. The reverse might be true for an ETC located in New England or Northern California.

For an ETC sponsored by suppliers—e.g., manufacturing companies, raw resource producers such as coal companies, agricultural concerns, and trade associations, among others—the ETC's product lines will be determined largely by the product lines of the sponsors. This will be the case for two reasons. First, the sponsors will want the ETC to focus on their product lines, since increasing their own international sales would presumably be one of the primary reasons these sponsors have entered into an ETC in the first place. Second, the ETC will likely have a comparative advantage in these product lines in terms of access to the specific products, knowledge of market demands, and knowledge of distribution channels. This is not to say that such ETCs will not also trade in unaffiliated product lines, for many in fact may do so, but rather that the sponsors' product lines will form the backbone of the ETC.

Not all ETCs, and maybe only a minority, will be sponsored by suppliers. Many will be sponsored by organizations unaffiliated with any specific product lines--e.g., banks, insurance companies, other financial institutions, and port authorities. For these, the choice of representing unlimited product lines will be both an advantage and a disadvantage. It is advantageous in the sense that these ETCs will be less constrained in their product choice, but disadvantageous in the sense that they will have less comparative advantage in selecting and selling their product lines.

The principles for choosing product lines will be the same for these ETCs, however, as they are for supplier-sponsored ETCs. Specifically, the ETC will need to examine both its access to products and overseas market demand to choose which product lines to export. In this case, product representation will not be determined by the products of the sponsor, but rather by the products of the geographic region, products suitable to the ETC's distribution system, or other factors that provide comparative advantage. For instance, if the ETC's bank sponsor already has substantial banking relationships with the manufacturers of certain major product lines, the ETC should consider including those product lines.

Other considerations can also affect the choice of products. An ETC must take into account the requirements of exporting particular products--for example, required servicing after sales. Technical expertise will play a role for some product lines, especially high technology products. Also, products sometimes will be dictated by countertrade and barter requirements once operations commence.

Finally, the ETC need not specialize in just one product line, or even just a few. To the extent that ETCs export multiple product lines, however, they should select, at least at the outset, product lines that have common attributes, such as similar transportation and distribution requirements.

The choice of product lines is important for at least two reasons. First, choosing marketable products is crucial to the ETC's success. Second, the choice of product lines will play a major role in determining the ETC's structure, including the markets that can be served and the ETC's infrastructure requirements.

Markets

The second important decision that an ETC must make concerns the foreign markets to which it will export. Markets in this regard can be thought of in terms of either geographic areas or types of buyers. From a geographic perspective, the ETC must decide whether to focus on one country, a few countries, a particular region of the world, or a group of countries with commonalities such as language (e.g., French-speaking countries). Alternatively, the ETC may decide to be truly worldwide.

Again, two factors will play an important role in this decision. First, and most obvious, the ETC must look for the markets that have the greatest potential. It would do little good to develop an elaborate strategy for marketing to an area of the world where there is likely to be little demand.

Abstracting from particular products for a moment, the markets in different regions of the world have both positive and negative characteristics. For example, Europe is clearly a large, established market, and the risks of doing business there are likely to be minimal. On the other hand, the European market as a whole is not likely to grow substantially in the years ahead (although the market for a given product may have considerable growth prospects). Moreover, the competition in Europe is already substantial.

The Pacific Rim may have the greatest potential. The countries of the region are generally prosperous, and are expected to become more so. As a result, the demand potential is clearly substantial. Language and customs pose greater obstacles there than in a western culture such as Europe, however. Again, the competition is fierce, especially from the Japanese, Koreans, and Taiwanese.

Looking to the future, many analysts believe that Latin America may be one of the most promising markets of all. The growth prospects could be substantial. In addition, the United States enjoys some particular advantages in Latin America relative to other exporting countries: geographic proximity, a similar heritage, less competition from other nations, and the fact that many U.S. firms already have a foothold there. At the same time, there are clear risks in doing business in Latin America, especially given the historic political instability of many countries in the region, and the countries of the region are by no means wealthy.

These are certainly not the only areas of the world; nor are they the only ones upon which an ETC ought to focus its marketing efforts. Some U.S. trading concerns have done well in the Middle East. Parts of Africa also may hold promise for certain export activities, although in general the markets there do not seem as promising as in other parts of the world.

In selecting markets, the ETC should of course look beyond general market characteristics to the market potential for its particular mix of products. Moreover, since many parts of the world may hold promise, the ETC needs to look for those markets in which it is likely to enjoy the greatest comparative advantage.

To determine where it has a comparative advantage, the ETC needs to look to its own strengths. For example, if the ETC's sponsors already conduct trade internationally, then the ETC ought to look first at those countries where its sponsors already have trade relationships. Large banks and other financial institutions, for instance, usually have offices or correspondent relations abroad, as do the larger ports in this country. Even "sister city" relationships could be beneficial. Although many are only ceremonial in nature, they nevertheless may provide a useful beginning.

Other possible strengths of the ETC deserve attention. For example, the ETC's key personnel may have particular language capabilities. Geographic proximity will likely play a major role in determining where many ETCs trade. West Coast-based ETCs are likely to have more success in the Pacific Rim than in Europe, other things equal, while ETCs from the Southeast are likely to fare well in Latin America.

Although it may be most common to think of markets in terms of geography, markets can also be categorized according to the type of buyer. For example, rather than focus on particular regions of the world, a food products ETC might focus on such different types of buyers as international hotel chains, fast food outlets, airline caterers, ship chandlers, or foreign military establishments. Many such buyers have centralized purchasing authorities, thus enabling the ETC to focus its marketing efforts on a limited number of high priority target clients. Of course, an ETC might develop a marketing strategy that focuses on a combination of geographic areas and types of purchasers.

Finally, it is important to recognize that the choices of markets and products cannot be decided in isolation from one another. Nor can we say that one will necessarily take precedence over the other. Some ETCs will be product-driven while others will be market-driven. Even in these cases, both products and markets have to be considered together. To illustrate, an ETC with supplier sponsors, such as a group of companies or a trade association, is likely to be product-driven--that is, the nature of its sponsor will determine what general product lines are to be exported. Given these product lines, it must then proceed to look for the most appropriate markets. Alternatively, a bank-sponsored ETC may have little predisposition toward the specific products to be exported, but might know the markets upon which it wants to focus. Given these markets, it must then search for the most appropriate products to export. Finally, some ETCs will have little predisposition toward either specific products or specific markets, and must thus examine the two simultaneously.

The point here is again twofold. First, selecting the markets for exporting is an important decision that the ETC must make. Second, the decision regarding the markets for export will have a substantial impact on the structure of the ETC. It will affect the products that can be exported, the ETC's infrastructure and export financing requirements, and the need for such operations as countertrade.

Domestic Scope

The third element that needs to be considered in structuring an ETC is its domestic scope--that is, the area of the United States from which the ETC will draw the products and services to be exported. In other words, the ETC must decide the geographic area it wants to serve.

To illustrate, an ETC might be structured primarily to serve a particular metropolitan area, such as the Philadelphia area or the Los Angeles area. Such an ETC would focus its primary marketing efforts on the chosen metropolitan area, although it may occasionally export products from other areas as well.

Alternatively, the ETC might take on a somewhat larger geographic region of the country. One can envision, for instance, regional ETCs serving such areas as the Northwest, the Southeast, the Ohio River Valley area, and so forth.

ETCs need not be limited to a particular state or region, either. Indeed, some nationwide ETCs probably will be developed. A nationwide orientation is especially likely for product-driven ETCs such as those sponsored by a trade association. A general trading company could also be nationwide. However, a truly national general trading company (not product or market specific) would need to be extensive, and thus expensive to develop.

Domestic scope will be a particularly important consideration for those ETCs without a natural product base. A bank or port authority may have little predisposition toward any particular products. The key to choosing products for such an ETC may lie in the way it defines its domestic scope.

Again, it is important to make domestic scope, or geographic coverage, an explicit part of the ETC structuring decision. The reason for this is that the extent of domestic geographic coverage will have an important effect not only on defining the ETC's domestic client base but also on such factors as domestic infrastructure requirements.

Activities

A fundamental part of the ETC structuring process consists of determining the activities in which the ETC will engage. Activities in this sense refer to the types of international trade that the ETC will conduct. There are many trading activities that the ETC can perform in addition to exporting. Indeed, the large general trading companies of other nations engage in the full spectrum of international trade activities.

The types of trade activities an ETC can pursue include the following:
- Export--i.e., exporting from the United States to another country
- Import--i.e., importing from another country into the United States
- Third-country trade--i.e., trading between two nations other than the United States
- Barter, or countertrade--i.e., taking payment in goods rather than currency
- Switch trade--i.e., using a third country's currency in the trade between two nations

Some ETCs--primarily Webb-Pomerene-type ETCs and other manufacturer-sponsored ETCs--will engage solely in exporting. Others, however, will engage in at least some of the other trade activities that large foreign trading companies pursue, either because they have to or because they want to. To decide which, if any, of these other activities will also be utilized, the ETC will need to consider two factors: (1) what activities will be required to make the ETC succeed, and (2) what activities are consistent with the ETC's objectives.

As an illustration of the former, ETCs that trade with the Communist bloc and certain less-developed countries are likely to find that countertrade is essential, given those countries' lack of foreign currency. For very large transactions, such as aircraft purchases, bartering may be required even for some developed countries. Indeed, countertrade is the reason why multinationals such as General Electric and Control Data entered into their trading companies in the first place: in order to sell their primary product in a specific country, the multinational had to accept payment in goods produced by the purchasing country. Most multinationals today have some sort of countertrade operation.

In other cases, an ETC will decide to pursue some of these other trade activities simply because of the business opportunities such activities create. A port-sponsored ETC may try to balance imports and exports to optimize the use of the port's facilities. Alternatively, consider the case of an ETC that develops a substantial foreign infrastructure for its exporting activities. The marginal cost of importing from or engaging in third-country trade with those countries where the ETC has facilities may therefore be very little. These ETCs may find themselves naturally drawn into these other activities.

More generally, if the United States is to spawn large general trading companies that can compete effectively with the world's other giant trading concerns, then at least some American trading companies will have to engage in the entire spectrum of trading activities. This is not to say that such American ETCs need to start out on such a grand scale, for few probably will. Rather, it is likely that the more successful ETCs will find themselves gradually drawn into activities beyond exporting, and will thereby begin gradual expansion, in terms of markets, products, domestic scope, and trading activities.

Finally, although many ETCs are likely to find themselves participating in some nonexport trade activities, those certified under the ETC Act must be mindful of the constraints imposed by the legislation. An ETC with a bank sponsor, for instance, must earn at least half its revenues from exporting. There are ways of overcoming this constraint, but they must be considered carefully when the ETC is structured. Alternatively, those ETCs seeking antitrust protection under Title III of the Act should be aware that only their exporting activities can be certified, not their importing or other activities.

It is thus important to consider carefully which trade activities will be pursued, and how the implications of these decisions will affect the structure of the ETC. Consider, for example, the case of countertrade. If the ETC wants to trade with countries short of hard currency, then it must structure itself to undertake countertrade. Alternatively, if the ETC makes a policy decision not to engage in countertrade, then it may limit its opportunities in those markets. In either case, this example illustrates how the activities the ETC decides to pursue (or not pursue) can affect its structural requirements.

Services

Just as the ETC must decide on the range of its activities, it will also need to decide on the scope of specific services to provide its customers. A full-service ETC, or "one stop shop," will provide the entire array of export services, including:
- Buyer identification
- Product adaptation and packaging
- Export financing
- Shipping, insurance, and customs documentation
- Warehousing, distribution, and after-sales servicing
- Market intelligence

Moreover, such an ETC must also be prepared either to take title to goods or to act as a commissioned agent for any particular transaction.

Not every ETC, however, will need to provide all these services. Some may instead choose to specialize in a subset of these services. For example, the "hub" or "core" ETC model consists of an ETC that specializes in the logistics of trade, including shipping, insurance, customs documentation, warehousing, and distribution. Such an ETC might be sponsored by a port authority, transportation company, freight forwarder, or other organization that is regularly involved in some aspect of international trade logistics.

Full-service ETCs will probably be more common than those specializing in just a few of the trade services, at least in the long run. Small- to medium-sized businesses making up the ETC's client base are typically inexperienced in most aspects of international trade, and thus require the full range of services. The ETC need not provide all these services itself, however. Smaller ETCs in particular are likely to specialize in marketing and sales, at least initially, while subcontracting to others such as freight forwarders for such export logistics as shipping and insurance. Although the needed export services would thus be provided by a number of organizations, the ETC would be the single point of contact for the client--in effect being a "one stop shop."

The same theme underlies this discussion as it has for other structural factors that have been described. The choice of which services to provide is itself an important decision and will have a substantial impact on the structural requirements of the ETC.

Distribution Network

The sixth element that must be considered in structuring an ETC is the ETC's distribution network. This includes (1) domestic transportation, collection, and warehousing, (2) overseas transportation, and (3) transportation and distribution within the foreign market. There are a number of components in the overall distribution network:

- Transportation--railroads, trucking, steamship lines, barge lines, air carriers
- Ports and riverports
- Warehousing
- Distribution
- Freight forwarding and customhouse brokering
- Wholesaling
- Retailing

The ETC must determine, first, which of these elements are needed and, second, how best to provide them.

Although having a smooth, well-functioning distribution system will be important to the success of most ETCs, in some cases this distribution system will be of special importance. For example, for products whose transportation costs are high relative to the price of the product itself, transportation considerations will assume greater importance. This will often be the case for bulk products such as raw resources, grain, and lumber. In the reverse case of light, high value-added items such as electronic components, transportation will be less important, but insurance and security will become more important.

A large, general-purpose ETC will need to develop a full distribution network, since it will presumably trade in products with a wide variety of characteristics. Again, however, not all ETCs will start on such a grand scale. In these instances, it may instead make sense for the ETC to focus on products with common distribution requirements.

For example, one can envision an ETC that specializes in bulk products. Such an ETC might focus primarily on coal, lumber, and grain--all products that are shipped in bulk. Other products with common characteristics that might be considered include perishable goods, high value-added items, and certain capital goods.

Although the distribution system will generally derive from the mix of products, markets, and domestic coverage, in some cases the transportation and distribution system will be a driving factor. For example, an ETC sponsored by a railroad should consider structuring itself to maximize the use of the railroad. In this case, the products for export and the domestic scope of the ETC would be determined mainly by the appropriateness of domestic rail shipment and the domestic reach of the railroad.

The distribution system is therefore not only the key to the eventual success of the trading company, but will in some cases even play an important role in the actual structuring of the ETC.

Investors

The seventh element that needs to be considered is the ETC's investors. Those who invest in the ETC may collectively be one of the most important determinants of the ETC's structure, since the objectives and constraints of the investor(s) will play perhaps the leading role in determining the broad mission and purpose of the trading company.

In the case of an ETC sponsored by a single investor, the objectives, constraints, and comparative strengths of the sponsor will be the most important determinant of what the ETC is and how it will be structured. The investor in this instance will need to evaluate his strengths vis-à-vis a trading company and build from these strengths in accordance with his objectives and constraints.

The same principles apply in the case of multiple investors. Additional investors can bring important new strengths to complement other investors' inadequacies. For example, additional investors could bring an export financing capability, products to export, or important parts of the required domestic or foreign infrastructure. They can also bring new contacts, thus adding market potential. The price of multiple investors, however, is dilution of profits and control of the trading enterprise. Investors must decide whether the benefits brought by additional investors are worth the cost. Note, however, that the corollary to these "costs" is risk sharing--that is, by bringing in additional investors, the risks of the trading venture can be shared.

Additional investors also bring something else: the possibility of redundancies and accompanying conflicts among the partners. For example, two banks from the same area could participate in an ETC, but are likely to bring similar strengths to the trading company. Redundancies such as these need not preclude investment in the ETC if they are dealt with at the outset. The charters, obligations, authority, and responsibilities of each partner in the joint venture need to be negotiated to define roles clearly for each partner before the ETC commences operations. This is important for all cases where more than one investor is involved, and especially so when the partners either have common characteristics or differ in their objectives.

Again, the nature of the investors can play an extremely important role in shaping the structure of the ETC. For example, an ETC with a foreign trading company as one of the investors will need to do far less in structuring an overseas marketing and distribution system than the ETC formed without a foreign trading company as one of the sponsors.

Financial Resources

The final factor that all ETC investors must consider is the amount of financial resources to be committed to the ETC. The level of financial commitment is important, not only because it sets the ETC's capitalization, but also because it will play a critical role in determining the size and scope of the ETC. For example, the number of markets the ETC can enter, the number of products it can carry, the domestic scope of the ETC, and the range of services provided by the ETC are all dependent at least in part on the financial resources available to the ETC. The level of financial commitment is thus one of the first issues that all ETC investors must resolve. Calculation of financial requirements will be discussed later in this paper.

Structuring the Trading Venture

Together, the eight factors just described will serve as the structural backdrop for all ETCs. Although trading companies will differ substantially in the particulars of these factors, every ETC will need to address these eight factors. Moreover, the decisions that are made regarding the particulars of these eight factors will end up shaping the basic structure of the ETC.

This fact raises two important points regarding how to structure a trading company. The first is that there are many different kinds of ETCs that can and will be developed. The prospective investor must therefore engage in a process of reducing the many options that he faces. In a sense, the vast variety of available options is both an advantage and a disadvantage. The advantage is that, by not having a prescribed format, investors can be innovative in the way they approach the problem. Adroit investors will thus be able to develop ETCs uniquely suited to their own needs and objectives. As a result, we are likely to see a variety of different ETCs, many of which have not yet even been envisioned.

The problem, of course, is that having so many choices complicates the decision process substantially. To some potential investors, the vast array of options may well be bewildering and discouraging. The fact that there is no prescribed format could thus discourage the formation of some trading ventures, at least at the outset.

The second important point to be gained from this discussion is that, although all eight factors must be considered in developing an ETC, different investors will be driven by different factors. A multinational corporation, for instance, will likely be driven largely by the broad product lines it presently sells and the foreign markets in which it now operates. A transportation company, on the other hand, will be driven more by the distribution network of the ETC, and will thus be more likely to select at least the ETC's primary products according to how well they conform to the transportation company's distribution system. A retailing chain is most likely to enter markets where it already has agents or operations, usually for import to the United States, and would be most likely to focus on consumer products. A port authority will probably rely most on the products that are traditionally shipped through the port. These examples illustrate only a few of the many ways that the various structural factors can come together to suggest specific ETC models.

Moreover, not all investors will be driven by a specific factor. Banks and other financial institutions may be the most notable example because they are likely to be among the least constrained of all investors in terms of the range of options they can consider. For instance, although a bank will probably rely mostly on products from its general geographic vicinity, for most banks the choice of products will still be quite large. Thus, in a sense, banks and other financial institutions will have the most difficulty in developing an ETC, simply because their options are so open. Yet, the potential opportunities will probably be sufficient to entice at least some banks to overcome this obstacle.

Developing the ETC

After the ETC has been broadly structured, as just described, the ETC's investors must resolve a number of specific issues in order to develop the ETC. Four in particular stand out: (1) filling out of the structure, (2) organization and control of the ETC, (3) financial requirements, and (4) growth strategy.

Filling Out the Structure

To be successful, a trading company must possess a number of important attributes. It needs financial resources as well as a credit issuing capacity. It needs access to products for export. It further needs support, transportation, and distribution networks, both domestic and foreign. An ETC must have experience in the actual conduct of international trade, tailored to the laws, customs, and business practices of its trading partners. It must know international finance, and it must have access to foreign market information and market needs.[2]

Few individual investors will have all these attributes themselves. This raises the natural question of how to best fill out the structure of the ETC. Sometimes this issue is cast in terms of "acquisition versus de novo"--that is, should the needed expertise be acquired or developed anew? In reality, the choice is more complex, and more interesting.

To begin with, the prospective investor must first undertake a careful inventory of his own organization's strengths and weaknesses relative to the ETC he envisions developing. He must them determine what his ETC will need.

---

2. For a more complete discussion of these aspects, see my Potential Investors (cited in note 1).

Finally, he must assess his organization's strengths and weaknesses in relation to the ETC's requirements, thereby identifying the gaps in the ETC's structure.

For those areas where the prospective investor(s) falls short of the ETC's needs, there are five options for filling out the structure:

- Develop in-house
- Hire
- Purchase
- Contract out
- Form a joint venture

For some areas, the investor may want to develop the required expertise in-house. Alternatively, when the missing element is something that can be satisfied by particular personnel, the ETC can hire the needed individuals. The expertise of certain trade specialists will often be obtained this way.

When the investor needs more than just personnel, he can consider acquiring an entire organization with the needed experience, capacity, and infrastructure. The most likely candidates for acquisition are export management companies (EMCs), since they are among the few independent institutions in the United States with considerable experience in international trade. At the same time, the acquisition of EMCs should be approached with caution, since EMCs sometimes offer little beyond the experience of the proprietors themselves. This not only can make it difficult to price the acquisition, but also can reduce the value of the acquisition once it is completed. The latter problem can be mitigated either by a lock-in agreement with the EMC's proprietors or by acquiring an EMC large enough to have value beyond the immediate experience and contacts of the proprietors.

Larger ETCs may also consider acquiring other parts of the needed support and infrastructure, such as key transportation, warehousing, and distribution facilities. Investors other than financial institutions may also acquire a finance company.

In many instances, the ETC will wish to contract out for the required services. This is particularly likely for part or all of the transportation and distribution system. Recent revisions in railroads' authority to negotiate rates, plus trucking deregulation, may make this approach particularly useful for domestic transportation. Other candidates for contracting out include banking, freight forwarding, domestic and foreign warehousing, and overseas sales agent service, among others.

Finally, though by no means a last resort, the investor can seek joint venture partners. Bringing in other investors can add to the strength of the ETC, enhance its domestic and foreign contacts, and enable the various investors to share the risks. The disadvantages are dilution of profits, dilution of control, and possible conflicts among the partners, although the last of these can be at least partially offset by careful planning before formation of the partnership.

One can envision a wide variety of partnerships, some of which have been previously mentioned. Another deserving particular attention is a partnership between the American ETC investors and a foreign trading company. Although this approach will not always be the most appropriate path to follow, in some instances it will have the substantial advantage of tying the American ETC investors to an organization very knowledgeable about foreign trade, one with extensive marketing knowledge abroad, and one with a foreign infrastructure already in place.

In sum, there are a variety of ways to fill out the structure of the ETC. The most appropriate choice will depend upon the particular circumstances and the needs and objectives of the investors.

Organization and Control

As with any business--or, indeed, with any organization--it is important to establish a proper organization and management structure. Although this may seem obvious, it is nevertheless useful to begin by specifying some basic management and control principles. There is a wide variety of generally accepted business management principles, so rather than address all of these, I will focus on those that have particular importance for a trading company.

First, roles and responsibilities must be clearly defined. Although this principle is of obvious importance to all organizations, it assumes a special importance in the case of an ETC sponsored by more than one investor. The failure to do this properly can result in needless redundancies, inefficiencies, and business conflicts.

Second, an ETC, more than most other businesses, must be extremely flexible. By the very nature of the business, traders must be opportunistic. They must recognize opportunities when they arise, and must be prepared to act quickly. Thus, while a trading company should seek to establish long-term customer relationships, much of its business may be derived from new opportunities, which will sometimes appear quite unexpectedly. This is likely to be true for ETCs in their early phases, before they have had the chance to develop an established customer base. The ETC's management structure must therefore be flexible enough to adapt to opportunities when they arise, and to act on them quickly before they evaporate.[3]

Third, the fact that trading is such a fluid business makes it all the more important to have the proper management controls in place and to maintain accountability. In other words, although the ETC must remain flexible enough to take advantage of opportunities when they arise, it must also maintain controls and accountability. Indeed, it is precisely because of the flexible nature of the business that controls and accountability are so important. These controls should not take the form of lengthy management evaluations, but must rather take the form of guidelines to which traders can respond.

---

3. ETCs that have been certified under the ETC Act need to be cognizant of exactly what has been certified, and how new opportunities that arise fit into the ETC's certificate. ETCs face a number of options in this regard. First, they can at the outset seek a broadly worded charter. If the certificate is awarded, the ETC needs to worry less about whether these new opportunities conform to its certificate. A broadly worded application, however, may have less chance of being accepted. Alternatively, an ETC can file for amendments to its certificate as the need arises. The problem with this is that, even though the government has promised a speedy review process, approval may still not come quickly enough to take advantage of opportunities as they arise. Third, the ETC can proceed to pursue new opportunities, irrespective of its certificate. In this regard, it is important to recognize that going outside the ETC's certificate is not in itself illegal. It only means that those activities are not covered, and are thus open to suit. When considering new trading opportunities, the ETC thus needs to compare the business potential of the opportunity with the risks of going outside its certificate. Fourth, the ETC can pursue a combination of the second and third options: namely, proceed without certification while simultaneously filing for an amendment to its certificate. Finally, it is important to remember that these comments apply only to those ETCs that have been issued certificates under the provisions of the ETC Act.

Finally, and again perhaps even more so for trading companies than most other businesses, the ETC will need a meaningful financial and management information system to properly measure the ETC's performance. With such a system in place, management can compare the ETC's actual performance with its objectives. Such a system must be tailor-made to the specific ETC's objectives; traditional financial and accounting measures, though still appropriate for such things as tax filings, may not properly reflect the ETC's performance relative to its basic purposes.

For example, consider the case of an ETC designed not to earn a profit on its trading operations, but rather to generate profits to the investor through increased sales of the investor's product or through reduced costs of the investor's international infrastructure. In this instance, management should not view the lack of profit from the trading operations itself as an indicator of poor performance, for the ETC was not designed to earn a profit in this sense. This does not mean that the ETC should not be held accountable. Rather, it means that the ETC should be evaluated in terms of its original objective--i.e., increased sales (or reduced costs) for the parent investor--and the management information system should reflect this. Similarly, banks may sponsor an ETC both to attract new customers for their traditional operations and to increase customer retention. The management information system for these ETCs should be developed accordingly.

Not all ETCs, however, will be developed for their indirect benefits. Some will be developed to earn a profit from their trading operations, and they should be judged accordingly. The financial and management information systems developed for these ETCs should therefore reflect this objective, plus any others that the ETC may have.

Having outlined some basic management principles that have special applicability to ETCs, we can now consider the organizational structure of an ETC. Many different organizational forms will undoubtedly be devised, but it is probably fair to assume that all will have at least three points in common: an operations function, a marketing function, and a finance function. The exact structure and organization of these functions will vary from ETC to ETC, but each will need these three functions.

Lastly, we can consider the legal structure of the ETC. An ETC sponsored by a single investor can be either incorporated as a wholly owned subsidiary or created as a division within the parent. The form used will depend primarily on the sponsor's preferences regarding organizational form, the amount of liability the parent is willing to assume, and the tax implications. For an ETC sponsored by more than one investor, there are several options: incorporation, subchapter S, partnership, and cooperative. And, within each of these, there are several possibilities, e.g., for a corporate form, the mix of common and preferred stock. As with the ETC sponsored by a single investor, the form adopted for a particular ETC with multiple investors will depend on such factors as the sponsors' preferences for the different alternatives, their willingness to assume the liability of a partnership, the cost of raising funds, and the tax implications for the investors.

Financial Requirements

One of the most frequently posed questions about ETCs is, "What will it cost?" Regretfully, there is no ready answer, as each situation will differ. It is conceivable, for instance, that a small ETC could be launched for less than $1 million. At the other extreme, it is not unreasonable to assume that a large-scale general trading company with operations around the world and throughout the United States would require well in excess of $100 million to initiate operations.

In general, the capital requirements for the ETC will depend on two factors. First, they will depend on the scale, or scope, of the ETC. A larger, more extensive ETC will, other things equal, obviously cost more. Second, capital requirements will depend on what the ETC already has in place. For example, an ETC that can tap into the existing international infrastructure of one of its investors will clearly require less initial capital than one not so equipped.

To determine the capital requirements for a given ETC, it is thus necessary, first, to assess the costs required to support the ETC's method of operations, products sold, markets served, activities and services, and infrastructure needs. Costs should represent the total cash expenditure requirements for annual operations at estimated sales volumes, less the cash flow generated from the estimated annual revenue stream. Total capital requirements can then be determined by subtracting from these costs any savings that arise from using facilities and services that the sponsors can make available to the ETC.

After determining the level of capitalization required, the ETC's sponsors next need to decide on the sources of capital--particularly the equity and debt mix. Again, though, there are no ready rules of thumb, since different ETC sponsors are likely to settle on different approaches. Some, in fact, may try to emulate the foreign trading companies, which tend to be highly levered. Debt/equity ratios of 20 to 1 or more are not uncommon among foreign trading companies, and these ratios sometimes run even higher.

I do not believe that the majority of U.S. ETCs will be so highly levered, however. For one thing, much of the foreign trading companies' high leverage derives from the unique relationships that they have with their banks.[4] For another, the high interest rates experienced during the late 1970s and early 1980s have made American businessmen cautious about being too dependent on debt. As a general rule, though, the nature of the trading business is likely to cause ETCs--especially those that take title to goods--to be more highly levered than traditional U.S. businesses. Bank participation in ETCs, as allowed under the ETC Act, should help to facilitate this. Nevertheless, American trading companies are not likely to be as highly levered as their foreign counterparts. Ultimately, each ETC will need to determine the optimal composition of its own capital structure, and assess its ability to attain that debt-equity mix. The answer to both questions will depend on such factors as the ETC's method of operations, whether it takes title to goods, the financing it provides to suppliers and purchasers, and its access to funds, as well as on the structural aspects of the ETC.

Thus, unsatisfactory though it may seem, we simply cannot say precisely what an ETC will require in the way of initial capital or what the composition of that capital will be. There are numerous options regarding possible ETC models, and each will have different financial requirements.

Growth Strategy

Finally, consider the development of a growth strategy for the ETC. Although it might at first appear premature to specify a growth strategy before the ETC is formed, I

---

4. For a more complete discussion of the trader-banker relationships abroad, see Richard V.L. Cooper, "ETC-Bank Relationships: Lessons from Foreign Trading Companies," remarks to the U.S. Chamber of Commerce Conference on Export Trading Companies, Washington, D.C., November 30, 1982.

believe that careful consideration of the possible alternative growth strategies can shed light on the best way for the ETC to initiate its operations.

Conceptually, we can think of three basic growth strategies. Under the first, the sponsor(s) would plan on a large general trading company, and begin full-scale operations as such from the outset.

The second strategy, even for those ETCs that eventually plan to become large general trading companies, calls for a more gradual development of the ETC. The ETC could begin operations by exporting to just a few countries, possibly even just one, or by exporting just a few major product lines. As the ETC gains experience, new countries and/or product lines could be added gradually, either as a result of taking advantage of new opportunities when they arise or as a result of preplanning. For example, the ETC might begin by exporting exclusively to the Philippines, then plan on adding other countries in the East Indies area, then other countries in the Pacific Rim, and, finally, countries in other parts of the world.

Under the third strategy, an ETC could begin by working on a deal-by-deal basis, acting more as a broker than a trading concern. As it gains experience, the ETC could gradually move more toward a trading operation, and then expand according to the second strategy.

The first strategy is clearly the riskiest of the three. It would require the greatest commitment of resources, and thus stands to lose the most if not successful. On the other hand, if successful, it could achieve the greatest market penetration and promises the greatest potential payoff. This route will be followed rarely, if at all, simply because of its risk. Those most suited to this approach are the major multinational corporations and, possibly, the large money center banks, especially if teamed with a foreign trading company. The reason for this is that these are the ones with the most infrastructure in place before the ETC is initiated. As a result, these ETCs would have to incur smaller incremental costs than those without an existent international infrastructure.

Just as the first strategy is the riskiest, the third is the least risky. It requires a relatively modest initial expenditure of resources, mostly in the form of time and expenses from the principals. This approach also promises less in the way of future payoff, since it could prove difficult and time consuming to translate its activities from brokering to trading. The ones most likely to follow this approach are the venture capitalists, although some banks and others also may employ it as well.

I believe that the second approach is the one that will be used most often. Although not without risk--indeed, no new or existing venture is--it provides a way of managing the risks, gradually developing the required trading experience and capability, and, perhaps most important, gradually developing the needed management expertise. In fact, the greatest difficulty in developing American trading companies is likely to be the lack of trading company management expertise in this country. The second growth strategy, although maybe without quite as much profit potential as the first, explicitly allows for gradual development of this essential ingredient.

Conclusions

Structuring and developing an ETC will not be an easy business--few new ventures are. This is especially true for ETCs, however, if for no other reason than the fact that the prospective ETC investor faces so many options. Add to this the complexities and intricacies of the trading business, and one must conclude that the prospective ETC investor faces a formidable task.

Yet, the obstacles will be manageable for many. With thoughtful planning, knowledge of one's own strengths and weaknesses, and a matching of these to the market needs and available resources, the prospective investor has a good chance of developing a viable and prosperous trading enterprise. The many potential options mean that some, perhaps many, investors will find the appropriate mix of ingredients for their circumstances to develop a successful ETC.

Acknowledgments

I am grateful to Roy Spiewak, Philip Odeen, and Catharine Findiesen for their many helpful suggestions and ideas.

INTERNATIONAL WOOD MARKETS: A EUROPEAN VIEW

John Wadsworth

The brief offered by the symposium organizers to this author was to review trends in world trade from a European point of view. A number of factors conspire to make such a view somewhat subjective. First, working for a British specialist consultancy company where 80 percent of our revenue comes from outside the United Kingdom means that we are subject to a number of international influences. It becomes increasingly difficult to separate one's own view from the opinions and commercial aspirations of the client companies (who are usually medium to large forest products producers). Their views naturally reflect their vested interests, but these vested interests are often in conflict. Take, for example, the sawmillers in one country who rely on imported logs and the log suppliers in another country who wish to establish sawmills and sell the first country added-value lumber products. Consider also the waferboard producer whose principal market is that served by his neighboring plywood mill.

In exporting, the frequent conflict is between a range of national suppliers of basically undifferentiated products. Most wood products fall into this category, such as sawn softwood (the majority of grades and species), medium-light tropical hardwoods, particleboard, and commodity grade plywood. To be a successful exporter, under these circumstances, requires a strong sense of purpose and a belief in the superiority of one's products. In dealing with these companies, the consultant can begin to feel like a Trilby faced by a polyglot collection of Svengalis.

Second, the distinction between an Englishman and a European is significant. To some observers, the group of mainland European countries cannot always be described as harmonious. Even when Europe reaches a uniform opinion (albeit temporarily), the British often disagree with them all. This is no recent phenomenon, as students of Roman, medieval, and modern history can confirm. As it is with political history, so it is with the forest products industry. Britain in particular has relied on imported lumber (and other wood products) for centuries. Europe tends to be more self-sufficient. Outside the U.S. we are the largest single softwood lumber importing country in the world, since we provide only about 12 percent of consumption from domestic sources. A very high proportion of our hardwood lumber originates from overseas, and we have found it difficult to establish a wood panel industry. (Again, outside the U.S. we are the largest plywood importer in the world.)

Thus an Englishman is well accustomed to the international trade in wood products, and his point of view will often reflect prevailing world views (at least the buyers' views). Nevertheless, many of his European colleagues in Paris, Amsterdam, Hamburg, or Milan are equally familiar with international trade. Together they deal with sellers from Finland and Fiji, Sweden and Switzerland, Singapore and South Korea, Brazil and British Columbia, India and the Ivory Coast, Moscow and Malaysia. With a few notable exceptions this familiarity is not evident in the U.S. industry.

Why this situation exists is not hard to discover. It lies with the size of the American domestic market. American forest product producers, for many years, have serviced the very large and growing market on their doorstep. The industry has long regarded the export market as a residual--something to fill spare capacity in times of local recession. Nothing better illustrates this point than the consumption and export of softwood lumber (see figure 1).

Figure 1 plots exports lagged one year against consumption. This has been done to take account of the time it takes (some) sawmillers to react to a poor domestic situation by increasing exports. This is the highest volume export product, and many organizations have, in addition to individual companies, worked very hard in such areas as tariffs, standards, and customer awareness to create a better trading climate. The graph shows clearly that the level of exports declines as domestic consumption rises, although the general trend in exports is upward. But even for those products that exhibit a steadily rising export performance, the total volume exported against production is a very low proportion. Softwood plywood is a good example of this. Figure 1 also suggests that the pattern of softwood export sales is likely to repeat itself in 1983.

There are a number of sound commercial and psychological reasons why the American industry behaves in this way:

- Hitherto the domestic market has been sufficiently large and growing steadily for the industry as a whole not to need export sales.
- The domestic market is reasonably homogeneous and supports "long-run" production of particular grades of product. This is a commercial fact which in turn creates a psychological barrier not to convert plant to short-run and varied production for exports when it will only have to be reconverted when the domestic market picks up.
- Physical distribution is (or was until recently) a problem in which West Coast mills--unless they organize their shipping--are expensively distant from European and Middle Eastern markets. The southern and southeastern mills did not have adequate port and shipping facilities until recently.
- Even with domestic markets in recession, the time taken for mills to create export sales could be too long to have any worthwhile effect before the next upturn appeared. Only those mills with existing sales and distribution facilities and knowledge of the customers can benefit in a reasonably short space of time. Mills that have never exported need to lay early plans; in the past this would have meant planning for exports when there was no immediate necessity. Few companies are blessed with managers of such prescience.

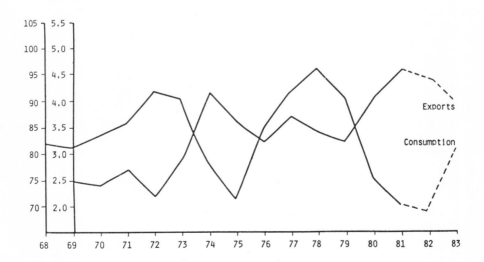

FIGURE 1. U.S. softwood lumber consumption and exports (million cubic meters).

Herein lie the themes of this paper: (1) some reasons why U.S. wood products companies should actively plan to start exporting (or export more), (2) some considerations when approaching export markets. In considering why U.S. companies might turn to exporting, it is not my intention to be unduly alarmist but rather to offer a series of opinions to stimulate discussion on a wider scale than perhaps has been done hitherto.

World Economies

The fearful world recession has finally caught up with just about every nation on earth. There are hints that 1983 and 1984 will be better. Indeed many eminent spokesmen are heralding economic recovery on a broad scale. But the question to ask is "better than what?" If they are only as bad as 1980, 1981, and 1982, mill closures will become permanent, and producers and users alike will be out of business.

The scale of growth required for most sectors of virtually all world economies to get us back to the position of the early and late 1970s is anywhere up to 40 percent. There is not the space in this paper to catalog all appropriate examples, but one very close to home will illustrate what I mean. The U.S. consumed 218 billion board feet of softwood lumber in the six years up to and including 1981. To match this you will need to sell 36.3 billion board feet in each and every year for the next six years to 1987. Your industry has managed to achieve that level in only six of the past sixteen years. Do you believe you can achieve these minimum levels every year--especially when they were missed in 1982 and probably in 1983?

Lower oil prices are seen as a stimulus to economic recovery. Certainly it is good news for Japan and West Germany. But I wonder what the reaction is in Texas? What about the U.S. banks with outstanding loans to Mexico? What about U.S. oil and construction companies operating in the Middle East? The amount of money these countries and others (like the U.K.) will have to reflate economies will be reduced. What will happen to a strong international recovery then?

Another significant factor to consider is inflation. Balanced budgets are rare events these days. Despite government expenditure cuts, they have enjoyed only limited success in restricting total spending. This is only one element of inflation. There are several others, each of which governments will attempt to come to grips with because inflation is an ever-present ogre and a key political issue.

__Exchange rate__ control is another sensitive issue for governments. Seesawing rates affect us all directly and indirectly. The strong dollar, for example, is a positive hinderance to world trade when so many deals are negotiated in that currency. On the other hand, is a weak dollar politically acceptable to President Reagan? I cannot predict a period of stable exchange rates. The oil price changes and the differing phases of recovery around the world will each contribute to this instability. In turn, this instability directly influences governments' actions to reflate economies.

One of the major socioeconomic problems facing most governments is __unemployment__. I do not see unemployment falling by very much, if at all, in the next two to three years. The developed economies are undoubtedly going through a period of structural change in the mass-employment industries like steel, shipbuilding, automobiles, and heavy engineering. It will take a considerable quantity of microchip and computer-game factories to take up the shake-out from these older industries.

To summarize, I expect there to be some sort of recovery in the U.S. and in other parts of the world. What is in doubt is the scale and duration of this recovery. There are many politicoeconomic "hot potatoes" for governments to juggle--inflation, unemployment, exchange rates--where attempts to keep them under control might well result in fiscal and economic policies that would cause any recovery to be short, jerky, or even stifled. Direct government action in the economy is now commonplace in Europe, and I see no reason why the White House should behave any differently. Regarding wood consumption, there are other factors to be considered that are not directly related to general economic recovery. It is perhaps easiest to deal with these factors under the headings of the main wood-using sectors.

Wood in Construction

A simple sum illustrates current and likely future trends: slower rates of dwelling starts __plus__ smaller units (because of cost of building) __plus__ use of graded lumber (for maximum utilization of yield) __plus__ use of wood panels __plus__ use of nonwood materials __equals__ less wood used.

The construction industry is the most important customer for wood producers, and it is now widely reported in the U.S., Europe, and Japan that new dwelling starts, in the medium term, are not expected to grow particularly quickly. Added to this is the fact that the higher costs of building are leading to smaller units, especially multifamily urban units using little structural lumber.

Machine stress rated lumber will doubtless penetrate the structural timber market, and the smaller dimensions generally pertaining will reduce the overall volume of timber consumed. The use of wood panels has already had an impact on timber consumption, and this will continue not only in Europe but in other countries. It takes, of course, a smaller volume of wood panels to replace a given volume of lumber. Timber and wood panels are also threatened by nonwood materials. Cement, aluminum, and plastics are gaining ground, as has plasterboard. In some parts of the world plasterboard is new to the market and has already eroded the use of wood panels.

A brief comment about timber frame housing, which, though enjoying considerable publicity, is making only apparent headway in the U.K. Elsewhere, the major timber frame users, the U.S. and Japan, have probably peaked. In any event, in the U.K., the use of timber frame offers greater benefits to the builder than the sawmill or plywood mill. The actual additional timber increment is quite small compared with our traditional methods.

The evolution in timber usage, and I hope to demonstrate later that there is an evolution in the use of timber materials, suggests that Europe has evolved further than the U.S. In effect, we are Homo sapiens, while the U.S. is the dinosaur. That makes the U.S. timber industry a very big target.

Furniture and Joinery

In addition to the increasing and more sophisticated use of panel products, there is another phenomenon that should be of direct concern to these industries both in Europe and the U.S. There is increasing emphasis among the developing countries on creating and extending export-oriented production facilities. Not only is this being developed indigenously, but wholesalers and retailers in the buying countries are encouraging these offshore industries. Countries like Taiwan, South Korea, Singapore, the Philippines, and eventually Indonesia and Malaysia will be looking even harder at U.S. markets. If their expansion is achieved at the expense of domestic producers, then the wood industry will lose domestic orders.

Packaging and Other Uses

Packaging is another customer of the wood industry that has switched to sheet materials, corrugated cases, and plastics. New warehousing and distribution methods call for less strength-critical transit packaging and lighter pack weight.

Individually these uses account for only small proportions of business, but collectively they are significant. Metal and plastics have replaced much of the timber used. Again this trend is not just restricted to Europe. The American market is also changing. that 36.3 billion board feet a year is looking more and more difficult to attain.

Solutions

Whether or not these changes are socially desirable, commercially appropriate, or politically necessary, there is an important timber industry in existence. That industry is a major employer, a major purchaser of other materials, a depository of billions of dollars of shareholders' funds. Therefore, it is required that the industry seek solutions. Initially, in the short and medium term there are really only three solutions:

1. <u>Reduce Output and Costs</u>. This solution is socially unacceptable if it involves redundancies. Furthermore, the shareholders' interests are not served either. Ultimately, part of the industry may have to adopt this solution, in the longer term.

2. <u>Develop New Products</u>. This is a costly exercise and certainly not a short-term solution; medium to long term is more likely. I do not decry those efforts to create new products. It is textbook marketing in mature markets. But considerably more thought needs to be applied to wood, its fundamental characteristics, and how they can be adapted to match consumer needs. Additionally, more research is required to understand why consumers like wood and yet are prepared to substitute so readily. I would challenge the ability of our industry to identify any new wood product developed in the last twenty years or more that has successfully created for itself a market as a substitute for cement, plastic, concrete, chrome, glass, et al. All new products introduced in this time have competed only with other wood products. Our industry is most incestuous. It is also the prime target for the nonwood material producers.

3. _Export_. This is not necessarily the soft option. It requires special skills, resources, and patience to be a successful solution. Nevertheless, in the remainder of this paper I would like to offer some helpful comments and suggestions which really only form the initial part of a corporate export plan.

The Scale of the Markets

In the first instance it is important to adjust the scale of your perception to market size. Other countries cannot match U.S. consumption, but it is instructive to measure their imports.

Figure 2 shows the volume of softwood lumber imported by certain countries. In fact they are the only countries that import more than one billion board feet. Some of them are growing--such as Japan, Italy, and the Middle East--and there are other countries not on this list that might make it in the next few years. The United States is the largest softwood lumber importer in the world (although Japan imports a greater quantity of softwood logs for domestic conversion). All other countries represent only a fraction of the U.S. total.

Regarding plywood, Western Europe consumes about 4.5 billion square feet, of which about 2.7 billion square feet are imported. Once this is apportioned among the principal countries it is quickly appreciated how small each market is.

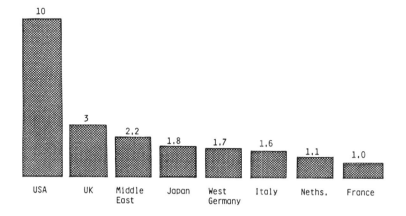

FIGURE 2. Softwood lumber imports, 1981 (billion board feet).

Taking both softwood lumber and plywood and dividing each market by the various species, grades, and specifications, some idea can be gained of the different approach to the market that has to be adopted. The scale of orders processed, the minimum order sizes accepted, the shipping and delivery methods must each be adapted to this trading environment, which is quite distinct from the North American experience.

The Nature of European Markets

I would like to offer some suggestions that exporters to Europe should incorporate into their strategy. But first, a brief explanation of some forecasting systems which BIS has developed in conjunction with Dr. Carl Smith, one of our professional associates. In the course of our work for forest products clients, there is need for reliable forecasts of demand trends. Regression models and even complex systems based on a whole series of regression models are not always satisfactory, and we have developed a relatively simple method of cross-checking.

By way of example, we take annual consumption of, say, softwood lumber, hardwood lumber, and the total of all wood-based panels, convert them to a standard unit of measure, and total them. Then the percentage share of that total enjoyed by each product is calculated and plotted on a graph, year by year. Figure 3 shows this method applied for the above mentioned products for U.K., France, West Germany, and the Netherlands (aggregated).

Note that these percent share curves are very smooth, unlike real consumption data, which oscillate through time. These curves are strong, constant--and measurable. The smoothness and strength of direction are surprising over the period of time indicated when you remember the boom and slump conditions in the market, the price fluctuations, oil shocks, and new products.

Figure 4 is another example of this approach where total European panel consumption has been split into its three basic components: particleboard, plywood, and fiberboard. Irrespective of what is happening to total panel consumption, the graph shows the relativities between the basic attitudes to the three products, converted into purchases.

The curves in both figures 3 and 4 can be extrapolated to produce a forecast. Regression models are used to obtain a forecast of total wood products consumption. This is relatively troublefree, because total consumption responds

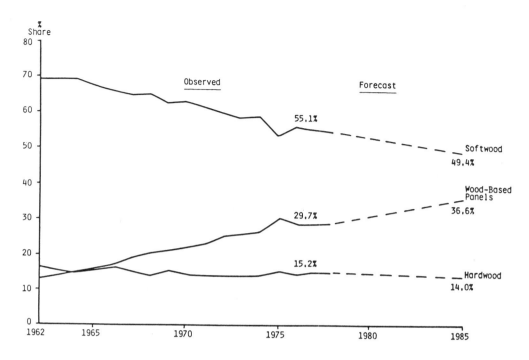

FIGURE 3. Trends in relative share of softwood, hardwood, and wood-based panels, 1962-1985 (U.K., France, West Germany, and the Netherlands).

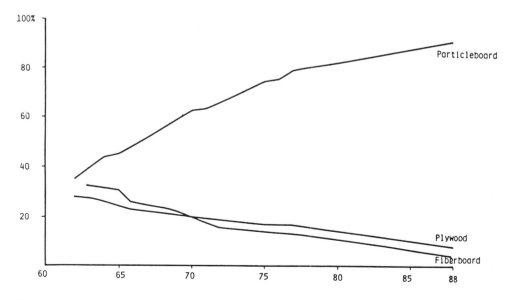

FIGURE 4. Panel market share of fourteen European countries (non-Comecon).

well to readily available forecasts of key variables such as housing starts or interest rates. By applying the percent share forecasts to the total forecast we can obtain an estimate of future volume consumption for each product.

Again we can use regression techniques for each product and compare the results. As a forecast of trends, the method tends to be fairly reliable. In 1979 we made some predictions about softwood lumber consumption in Europe to 1985-86. So far in 1983 they are still on track: static to slight decline over the period.

This technique has revealed an evolutionary pattern in the consumption of wood products. The basic pattern revealed in these graphs--softwood lumber's share declining and panel's share growing (but particleboard growing at the expense of plywood and fiberboard--is repeated in just about all European countries investigated. There is no difference in the shape of the curves despite the significant differences in the building methods, strength of the domestic products industry, reliance on imports, wood costs, and many other factors.

It is possible, however, to plot the changing share over time for a particular product (say particleboard) for all countries. They all fit on the trend line, only spaced out at different intervals. One conclusion to draw is that the only difference between the relative market shares of the products in each country is time. Certain countries, such as West Germany, Austria, and Switzerland, have evolved further than Sweden and the U.K.--that is, their share of softwood lumber, plywood, and fiberboard is lower. In the U.S., plywood and softwood lumber still enjoy relatively high shares in total consumption, but early evidence suggests this is beginning to change.

Potential exporters can draw several conclusions from these patterns. The visually obvious one is that when total consumption of wood products is forecast to grow slowly or is static, these upward and downward curves indicate real increases or decreases in the volume of each product consumed. Therefore, for two important items of U.S. forest products exports, softwood lumber and softwood plywood, figures 3 and 4 suggest a static market over the medium term. Consequently, the strategy of an exporter has to be oriented to winning market share from competitors as opposed to participating in a naturally expanding market. It is a very different set of tactics required to win or maintain market share. For example, a supplier's image and reputation become very important. This involves how your customers and potential customers perceive your product strengths, your range of commercial services like delivery, its speed and reliability, how you respond to inquiries, your claim handling procedures, and so on.

These aspects are additional to the basic matters of meeting official norms and standards and providing the appropriate grades and specifications. The problem with image is that it is not what you know to be true but what others (your customers) think is true. Refer back to figure 1 and it is plain why U.S. sawmills are regarded as occasional suppliers, despite the underlying upward trend shown by the graph.

BIS is undertaking an increasing number of these types of investigations for forest products companies that already perceive this to be an effective approach. There are a number of other factors that can be cited to encourage forest producers to consider exporting.

Household Formations

It has been said previously that housing starts are not forecast to grow particularly quickly. The observation is made having taken into account assumptions about slow population growth, adequacy of existing housing stocks, and limited expenditure by the public sector. Other commentators have pointed out that perhaps household formations have been underweighed in the forecasts. Household formations have been taken into account, but insufficient emphasis has been given to them. Any number of factors contribute to an increase in household formations, such as longevity of life, single parent families, divorced persons, single young persons (with the income to afford a home), second family homes, and so on. These are especially relevant for the Western world.

In the East, multigeneration family household units are common. But there is some evidence to suggest that this may be breaking down. Apart from giving a general impetus to housing starts, the change is essentially a middle-class phenomenon where the financial means pertain to satisfy the desire for a separate home. There is an additional bonus for the forest products industry in that new houses require furniture and electrical goods.

Hardwood Versus Softwood

The extent to which softwood could substitute for hardwood is a complex issue which is not fully explained or resolved. This is likely to be a significant factor in the

Asian markets during the coming decade. Recent work we have completed shows that, for several reasons, substitution is taking place which will afford attractive opportunities for softwood sawmills. The plywood situation is more problematical. Although the potential exists, softwood plywood will not fulfill it to any great extent in Asia.

Plywood Production Costs

Following on from the last point, it should give U.S. plywood exporters great heart to realize that they are probably the lowest-cost producers in the world. For structural grades this is certainly true. While freight costs may offset some of this advantage, knowledge of the fact is an invaluable asset. As waferboard capacity comes on stream and releases more veneer-based product, producers should grasp the opportunity.

Distribution Systems

These remarks really apply to Europe, but similar trends are discernible elsewhere. The multistage distribution system which has been traditional in Europe--shipper to agent to importer to merchant to user--has been changing. The conditions are suitable for suppliers to shorten the selling and distribution chain, set up their own sales offices, and generally get closer to the user. In this way the supplier can better educate the user as to the nature and benefits of his products, species, and grades.

Summary

In summary, there are four points to remember:
1. The future prosperity of U.S. forest products companies solely concerned with the domestic market is uncertain. Economic growth is not guaranteed, and furthermore the nature of wood usage is changing.
2. Exporting is a viable solution because it spreads the risks associated with this uncertainty.
3. Market research is an essential step to more clearly identify the opportunities, define their relevance to the supplying mill, and describe their idiosyncracies.

4. Potential suppliers need to adjust their perceptions and attitudes regarding market size and marketing strategies to exploit these opportunities. This point is important because many suppliers would be dissuaded from exporting if the markets appeared either too small or too alien. Much time can be wasted in trying to identify markets that are large in terms of volume and tend not to be too dissimilar to the U.S. in terms of product specifications.

I hope this paper will go some way to helping potential exporters recognize that export markets will be different in both scale and products required but nevertheless the opportunity is worth the extra effort. In 1960 Theodore Levitt said in Marketing Myopia, "In truth there is no such thing as a growth industry, there are only companies organized and oriented to create and capitalize on growth opportunities." Too often this fact is forgotten regarding exporting. But be assured, in the 1980s, that those forest product companies with the desire to organize and orient themselves to international trade are more likely to be the commercially successful ones.

NEW MARKETS: PEOPLE'S REPUBLIC OF CHINA

Vivien F. Lee

The initial exuberance over the unlimited potential of foreign trade with the People's Republic of China (PRC) has worn off and the development of the Chinese market has settled into a steady pace. Just how steady is this pace in the area of forest products?

That China truly needs forest products cannot be overstated. Her modernization program presently emphasizes communications, transportation, and infrastructure. This requires wood for electrical poles, railroad ties and trestles, marine pilings, highway bridges, bulkheads, and pitprops. To generate foreign currency to finance the modernization, there is a steady rise in her exports of light industrial products. This requires an equally steady rise in Chinese imports of packaging materials and paper. To improve the quality of life in China, the building industry is kept busy planning and constructing new dwellings. There is a shift away from heavy reliance on energy-intensive materials such as metals, cement, and bricks; and wood is the logical substitute. Wood and wood products are needed for doors and windows as well as for furniture and millwork. Education is once again emphasized and wood is required for desks and tables, while pulp and paper products are required for books and notepaper. Wood as a fuel source is another important consideration.

There is available data on the China market regarding her resources, her production, her consumption, and her trade. The data, however, can be contradictory depending on the source. Some of the numbers are quoted here. They are not to be taken as gospel, but are to be used as a starting point to be confirmed or refuted over time.

China's Forest Resources

China herself currently produces nearly 50 million cubic meters of logs annually (China's Agricultural Yearbook 1980). The provincial log production figures are shown in

Table 1. Log Production by Province.

Province	Production (thousand cubic meters)
Beijing	--
Tianjin	--
Hebei	62.0
Shanxi	185.5
Nei Monggol	3,423.5
Liaoning	375.5
Jilin	5,132.0
Heilongjiang	13,839.9
Shanghai	--
Jiangsu	--
Zhejiang	618.0
Anhui	468.2
Fujian	3,812.7
Jiangxi	2,649.1
Shandong	11.3
Henan	73.9
Hubei	661.6
Hunan	2,414.1
Guangxi	1,612.6
Guangdong	2,218.0
Sichuan	4,035.5
Guizhou	831.7
Yunnan	2,341.0
Xizang	194.8
Gansu	638.0
Qinghai	40.3
Ningxia	6.6
Xinjiang	52.2
Total	46,715.8

table 1. The total area of forest in China is about 122 million hectares, or 12.7 percent of the total land area: 98 million hectares (80.3 percent) are timberland; 8 million hectares (6.6 percent) are shelter forests; 8.6 million hectares (7.1 percent) are economic forests (fruit, coconut, rubber, tung, and lacquer trees); 3.3 million hectares (2.7 percent) are fuel forests; and the other 4 million hectares are bamboo forests or forests for other purposes. In recent years there has been decreasing log production, from about

Table 2. Log Production (cartload of wood; million cubic meters).

Year	Log Production	Year	Log Production	Year	Log Production
1949	5.67	1960	41.29	1971	40.67
1950	6.64	1961	21.94	1972	42.43
1951	7.64	1962	23.75	1973	44.67
1952	11.20	1963	32.50	1974	46.07
1953	17.54	1964	38.00	1975	46.26
1954	22.21	1965	39.78	1976	48.73
1955	20.93	1966	41.92	1977	51.62
1956	20.84	1967	32.50	1978	51.60
1957	27.87	1968	27.91	1979	54.39
1958	35.79	1969	32.83	1980	53.39
1959	45.18	1970	37.82	1981	49.42

Sources: China's Agricultural Yearbook 1980 (for 1949-77); State Statistical Bureau Announcement, April 30, 1979 (for 1978); SSB, April 29, 1980 (for 1979); SSB, April 30, 1981 (for 1980); SSR, April 30, 1982 (for 1981).

54 million cubic meters in 1979 to 53 million cubic meters in 1980, to 49 million cubic meters in 1981 (table 2).

Production, Consumption, and Trade

The Ministry of Forestry has announced that domestic production of timber has grown steadily over the past three years, from 46.68 million cubic meters in 1980 to 47.16 million cubic meters in 1981 to a probable 48.8 million cubic meters in 1982.[1] According to the Chinese Timber Corporation, total domestic consumption has also increased from 48.86 million cubic meters in 1981 to 52.2 million cubic meters in 1982. A UN report says China's timber consumption is 200 million cubic meters a year, including

---

1. Home Demand for Timber Still Outstrips Production," China Daily, October 22, 1982. (This source is inconsistent with the information in table 2.)

firewood used by local people, but Chinese officials have neither confirmed nor rejected it. The official figure for annual consumption in the national economic plan is 50 million cubic meters, of which 30 million are purchased and used by the state. The rest is left for the provincial governments. Approximately 24 million cubic meters, nearly half of the annual consumption, are used for industry; 1.6 million for agriculture; 1.5 million for housing and furniture; and the remainder is set aside for disaster relief and foreign aid. Despite the ambiguity, it is generally acknowledged that domestic production has not been able to meet domestic demand.

To meet its forest products needs, China has been importing from the United States and other nations. For example, it is known that United States earnings from softwood log exports have grown rapidly over the past three years. Sales to China were worth $41.5 million in 1980, reached almost $90 million in 1981, and surpassed $200 million in 1982. Aside from timber, China also imported other forest products from the U.S., including lumber, pulp, and paper (table 3).

The Chinese Market

Assessing the true size of the China market based on the U.S. export performance of the past several years may prove confusing. For example, the surge in volume purchase of kraft linerboard or pulp in 1980 may have been planned to cover needs for more than a year; moreover, the current infrastructure means longer transit lines for the goods to reach consumers, and there have been delays due to lack of availability of Chinese vessels to pick up goods.[2] In addition, the perennial shortage of foreign exchange is a very real deterrent to the projection of market trends. Although buying surges may occur from time to time, one should not anticipate regular annual purchases to increase quickly. It will come, but slow and steady is more realistic, as China makes progress with her industrialization program and her need for imports of forest products gains momentum.

Aside from the United States, China has also imported forest products from the Soviet Union, from Japan and the

---

2. "China: Promising Outlook in the 1980s," Pulp and Paper International (Brussels), August 1981, p. 70.

Table 3. Forest Products: U.S. Exports to China, by Product, 1978–1982.

Item	1978	1979	1980	1981	1982
	(in thousands of dollars)				
Logs, lumber, and other rough wood products:					
Logs:					
Softwood					
Southern yellow pine	-0-	-0-	1	-0-	1,995
Douglas-fir	-0-	-0-	34,286	63,978	161,959
Western hemlock	-0-	-0-	5,640	17,629	42,476
Other	-0-	-0-	1,508	7,604	5,423
Hardwood	-0-	-0-	3	-0-	45
Lumber:					
Softwood					
Spruce	-0-	-0-	-0-	2,215	3,387
Douglas-fir	-0-	-0-	-0-	2,530	581
Western hemlock	-0-	-0-	-0-	4,672	1,016
Other	-0-	-0-	5	52	3
Hardwood	-0-	-0-	6	480	40
Other	-0-	-0-	-0-	-0-	39
Wood manufactures:					
Plywood:					
Softwood	-0-	-0-	2	-0-	1
Hardwood	-0-	-0-	-0-	-0-	1
Building boards	-0-	-0-	28	-0-	29
Other	15	-0-	1,402	55	58
Woodpulp:					
Sulphate and soda	-0-	630	11,779	24,927	4,777
Chemical, special alpha	2,493	1,904	15,024	13,524	6,072
Sulphite, bleached	1,558	-0-	32,846	23,757	64
All other	-0-	1,350	7,210	6,418	7,067
Paper and paperboard:					
Unbleached kraft linerboard	438	2,535	93,327	49,542	32,245
Other	42	932	37,111	11,214	3,948
TOTAL FOREST PRODUCTS	4,556	7,351	240,178	228,597	271,226

Source: Compiled from official statistics of the U.S. Department of Commerce.

Southeast Asian nations, from Chile, Canada, and countries in North Africa. China has also signed timber import contracts with Liberia and Ivory Coast and will continue to open timber markets in South America and North Africa in the coming years.

Despite the international competition, there are reasons why the U.S. can be optimistic on counting China as a dependable market. Costs are a basic and probably enduring reason for export opportunities. The U.S. has some of the world's lowest-cost wood; domestic paper companies have an abundance of wood wastes which they can burn as an economical fuel; and they operate large efficient mills. Japan's lack of wood and fuel resources has made it increasingly uncompetitive as a traditional source of packaging materials for China. In the northern climates, such as the Soviet Union, Scandinavia, and Canada, tree growth to maturity can take a long time compared to the 30 years or so necessary for tree growth in the southern U.S. Many of the developing countries have inadequate infrastructure and have not yet reached their exporting potential. In addition, a number of countries have been closing plants with uneconomical, inefficient processing capacity, further opening market opportunities for the U.S.

China's Conservation Efforts

As mentioned, China does acknowledge that her consumption far outstrips her production. She is conserving her limited supply of wood and fiber, and has actively pursued a program of reforestation and afforestation. According to the "General Treatise of Agricultural Geography of China," 28 million hectares of land were brought under forest during the period from 1949 to 1977. The effort continued in 1978-1981 (table 4).

In February 1979, forestry laws were passed by the Standing Committee of the People's Congress which stipulated that the nation should bring 30 percent of the Chinese land under forests. In mountainous regions, forest cover should be 40 percent or more of the total area of the region, whereas on plains the percentage should be no less than 10 percent. No date has been given for completing the program.[3]

---

3. Attaché Report filed November 2, 1982, by Ed Heskop, Agricultural Attaché, Hong Kong; Foreign Agricultural Service, USDA.

Table 4. Afforestation in China, 1978 to 1981.

Year	Afforested Area (million hectares)
1978	4.49*
1979	4.49
1980	4.45
1981	4.11
Total	17.54

*Derived from the 1979 figure.
Sources: State Statistical Bureau Announcement, April 29, 1980 (for 1979); SSB, April 30, 1981 (for 1980); SSR, April 30, 1982 (for 1981).

After a national conference on forestry in February 1981, attended by Chairman Hu Yaobang and Premier Zhao Ziyang, a joint decision by the party central committee and the State Council was issued outlining the following policy guidelines:
. Stabilize ownership rights of forest lands and implement responsibility systems in forestry.
. Unify local timber management, establish quotas and rotating zones for logging, and ban timber from free markets.
. Develop multipurpose uses of wood by-products and waste, such as particleboard, and promote use of timber substitutes in construction (including metal pitprops in mining, cement railway sleepers, and metal door and window frames).
. Reorganize financing for the forestry industry and establish funds for growing and cultivating saplings.
. Develop forestry science and education.[4]

Since 1949 there have been numerous campaigns to encourage tree planting and to control logging. Regeneration needs a great deal of care, especially in areas that have been deforested and have suffered soil erosion. Commercial foresters generally use timed applications of

---

4. *Far Eastern Economic Review*, August 20, 1982, p. 60.

herbicides to keep down fast-growing shrubs and ground cover until the young trees have established themselves. Hybrids developed for commercial tree harvesting grow quickly, but they must be carefully matched to the microclimate and soil characteristics of each site. So far, modern commercial tree farming technology is not being applied on a wide scale in China.

Adapting the responsibility system to forestry (by making farmers into quasi-private producers) may improve the survival rate of seedlings, but it is not clear how a family-based forestry management system can be made economically feasible. Assuming that a newly planted tract of fast-growing commercial trees could be harvested in 20 years, each household would require 20 individual tracts, each big enough to provide one year's income.

As most trees take years to grow to commercial size and most forests take decades to mature, it is unlikely that China will be able to achieve the goal of having 30 percent of the land under forest before the end of this century. Until then, China will still have to import logs and other forest products to supplement domestic production.

Market Access

It is all very well to look at numbers and statistics, to study five-year plans and plot growth curves, and to recognize that China's domestic demand for timber outstrips production and as modernization proceeds and demand continues to outstrip supply there will be even greater needs for forest products. It is another matter to deal with the realities of accessing the PRC market.

Possibly the most important starting point in any business transaction is contacting the right person(s) to do business. Dealing with the vast Chinese bureaucracy, it is essential to approach it with some knowledge of the system, with an established credibility among the Chinese, and an introduction to the decision maker(s). An intermediary who meets these requisites and can facilitate access to the proper party is well worth his fees and can help a company get under way with the Chinese.

It should be understood that with whomever one deals, the Chinese counterpart is not negotiating as a private citizen. He is a representative of the Chinese government, a member of an institution authorized to deal with foreign firms. These institutions may be created by the State Council or they may be under the Ministry of Foreign Trade--as foreign trading companies--or under other ministries or government organizations. Some provincial and municipal governments have also assumed authority in trade matters and can execute agreements in their own names. A number of special economic zones have also been established. Because they are all representatives of the same Chinese government, there is often competition among themselves to be the agency to bring in the foreign exchange or to make the purchase. It is therefore very important for the U.S. party to be dealing with the "right people" and/or the end user.

For a company to deal with the Chinese in China, it must receive an invitation. The issuer of the invitation is the host. Negotiations on large-scale projects with the national government are preferably conducted in Beijing, that city being the seat of government. But depending on the nature of the transaction and the type of sales envisioned, negotiations can take place in provinces or with other entities with a fair degree of autonomy. China wants products and/or technologies that are "truly advanced and appropriate to China's needs." Companies pursuing PRC business should prepare a proposal offering a mutually beneficial relationship that appears commercially viable; it should have a brief outline introducing the firm, describing its products, and--if technology is involved--emphasizing the performance. The proposal should then be forwarded to the Chinese host for review. If the Chinese are interested, they will request a more detailed outline, and will schedule a meeting with the right contacts.

After the preliminary matters are cleared and the Chinese are satisfied that one's products, services, and technologies are beneficial to the PRC, they will issue the necessary invitation to the company to visit China. Without an invitation, one is powerless to go into China except as a tourist. With that invitation one has access, and the process for negotiating the sale can begin.

FINANCING EXPORT SALES FOR THE SMALL- AND MEDIUM-SIZE
FOREST PRODUCTS FIRM

Robert M. Ingram III

Introduction

In March 1982, when the recent forest products depression was about at its deepest point, Governor Evans of Idaho held a Governors' Conference on Lumber Export. The conference, attended by a majority of the Idaho sawmill owners, was designed to explore the steps needed to be taken in order for Idaho to capture a larger share of the lumber export market. I attended that conference to discuss export financing. During my presentation I was asked a question which puts in perspective the relationship between many small- and medium-size firms and the export market. That question was, "Don't banks consider those sawmills that sell to the export market as having a higher lending risk than those sawmills that sell only to the domestic market?" My answer to the sawmill owner was, "Not at all. In fact, I often consider those sawmills that do not tap the export market as having the higher lending risk." After a few questionable stares I went on to say, "Those sawmills selling export often have better quality receivables, assuming the export sales are done under an irrevocable bank letter of credit. More importantly, those sawmills selling export at least have some orders which would help assure their survival through the 1980 to 1982 forest products depression."

Another way to put the export issue in perspective is to look at the color of my hair. At Rainier National Bank I manage the Forest Products Commercial Banking Center, where we handle all of the banking requirements of Rainier's forest products corporate customers. In 1979 my hair was black, today it has streaks of "banker" gray; however, if most of my customers had not been selling to the export markets during the recent depression, my hair would probably be "bankrupt" gray today. Selling to the export market literally saved the bacon for many companies during the last three years. Now that you know where I stand regarding the export market, I want generally to discuss those services available from your bank to help you in tapping the export market, and, finally, I will conclude my remarks with a few words about the recently passed Export Trading Company Act.

International Banking Services

We all know that the survival of the Pacific Northwest forest products industry, as it exists today, depends on the industry's successfully increasing its share of the export market. This is why we are here at this symposium this week and this is why, all over the Northwest, we are seeing a growing number of export conferences. The major forest products firms have long known the importance of the export market and have been exporting their products for years. They all have international departments which are well versed in marketing and financing export sales. Unfortunately, the small- and medium-size forest products firms often shy away from this important export market because of its perceived complexities. These firms need not, and indeed should not, shy away from selling into the export market, for they all have extensive international departments available to them at their local regional banks. It is to these nonexporting small- and medium-size firms that my remarks are primarily addressed.

If you are considering exporting for the first time, or want to expand your export business, get in touch with the people in your bank's international department. All regional banks have full-service international departments. If you bank with a local bank, that local bank will have a correspondent relationship with one of the regional banks, to whose international department you can be introduced. When the nonexporter first approaches the international banking department, he is first educated, encouraged, and assisted in the mechanics of exporting before export financing is tailored to meet his particular needs. I contend that the fear of selling to foreign buyers and the fear of export documentation are two of the greatest barriers to tapping the export market.

Your international banker will first help to educate you regarding the special documentation requirements of exporting. This process probably will include supplying you with any number of pamphlets explaining export documentation and operations, including the "Uniform Customs and Practice for Documentary Credits," which is the international trade law governing export transactions. This education process may also include attending frequent bank-sponsored export seminars and special training visits to your firm by bank personnel. Your bank can also supply you with names of export brokers, freight forwarders, and others who will assist you with your export requirements.

Once you become familiar with the documentary aspect of exporting, your international banker can provide you with vital information concerning the credit-worthiness of potential foreign buyers along with the laws, accepted credit terms, and credit risks of foreign countries. This information is accurately and quickly obtained through either the bank's own overseas offices or the bank's correspondent relationship with foreign banks. This network of overseas branches and foreign correspondents is also a valuable source of market information.

The assistance you can expect to receive from your bank's international department can probably best be illustrated by relating to you the following four examples:

1. Two years ago a medium-size forest products firm asked me for help in exporting its products directly rather than through export brokers, which it was then currently using. After going through the process of educating, encouraging, and assisting this customer in the mechanics of exporting, which included individual training of the customer's employees by the bank's international personnel, this customer is today a confident and successful exporter.
2. Throughout the year I receive numerous letters and telexes from our overseas offices and foreign correspondent banks requesting the names of customers who can supply logs and lumber to various foreign buyers. These requests are passed on to the bank's customers who might be potential suppliers.
3. In the fall of 1981, when log buying by China was decentralized, allowing the provinces to buy directly, usually through Hong Kong brokers, we met with many of our customers and prospective Hong Kong brokers to determine, for our customers, which of the brokers were capable of performing; thus, we protected our customers' interests.
4. When our forest product customers take marketing trips overseas, our overseas offices or foreign correspondents are requested to extend any assistance necessary to ensure the trip is a success.

I will now turn my attention to the typical export transaction. The small- or medium-size forest products firm can either export directly or through an export broker. If the sales are through an export broker, which in many cases may be another forest products firm well versed in the mechanics of exporting, the sale is handled as any other domestic sale. If the export is direct, most terms of sale usually call for payment in U.S. dollars and use either an irrevocable sight letter of credit or documentary collection. By calling for payment in U.S. dollars, foreign exchange risk is eliminated.

As a rule, forest products export sales to the Far East are made under a sight letter of credit. This is an irrevocable, written commitment of a foreign bank to make payment to the exporter, via the exporter bank, upon presentation of certain shipping and other documents as specified in the letter of credit. Payment is made either when documents containing no discrepancies are negotiated at the exporter's bank or when they are negotiated at the foreign bank.

Forest products export sales to Europe generally are handled under a documentary collection. Here the exporter's bank forwards the title and shipping documents to the foreign bank of the buyer. The foreign buyer's bank then releases the documents to the buyer only after payment is made. In this case the bank acts solely as an agent and does not guarantee payment, as is the case with an irrevocable letter of credit. If the foreign buyer decides not to pick up the documents, the exporter is left with unsold goods sitting in a foreign port.

Financing Export Sales

How are these export transactions financed? Again, please remember that I am limiting this discussion to the most common type of export financing used today by the small- and medium-size forest products firms. The most common forms of export financing are the firm's own working capital line of credit and preexport banker's acceptance financing.

For the occasional exporter, the firm's own working capital line of credit provides the necessary export financing. This line of credit, usually secured by accounts receivable and inventories, is priced at anywhere from prime to about prime plus 2 percent. The firm will use funds from its line of credit to accumulate and process the export order, which will be sold in U.S. dollars against a sight letter of credit or a documentary collection. If funds are not received immediately, as they would be under a sight letter of credit payable at the exporter's bank, the exporter's bank will often provide advances against the letter of credit or documentary collection, should funds be immediately needed for working capital purposes.

For the frequent exporter, the firm will most likely use preexport banker's acceptance financing to provide funds for the accumulation and processing of export sales. The Federal Reserve Act of 1913 created banker's acceptance financing for the purposes of stimulating exports. Banker's acceptances are negotiable time drafts drawn to finance the export, import, shipment, or storage of goods, and they are termed "accepted" when a bank assumes the obligation to make payment at maturity. Basically, in preexport banker's acceptance financing, the exporter gives his bank a copy of the export order and a time draft in the amount of the order. The time draft is a promise to pay his bank in 30, 60, 90, 120, 150, or 180 days. The bank will then "accept" the time draft and discount it, providing funds for the exporter to accumulate, process, and ship his export order. The maturity of the banker's acceptance time draft corresponds to the anticipated receipt of funds from the export order. When the exporter receives funds for the export order, he simply pays off his bankers acceptance time draft. The advantage of banker's acceptance is a favorable discount rate, which is usually somewhat below prime. The reason rates are lower for banker's acceptances than they are for lines of credit is because the bank can sell the acceptance in the national acceptance market, freeing its funds for other uses.

Whether a firm finances its exports by its own line of credit or through preexport banker's acceptances, the important thing to remember is that the export firm's bank is relying on the ability of the firm to perform and on the firm's financial stability. A good relationship between the exporter and his bank is extremely important. Good communication is the key to this relationship. Good communication means making sure your banker knows your company and its management and is kept up-to-date on all major developments. Make sure your banker routinely receives your monthly or quarterly financial statements. By keeping your banker updated on your firm you will be a few steps ahead on arranging for another important, but sometimes overlooked, type of export-related financing. This is financing, usually in the form of a secured term loan, to allow your firm to convert machinery to saw metric-size lumber, to build special export packaging facilities, or to add an export department. When thinking of entering the export market, make sure you consider these costs in your overall export finance planning.

Export Trading Company Act

Finally, I would like to make a few brief comments regarding the Export Trading Company Act of 1982. This act, designed to further stimulate exports, allows export trading companies certain exemptions from antitrust laws, and allows bank holding companies to acquire equity in export trading companies. Currently there are over 200 export trading companies in Washington State, if you define an export trading company as one involved in the exportation of goods and services.

This act will increase these companies' protection from antitrust laws and allow them to acquire a bank partner. It will also mean that these companies will have more competition as more export trading companies are formed as a result of this new act. The immediate benefit of this act to the small- and medium-size forest products firm appears to be one of economies of scale, as these firms can now form an export trading company with a common export department.

What are banks doing in regard to the Export Trading Company Act? For the most part, banks are all studying the act and deciding whether they should join an export trading company or simply continue to provide their international banking services to them. Whether a bank joins an export trading company or simply acts as the banker to an export trading company, the financial and international banking services it brings to the table will be the same. Because banking is a competitive and service-oriented industry, we must and will supply all of our services to all of our customers.

Summary

Selling into the export market may very well be the key to the future of the small- and medium-size Northwest forest products firm. This fact is recognized by banks, and their international departments are available to assist these firms with their export sales. Most forest products export sales call for payment in U.S. dollars and are handled by irrevocable sight letters of credit if to the Far East and by documentary collection if to Europe. The most common forms of export financing are the firm's own working capital line of credit and preexport banker's acceptance financing.

If you are considering exporting, do not forget to consider financing needs for such items as mill modifications. To facilitate all financing, it is important to keep your banker well informed. Finally, banks will continue to provide all international financing and banking services to all firms equally, regardless of whether or not the firm or the bank is a member of an export trading company.

MARKET PROSPECTS FOR TROPICAL HARDWOODS FROM SOUTHEAST ASIA

Kenji Takeuchi

The Importance of Tropical Hardwood

In terms of export earnings of developing countries, tropical hardwood is one of the most important primary commodities. In 1980, developing countries' exports of broadleaved (hardwood) industrial roundwood amounted to more than $3.5 billion, while exports of coniferous (softwood) roundwood amounted to only $100 million or so. If the derived products[1] are taken into account, developing countries' export earnings from tropical hardwood were worth double in value. Virtually all of the broadleaved roundwood exports from developing countries are in the form of sawlogs, veneerlogs, and logs for sleepers (logs, hereafter), with only a small fraction being pulpwood and other industrial roundwood.[2]

In volume, tropical hardwood logs are surprisingly only a modest component of total world wood production. In 1980, world wood production amounted to 3 billion cubic meters, of which 46 percent was industrial wood, the remainder being fuelwood. Within industrial wood, logs are the most important subcategory, accounting for 60 percent; but only 29 percent of logs are from broadleaved species, and only one-half of broadleaved logs are tropical hardwood. Thus, tropical hardwood logs account for only 4 percent of total volume of all trees harvested.

Within the "logs" category, however, the relative importance of tropical hardwood in total log supply has been increasing rapidly. During 1961-80, for example, production

---

1. "Derived products" refer to the products of first-stage mechanical processing--namely, sawnwood, sleepers, veneers, plywood, and other wood-based panel products.
2. "Other roundwood" includes pitprops, poles, piling, scaffolding, and formwork roundwood.

of tropical hardwood logs increased at 4.8 percent per annum compared with the 1.2 percent and 0.7 percent per annum growth in production of softwood logs and temperate hardwood logs respectively. Furthermore, while softwood production is dominated by industrial countries and centrally planned economies, tropical hardwood, which accounts for almost one-half of world production of hardwood logs, is available only from developing countries (except for some Asian centrally planned economies). Developing countries' exports of tropical hardwood logs and their derived products have risen rapidly over the last three decades, amounting to some $5.8 billion in 1979. This sharp rise has resulted primarily from the shortages of temperate hardwood supply in industrial countries.

Tropical hardwood comes from the moist forests in three tropical regions:

1. West and Central Africa (mainly Liberia, Ivory Coast, Ghana, Nigeria, Cameroon, Gabon, Congo, Central African Republic, Zaire, and Equatorial Guinea).
2. Southeast Asia and Tropical Oceania (mainly the Philippines, Malaysia, Indonesia, Thailand, Burma, Papua New Guinea, and the Solomon Islands; also Kampuchea, Loas, and Vietnam).
3. Tropical Latin America and the Caribbean (mainly the Amazon region of Brazil; also scattered throughout Central and South America as well as the Caribbean Islands).

While the tropical hardwood species of these moist tropical forests are, theoretically, renewable resources, how long it takes for them to grow into sizes suitable for sawlogs and veneerlogs is not at all clear. It is believed to take from a minimum of 40 years to more than 100 years, but very little is known about the natural regeneration of the commercial species. Large concession forests in tropical regions are "managed" on a 25 to 35 year cutting cycle. It is assumed that in that time, after the first "creaming" of the forest, the trees that were too small to be removed as sawlogs or veneerlogs will have grown large enough to be cut. This is not much more than a theoretical assumption, however.

Plantation growing of preferred species has been successful for only a few species in limited locations. Under plantation conditions, some "white wood" species in West and Central Africa have been grown to a size of 50 centimeters in diameter at breast height (the minimum size for veneer production) in 20 to 30 years. We all know that teak has been grown to a harvestable size in Burma and

Indonesia in 40 years. At this juncture, however, tropical hardwood may well be considered a quasi-nonrenewable resource.

The Pattern of Trade

Broadly speaking, most of the world trade in tropical hardwood flows from the three major producing areas--West and Central Africa, Southeast Asia, and Latin America--to three main market areas--Western Europe, Japan, and North America. For <u>logs</u>, the close traditional relationship between origin and destination and the transport cost factor have strongly influenced the pattern of trade. The major trade flows in logs have been from West and Central Africa to Western Europe, or from Southeast Asia to Japan, Republic of Korea, Taiwan (China), Singapore, and Hong Kong. The United States, Canada, Australia, and New Zealand import only small quantities of tropical logs. Log exports from Latin America have been negligible since the early 1970s because of the policies of no log exports in the major producing countries in that region.

For <u>sawnwood</u>, the pattern of trade flows is more diversified than for logs. Major sawnwood trade flows have been from West and Central Africa to Western Europe; from Southeast Asia to Western Europe, Japan, Australia, and North America; and from Latin American to North America. Exports of sawnwood from Southeast Asia to Western Europe, and recently to the Middle East, have been increasing; however, exports from Asia to Japan have been sluggish, as consumption of sawn hardwood in Japan tended to decrease during the 1970s.

World exports of <u>tropical hardwood plywood</u> originate mainly in Asia with two types of exporters--log-producing countries (Malaysia, the Philippines, etc.) and log-importing countries (the Republic of Korea, China, Singapore, etc.). From the 1950s through the mid-1960s, Japan was the largest exporter of tropical hardwood plywood and the United States was the main importer. As other in-transit processor exporters expanded their trade and as Japanese domestic consumption increased, Japan's exports of tropical hardwood plywood declined. The major trade flows in tropical hardwood plywood are from West and Central Africa to Western Europe; from East and Southeast Asia to the United States, Western Europe, and the Middle East; and from Latin America to the United States, Canada, Australia, and Japan.

World trade in <u>tropical hardwood veneers</u> is relatively small compared with trade in logs, sawnwood, or plywood. Unlike plywood, exports of tropical hardwood veneers come mainly from the log-producing countries. The major trade flows are from West and Central Africa to Western Europe, from Southeast Asia to the United States, and from Latin America to the United States. One notable feature of tropical hardwood veneer exports is that they have been rather stagnant since the mid-1960s, except for the surge in the period from 1971 to 1974.

The Market Structure

The world market for tropical hardwood logs is fairly competitive. There are many producers, sellers, buyers, and consumers of tropical hardwood logs; however, competition in the tropical hardwood log trade is less than perfect. First, because of the heterogeneity of tropical hardwood species and the vast variety of products (different sizes of sawnwood and plywood), there is no organized market that deals with tropical hardwood logs or tropical timber products. Information regarding "world market prices" for tropical hardwood logs, sawnwood, and plywood is not readily available; only indicative price quotations for some products in some national markets and c.i.f. and f.o.b. prices for some countries are regularly published. Second, trade in logs has tended to be geographically concentrated, partly because of the high transportation costs (relative to the value of the commodity). Third, companies importing South Sea logs in Japan are highly concentrated.[3] In 1977, when a total of 158 companies imported 21 million cubic meters of South Sea logs into Japan, the "top ten" firms accounted for 52 percent of the total imports and another ten firms for a further 24 percent. Considering that many of the other important firms are affiliated with the largest 20, the degree of concentration of market power in South Sea log importing in Japan is high, although there is no evidence of collusive market distortions by importers.

On the supply side, a large number of producers are engaged in tropical hardwood log production, but the effective number of independent companies engaged in log production in major log-exporting countries is considerably smaller than the numbers of logging licenses and operations

---

3. Tropical hardwood species (excluding teak) produced in Southeast Asia and South Pacific Islands are referred to as South Sea timber in Japan.

suggest because many companies are engaged in logging operations at more than one place and there are interlocking ownership relationships. The involvement of foreign companies in logging operations is substantial. In addition to outright ownership, dependence of local logging firms upon foreign firms through suppliers' credits provided for initial purchase of equipment and associated marketing arrangements for log output is pervasive.

Supply Outlook

During the 1970s, world production of temperate hardwood increased only marginally, at 0.3 percent per annum, because of the slowdown of world economic growth and the relative shortage of supply. In contrast, during the same period, production of tropical hardwood increased at a more rapid rate--3.1 percent per annum--although this growth rate was markedly down from the 5.9 percent experienced in the 1960s.

World production of temperate hardwood logs is projected to continue increasing only slowly (at 0.3 to 0.5 percent per annum), as the supply is constrained by the allowable cut based on the principle of sustained yield. Until recently, the rapid growth in production of tropical hardwood filled the potential gap in the total supply of hardwood logs, but in recent years concern with possible overcutting of tropical forests has been growing. This concern has become especially serious in some of the major producing countries in Southeast Asia and West Africa. Partly because of this concern, three major traditional Asian suppliers--Indonesia, Malaysia, and the Philippines--have taken decisive steps in the last few years to reduce log exports. The measures involve quotas (or outright bans) and/or increased government charges (royalties, export taxes, and so forth) on log exports. The objectives of these actions are threefold: to conserve these quasi-nonrenewable resources, to collect higher resource rent from the forest resources which are owned by governments, and to secure benefits from increased local processing of logs.

The recent actions taken by major supplying countries in Southeast Asia to restrict exports of logs have dramatically changed the future prospects for the world supply of tropical hardwood, because until recently these countries accounted for 80 percent of world exports of tropical logs, and indeed they own a predominant portion of the world's "richest" tropical forest resources.

In the Philippines, the overcutting problem has been most acute, resulting in serious environmental deterioration and forest denudation. Back in 1972, the Philippine government decided to ban all log exports in principle and announced a program aimed at phasing out log exports in four years. Due to recurring balance-of-payments difficulties, however, the government has repeatedly postponed the complete cessation of log exports. In addition, due to unrecorded log exports, forest resources in the Philippines have been far more depleted than the official statistics indicate. Unless a policy for conservation and rational utilization is rigorously implemented, the forests will not be able to sustain even the present level of wood processing activity beyond the next few years.

In Peninsular Malaysia, also, the policy of restricting log exports was established in 1972 when a ban on log exports was imposed, first on eleven species and later on sixteen species. Although the ban was somewhat relaxed in 1977, Peninsular Malaysia's logs exports declined steadily to nil by 1979. Indeed it has now become a log-importing region.

In the state of Sabah in Malaysia, exports of logs peaked at over 12 million cubic meters in the period from 1976 to 1978. In 1979 the state government established the policy of reducing log exports by applying export quotas and increasing the royalty rates sharply. If log production in Sabah continued at the rate of over 10 million cubic meters per annum, commercially attractive forest resources would become very scarce within the next five to ten years. Therefore, Sabah wishes to phase out log exports completely in the next few years, but the state is likely to continue to export logs, perhaps at a reduced rate, because of the heavy reliance of the state's fiscal revenue on log exports.

In the state of Sarawak, log exports have been rising rapidly, since there has not been any restriction and the government charges on log exports have been exceptionally low--around 10 to 15 percent of f.o.b. prices. Commercially attractive virgin forests in Sarawak are estimated to last for another twenty years at the current rate of exploitation, and the second round of cutting is expected to yield substantially less than the first round.

In Indonesia, log exports declined from 18 million cubic meters in 1979 to 4 million cubic meters in 1982. Although the main reason for the decline was the sharp decrease in demand resulting from the housing construction depression in the United States and other industrial countries, the tightening of the log export policy was also an important factor. A joint decree of four directors general in three different ministries--DG Forestry, DG Multifarious Industry, DG Domestic Trade, and DG Foreign Trade--issued in April 1981, and the modifications to the joint decree announced in February 1982, firmly established a new timber policy. Essentially, the policy is to phase out log exports completely by 1985 and establish an export-oriented, integrated wood-processing industry with plywood production as its core.

The policy has been successful in encouraging the growth of a plywood industry and processed product exports. Indonesia's plywood production capacity increased from 0.5 million cubic meters in 1977 to 1.5 million cubic meters in 1982, and its plywood exports expanded from the mere 30 thousand cubic meters in 1978 to 600 thousand cubic meters in 1981 and an estimated one million cubic meters in 1982. Indonesia's plywood production capacity is expected to increase further; however, at least for the next few years, Indonesia's exports of processed products are not likely to increase fast enough to compensate for the losses of log exports. It is important to note that the government maintains a liberal log export policy for Irian Jaya, where forests are not as commercially attractive as in the rest of the country.[4]

In Papua New Guinea, log exports peaked around 1974 at 0.4 million cubic meters per year and have remained at that level since. In early June of 1979, the Ministry of Forestry issued the Revised National Forest Policy, which reversed the previous practice of restricting log exports in favor of local processing. Under this policy, the government intends to "concentrate its efforts over the next few years in seeing to the efficient utilization of existing (and firmly proposed) timber processing capacity, and on the formation of a number of Papua New Guinea-owned export logging enterprises."[5] Although Papua New Guinea is believed to have a vast potential, because of the relative high cost of exploitation in its forests due to difficult terrains and a low incidence of commercially attractive species per hectare, log production would not rise

---

4. For a detailed discussion of the recent development of Indonesia's timber policy and export projections for 1983-1990, see Kenji Takeuchi, "Export Prospects for Forest Products in Indonesia, 1983-1990" (World Bank Commodities Division Working Paper No. 1983-1, January 1983).

5. Papua New Guinea, Office of Forests, Facts and Figures 1980 Edition, p. 1.

dramatically unless world market prices rise sufficiently to make it worthwhile.

Other countries in the region are only minor log exporters. Thailand has become a net importer. Burma could potentially increase log exports but is likely to do so only gradually. Vietnam is a wood deficit country. Thus, with the exception of Sarawak, Papua New Guinea, Burma, and possibly Laos, countries in the region are likely to continue to reduce their exports of logs significantly in the future.

In _Africa_, among the traditional suppliers, Ivory Coast, Ghana, and Nigeria have reached the limit of their production, and their future production levels are not likely to be higher than current levels, even if world market prices increase substantially in real terms.[6] On the other hand, Liberia, Cameroon, Gabon, and Congo do have the potential to increase production in varying degrees. In the long term, Angola, Mozambique, Equatorial Guinea, and Zaire also have potential to increase production significantly.

Finally, in _Latin America_, the resources in the Amazon area are believed to be very large. Nevertheless, considering the investment costs and risks involved, for a large-scale production in the Amazon area to be profitable, world market prices would have to be significantly higher than what they are today.

Tropical hardwood is likely to be in short supply in the long term. Limited additional exports can be expected from the established producing areas for the next few years, but if the projected demand beyond 1990 is to be satisfied, hitherto unexploited resources will have to be opened up on a large scale. Production of hardwood logs in 1970 and 1980 and projections for 1995 are shown in Table 1. The production costs of supplies from new areas--for example, the Irian Jaya region of Indonesia, Papua New Guinea, Amazonia in Brazil, and currently inaccessible areas in Central Africa--would, at least initially, be 50 to 100 percent higher than those in the traditional supplying regions. Since it takes 25 to 50 years for most of the marketable hardwood species to grow to minimum sizes suitable for veneer production, reforestation measures could have no significant impact on tropical hardwood supply of decorative quality until beyond the turn of the century.[7]

---

6. Nigeria recently became a net importer of timber.
7. For the latest estimates of forest resources in the tropics, see J. P. Lanly, "Tropical Forest Resources," FAO Forestry Paper No. 30, FAO, Rome, 1982.

Table 1. Nonconiferous Sawlogs and Veneerlogs--Production.

	Actual		Projected	Growth Rates	
	1970	1980	1995	1970-80	1980-95
	(million m^3)			(% per annum)	
Industrial Countries	72.7	71.6	77.0	-0.4	0.5
North America	38.9	42.3	46.5	0.6	0.6
EEC-10	17.6	16.5	17.5	-0.6	0.4
Oceania	7.0	6.1	7.0	-1.9	0.9
Centrally Planned Economies	36.4	35.5	38.3	-0.3	0.5
Developing Countries*	98.1	134.1	160.7	3.1	1.2
Asia	60.4	81.8	82.9	3.2	0.1
Malaysia	18.7	31.5	22.0	6.0	-2.4
Indonesia	10.7	21.2	25.0	5.9	1.1
China	8.6	13.8	18.2	5.1	1.9
Philippines	10.7	6.4	6.0	-4.5	-0.4
Africa	15.2	19.9	25.0	1.6	1.5
America	17.4	25.3	42.0	4.0	3.4
Brazil	7.5	13.3	26.0	7.4	4.6
World Total	207.2	241.2	276.0	1.4	0.9

*Countries include Southern Europe, South Africa, and China, in addition to the developing countries as defined by the United Nations.

Sources: FAO Yearbook of Forest Products; projections by World Bank, Economic Analysis and Projections Department.

Demand Outlook

As stated earlier, world consumption of tropical hardwood had grown rapidly over the period from 1950 to 1980, compared with consumption of softwood logs and temperate hardwood. Annual growth rates over the 1961 to 1980 period for tropical hardwood, temperate hardwood, and softwood were 4.8 percent, 0.7 percent, and 1.2 percent, respectively. The main explanation for the differential growth rates is the substitution by tropical hardwood for temperate hardwood and softwood. This substitution took place during this period mainly in the industrial countries, all of which are located in the temperate zone. It, in turn, was prompted by the shortage of supply of temperate hardwood and softwood logs and the rapid increase in the availability of tropical hardwood.

Between 1970 and 1980, when world economic growth slowed down, supply of temperate hardwood logs actually declined and consumption of softwood logs increased at only 0.3 percent per annum, whereas consumption of tropical hardwood logs increased at 3.1 percent per annum. World consumption of hardwood logs, and tropical hardwood in particular, declined in 1981, however, and appears to have remained at a low level in 1982. The recent decline is mainly attributable to the particularly depressed conditions of the building industry in industrial countries.

On the basis of the expected resumption of growth in economic activity and in building activity in industrial countries, demand for hardwood logs could be expected to grow at a reasonably high rate if timber prices would remain constant in real terms, but because of tight potential supply for both temperate hardwood logs and softwood logs, log prices are likely to rise in real terms unless tropical hardwood logs are readily available. Since the exportable supply of tropical hardwood logs from major producing countries is likely to be restricted, potentially significant rises in the demand for hardwood logs are likely to be choked off by sharply rising log prices, resulting in acceleration of substitution of nonwood materials for hardwood.

In certain end uses, past trends in the substitution of cheaper wood for high-quality wood, and in the substitution of nonwood materials for wood, are expected to continue. In several construction uses, for example, wood is being displaced by aluminum, plastics, cement, and steel. The ongoing substitution of aluminum and plastics for wood is not likely to decelerate materially.

World consumption of hardwood logs (which is assumed to be equal to production) is projected to grow at only 0.9 percent per annum during the 1980 to 1995 period. The annual growth rate in this period for consumption of temperate hardwood logs is projected to be only 0.5 percent per annum; that for consumption of tropical hardwood, 1.4 percent per annum. (Consumption of tropical hardwood in 1970 and 1980 and projections for 1995 are shown in table 2). These projections reflect the price increase that will bring consumption growth in line with the constrained supply of hardwood.

Consumption of tropical hardwood in producing countries is projected to grow faster than that in importing regions. Between 1980 and 1995, consumption in the producing regions is projected to grow at 1.7 percent per annum, while consumption in the temperate developing countries, centrally planned economies, and industrial countries is expected to grow at 1.7 percent, 0.5 percent, and 0.7 percent per annum, respectively. Consequently, the share of producing

Table 2. Tropical Hardwood—Consumption in Roundwood Equivalent.

	Actual		Projected	Growth Rates	
	1970	1980	1995	1970-80	1980-95
	(million m^3)			(% per annum)	
Developing Countries*	44.1	74.9	96.0	5.4	1.7
Producing Areas	39.5	63.0	80.6	4.8	1.7
Tropical Asia-Oceania	18.8	28.3	35.7	4.2	1.6
Tropical Africa	6.7	12.6	16.0	6.5	1.6
Tropical America	14.0	22.1	28.9	4.7	1.8
Importing Areas	4.6	11.9	15.4	10.0	1.7
Southern Europe	1.0	2.5	2.9	9.6	1.0
Temperate America	0.4	0.5	0.5	2.3	0.0
Asia	2.5	8.3	11.4	12.8	2.1
South Africa	0.7	0.6	0.6	-1.5	0.0
Industrial Countries	38.5	37.1	41.5	-0.4	0.7
United States	7.5	3.4	5.0	-7.6	2.6
Western Europe	9.3	13.4	14.2	3.7	0.4
Japan	20.1	20.1	20.0	0.0	0.0
Other Industrial†	1.0	2.0	2.3	7.2	0.9
(Residual)‡	(0.6)	(-1.8)	(0.0)	-	-
Centrally Planned Economies	1.0	1.4	1.5	3.4	0.5
World Total	83.6	113.4	139.0	3.1	1.4

*In addition to undesignated developing countries, Southern Europe, South Africa, and China are included in the developing countries.
†Canada, Australia, and New Zealand.
‡Errors and omissions.

Sources: FAO Yearbook of Forest Products; ECE/FAO Timber Bulletin; U.S. Forest Service data; Japan Forest Agency; projections by World Bank, Economic Analysis and Projections Department.

countries in world consumption of tropical hardwood will increase from 55.6 percent in 1980 to 58 percent by 1995.

Trade Outlook

Until recently, the bulk of tropical hardwood exports from the log-producing developing countries has been in the form of logs, with the remainder in sawnwood, veneers, and plywood. The share of processed products in the total tropical hardwood exports of log-producing countries had been relatively low. It remained constant at 24-25 percent in the 1960s. The share, however, increased during the 1970s, rising to 40 percent in 1980. Given the determined efforts on the part of tropical hardwood-producing countries to coordinate their export policies and to restrict log exports, it is likely that an increasing proportion of future exports will be in processed form.[8] Thus, the volume of exports of tropical hardwood logs is expected to decline, while exports of processed timber (in roundwood equivalent terms) are expected to increase at 3-4 percent per annum.

Price Outlook

Prices of tropical logs in real terms are projected to rise substantially, to close the gap between potential demand (at constant real prices) and available supply. The expected price increases are based on the notion that, as production moves to the untapped forest resources in the interior regions of Amazonia, Papua New Guinea, Zaire, Cameroon, and Central African Empire, production costs will rise considerably. The price of Lauan logs in Japan (the national average wholesale price as reported by the Forestry Agency) and the price of Sapelli logs (the best quality, f.o.b. Cameroon ports, as reported by the Marché Tropicaux et Méditerranée) are used as the indicator prices for the East Asian and West African markets, respectively. These indicator prices in 1981 constant dollars are expected to be 30 percent higher in 1995 than in 1981 (table 3).

Similarly, prices of sawn tropical hardwood are projected to rise in the projection period. The price of Malaysian red Meranti sawnwood (select and better quality, standard density, c.i.f. French ports) is used as the indicator price. In the long-term, the indicator price in 1981 constant U.S. dollars is projected to rise by 20 percent between 1981 and 1995.

---

8. An analysis of the issues and prospects of mechanical processing of tropical hardwood in the Asia-Pacific region is provided in Kenji Takeuchi, "Mechanical Processing of Tropical Hardwood in Developing Countries: Issues and Prospects of Plywood Industry Development in the Asia-Pacific Region," World Bank Commodities Division Working Paper No. 1982-1, January 1982.

Table 3. Timber--Prices, 1955-81 (Actual) and 1982-95 (Projected) ($/cubic meter).

	Lauan Logs*		Sapelli Logs†		Sawnwood‡	
	Current $	1981 Constant§	Current $	1981 Constant§	Current $	1981 Constant§
Actual						
1955	25.3	98.1	NA	NA	NA	NA
1956	24.4	92.4	36.0	136.4	NA	NA
1957	23.5	84.8	33.2	119.5	NA	NA
1958	22.4	76.7	33.2	113.7	74.3	254.4
1959	26.4	94.9	29.5	106.1	68.0	244.6
1960	29.0	101.7	34.8	122.1	84.9	297.9
1961	30.0	105.3	39.3	137.9	68.6	240.7
1962	32.9	116.7	37.5	133.0	71.8	254.6
1963	32.1	112.6	39.1	137.2	76.5	268.4
1964	27.4	94.8	39.5	136.7	82.5	295.5
1965	31.7	107.4	39.5	133.9	81.4	275.9
1966	33.3	110.3	38.0	125.8	73.1	242.1
1967	35.6	116.3	37.5	122.5	76.9	251.3
1968	36.5	127.2	42.0	146.3	84.1	293.0
1969	35.2	121.8	49.6	171.6	90.2	312.1
1970	37.2	116.6	43.0	134.8	92.9	291.2
1971	38.0	109.8	44.5	128.6	92.5	267.3
1972	37.6	98.9	52.5	138.1	109.5	288.1
1973	65.6	144.5	133.6	294.3	156.1	343.8
1974	78.6	139.8	120.5	214.4	143.1	254.6
1975	59.3	92.1	126.6	196.6	166.4	258.4
1976	79.6	121.5	142.3	217.2	168.1	256.6
1977	89.8	126.3	158.8	223.3	154.1	216.7
1978	91.8	109.0	191.3	227.2	205.4	243.9
1979	160.2	170.2	211.5	224.8	339.1	360.4
1980	192.9	186.2	251.7	242.9	365.1	352.4
1981	144.5	144.5	212.8	212.8	314.1	314.1
1982	144.7	144.0	175.9	175.0	302.0	301.0
Projected						
1983	160.0	150.0	215.0	201.0	335.0	313.0
1985	206.0	164.0	299.0	238.0	425.0	339.0
1990	286.0	170.0	421.0	251.0	587.0	350.0
1995	426.0	190.0	606.0	270.0	843.0	375.0

*Lauan for plywood and veneers, length over 6.0 m and diameter over 60 m, average wholesale price in Japan.
†West African, high quality (Loyal and Marchand), f.o.b. Cameroon.
‡Malayan dark red Meranti, select and better quality, standard density, c.i.f. French ports.
§Deflated by the World Bank unit value index of manufactured exports (SITC 5-8) from developed to developing countries (i.e., bilateral weight) on a c.i.f. basis. The source is the U.N., Monthly Bulletin of Statistics, for the historical period.

Sources: Japan, Ministry of Agriculture, Forestry and Fisheries: Nihon Keizai Shimbun; Marché Tropicaux et Méditerranée (actual); World Bank, Economic Analysis and Projections Department (projected).

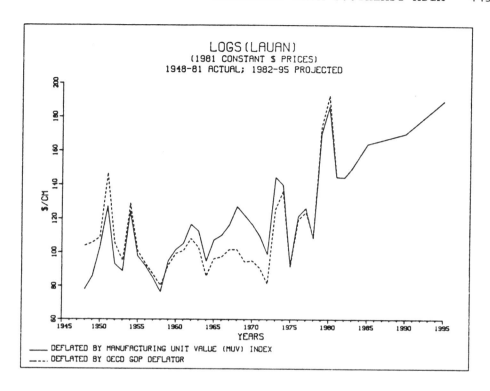

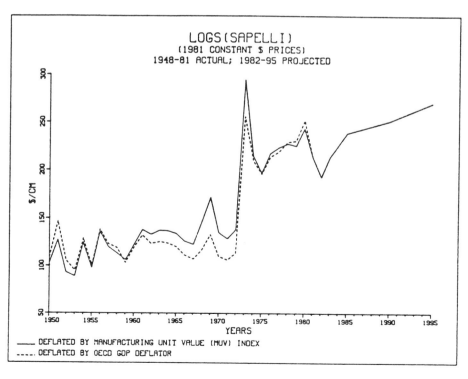

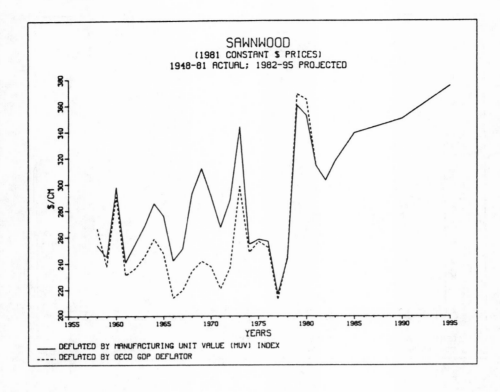

EXPANDING CORPORATE HORIZONS FROM DOMESTIC TO WORLD TRADE

Charles L. Spence

All the statistics show that there is a growing demand for forest products throughout the world. With the exception of such periods as the 1981 and 1982 recession (and perhaps 1983), this should prove to be true over the coming years. There is no question that the world has a pent-up demand for housing, and with the gradual return of stable economic conditions in various foreign lumber markets, our overseas business will greatly improve.

At the present time we have several very negative factors affecting our trade, and these are making it difficult to market forest products. The negative factors are high inflation in some countries, a strong U.S. dollar against a weak currency in many of the wood-consuming areas, and political instability and unrest.

Perhaps we should take this by country or area in order to understand the situation. England, a large consumer of softwoods, has been faced with economic problems for the past several years, along with a currency that has been slipping down against the dollar. The British are still buying some wood products, but until their economy improves and our dollar weakens it becomes an increasingly difficult market. Italy, France, Holland, Belgium, Greece, and Germany are now facing very difficult times and, as you know, have faced devaluation of currency, high unemployment, and political unrest. All of these countries are consumers of West Coast forest products in the form of sawn lumber, plywood, and other manufactured wood products.

In another area of the world is Australia, a large consumer of lumber from the Pacific Northwest. They also are suffering the hard times and devaluation of their currency, but the future is there, and Australia will continue to be a good customer.

The OPEC area countries have been importing in fairly steady quantities, but now with the oil surplus, we anticipate a big reduction in their use of our wood products until oil prices begin to rise.

The bright spots of the export world are the Far Eastern countries, Japan, Korea, China, Taiwan, and some Malaysian countries. Japan is the largest consumer of forest products. Logs have been a major import item for years, as have large quantities of sawn lumber.

In past years Japanese lumber size and length requirements were not attractive to the U.S. domestic mills. Also we had not been able to be competitive against the Canadians with their tidewater production and cheaper freight rates. But there has now been a radical change. The U.S. mills have modernized and installed capabilities to cut the sizes and lengths needed by the Japanese market. The Japanese people are moving away from the traditional house and are beginning to favor the two by four construction type home. This has now opened the door for American mills. Plywood may also become a major item if the duty can be reduced, and if producers and customers can resolve the three-by-six-foot size problem. We have also become price competitive with ocean freight.

One problem now affecting the sales of forest products to Japan is the strength of the U.S. dollar. The weakening of the yen against the dollar has made trading much more difficult. If the yen ever strengthens, we will see a much improved Japanese market.

China is the great unknown. It is a country that will need large quantities of forest products. Millions of people must be kept working in that country, and they prefer to purchase raw logs from the U.S. and Russia. Lately they have been showing more interest in buying products such as linerboard, plywood, and sawn timbers. As the Chinese enjoy more contacts with the western world, they will need more forest products to improve their ports, bridges, cities, and housing. This area will require much patience and understanding by anyone wanting to trade there.

Korea has been a fairly active buyer of logs, and Korea and Taiwan may ultimately require small quantities of sawn lumber.

To service the export market of the future in all the countries mentioned, we must have well-trained wood-oriented people who have a familiarity with the requirements of the world markets. Every country has its own particular specifications. There must be knowledge of the species required, such as hemlock, redcedar, Douglas-fir, redwood, spruce, and others. We need an understanding of all U.S. wood products grading rules.

We must determine the best method of developing a foreign market. In doing this we must study and understand countries that are actively buying U.S. forest products.

Information concerning markets can be found by obtaining leads through private companies, banks with international departments, and public agencies such as the U.S. Department of Commerce, state international trade divisions, state agriculture departments, and port authorities.

Banks with active international overseas departments are excellent sources of information for doing business in individual countries. They handle the payment mechanisms and offer contacts with foreign correspondent banks.

The methods used for selling our products overseas are very important to our industry. We have several options:
1. Working through foreign agents in each country, who work on a fee or commission basis. If a good reputable agent can be found, this is probably the best and safest way to operate.
2. Selling directly to foreign consumers, which is often not practical.
3. Establishing an overseas office with a staff. This is very expensive and, unless you have a large volume, impractical for the small exporter.

In making arrangements for sales in various countries, keep in mind the language barriers which will perhaps determine the selling method. Trying to sell directly to a customer in Hiroshima who speaks only Japanese becomes a large obstacle in direct business.

Finance is of major importance, for we must often carry the load from the mill to the customer. Letters of credit are of help, but you will find that your competitors in many countries sell on sight draft or other risk-type payment procedures. This often involves carrying accounts from mill to consumer for periods of 45 to 60 days. A good credit rating with a banker is a must for obtaining needed loans.

We need to be knowledgeable concerning the problems of overseas transportation by being familiar with the loading ports and the most competitive freight rates to importing countries. Know the requirements of an ocean bill of lading, with technical descriptions, credit requirements, and protection of goods. Marine insurance is a must, and should give you full protection until paid.

Here on the West Coast our biggest competitors overseas are the Canadians, so be familiar with the freight advantages they often have in various markets.

As a world trader you will become aware that the cost of lumber is only a part of the price to the consumer. We have lumber cost, transportation costs, dock charges, brokerage charges, freight, insurance, and agency commissions.

In addition to all of the aforementioned considerations, we must be familiar with the political problems, the rate of exchange with the U.S. dollar, and the people and their customs in the foreign market.

The exporting business is not a simple business. It is one that requires hard work and long hours. It is not all a life of exotic overseas trips, but rather one of expensive travel costs, language barriers, and hard work. On the plus side, it is a very challenging business. Doing better than the competitors, meeting and developing friendships, and making contacts with many fine people make it worthwhile.

If you are willing to accept some of my previous remarks, and are not afraid of hard work, you could become a forest products exporter. Good luck.

INTERNATIONAL JOINT VENTURES
AND THE PACIFIC RIM WOOD INDUSTRY

J. Frederick Truitt

Foreign Trade or Foreign Investment?

Pacific Rim Wood Trade as an Important Part
of World Wood Trade

Pacific Rim countries produce, trade, and consume a lion's share of the world's wood product, and trade among the Pacific Rim countries is the single most important piece in the mosaic of world wood trade: (1) The world's largest national economies and largest producers and consumers of wood--the U.S., Japan, and the U.S.S.R.--are Pacific Rim countries. (2) The world's largest coniferous forests, in Canada, the U.S., and the U.S.S.R., and nonconiferous forests, in Indonesia, Malaysia, and Papua New Guinea, are located in Pacific Rim countries. (3) Some of the largest and most extensive reforestation, 35 million hectares in China, 10 million hectares in Japan, 4 million hectares in Korea, and most exciting plantation forestry--in Malaysia, Philippines, Australia, Chile, and New Zealand--is taking place in the Pacific Rim (Sedjo 1982). (4) From their extensive natural and artificial forests, Pacific Rim countries produce 77 percent of the world's output of coniferous roundwood and 67 percent of its nonconiferous roundwood output (Aird and Calow 1982). (5) Sixty-eight percent of the world paper and paperboard capacity and 73 percent of world woodpulp capacity were located in Pacific Rim countries in 1978. (6) The world's largest importers and exporters of wood and wood products are located on the Pacific Rim, and Pacific Rim countries import 44 percent and export 52 percent of all forest products traded on world markets.

The largest and most proficient trading nations of the world--the U.S. and Japan--are Pacific Rim countries, and the presence of these two countries in any set of economic statistics shifts the center of gravity of just about anything to the Pacific. The Pacific Rim, however, is also the location of four of the most dynamic newly industrializing countries (NICs), the "chopstick economies" or the "Gang of Four." In three of these four countries--Taiwan, Korea, and (to a lesser extent) Singapore--the conversion of Pacific Rim logs into semifinished or finished product for export to Organization for Economic Cooperation and Development (OECD) and Middle East markets has played a role in the industrialization and development of these resource-poor but industry-rich Asian countries.

What Role for Foreign Investors?

We know from experience and research that wood and wood products are highly tradable.[1] In the wood industry the cost of moving the key factor of production, the forest land, across national borders is infinitely high but the cost of moving the commodity made from that factor, wood, is relatively low. These two conditions generate a high level of commodity movement in international trade--a classic case where international trade theory in its simplest form does a good job explaining trade flows.

If the only factors of production involved were forest land and chainsaw labor, and the only product were two by fours, and government policies did not enter the picture, we might be able to halt the explanation here. But large amounts of capital, technology, energy, and diverse and influential government policies very definitely do enter the trade and investment picture, especially when the discussion of product includes plywood, pulp, and paper. To find an explanation for the patterns of the trade <u>and foreign investment</u> in the wood industry we must extend our vision beyond the simple theory of comparative advantage to location theory and various theories of foreign direct investment.

---

1. Lawrence G. Franko (1971) developed a measure of "tradability" of products in the joint ventures he examined in his study. The measure was the sum of U.S. exports plus U.S. imports divided by U.S. production in the same three-digit SIC industry. In the 85 three-digit categories studied by Franko only 10 exceeded lumber in tradability and only one, "smelted and refined nonferrous rolled metals," exceeded "paper mill products" in tradability.

Two Important Strands of FDI: U.S. in Canada and Japan in North America and Southeast Asia

Two important strands of foreign direct investment (FDI) in the Pacific Rim wood industry were already well established before the disruptive and disappointing events of the 1970s. First, U.S. FDI in Canada was already a major feature of the North American forest products industry. the 1972 Gray Report, intended to warn Canadians of the magnitude of foreign ownership and control over the Canadian economy, put foreign control of assets, sales, and profits of the Canadian wood sector at 31, 22, and 24 percent, and of the paper and allied products sector at 40, 41, and 41 percent, respectively. Foreign ownership and control in the wood and paper sectors was significantly below the levels of foreign (mostly American) ownership of all Canadian industry: 58 percent of assets, 55 percent of sales, and 64 percent of profits.[2]

In recent years, most FDI in the Canadian forest industry has been for expansion of existing facilities or greenfield plant construction by existing subsidiaries and affiliates. There have been few takeovers of existing Canadian companies in the forest products sector. The share of foreign ownership in Canada remains high, even after a decade of increased hostility to foreign investment. In 1976, information provided by the Corporation and Labour Union Returns Acts indicated that foreign-owned companies accounted for 44 percent of sales in paper and allied industries and 26 percent in panel products.[3] More recently the <u>Foreign Investment Review</u>, Autumn 1980, reported that 37 percent of Canadian pulp and paper manufacturing capacity is owned by non-Canadian companies, and in British Columbia companies with majority foreign ownership hold 29 percent of sawmilling capacity, 37 percent of pulp capacity, and 43 percent of veneer and plywood capacity, but only 18 percent of paper capacity.

American investment in the Canadian forest products industry (about three-fourths of FDI in this industry is from the U.S.) can be explained (but not here) by the

---

2. <u>Foreign Direct Investment in Canada</u> (Ottawa: Information Canada, 1972).

3. Department of Industry, Trade and Commerce, <u>Review of the Canadian Forest Products Industry</u> (Ottawa: Minister of Supply and Services Canada, 1979), p. 140.

relative openness of the Canadian economy to FDI in the pre-FIRA era and various theories of FDI.

The second major strand of FDI in the Pacific Rim forest products industry is Japanese investment in North America and Southeast Asia. Japan's "outward dependency" on natural resources was already recognized and in the process of being alleviated by Japanese government policies, the general trading companies (GTCs), and process industries before the shocks and disappointments of the 1970s. The superrapid growth of the Japanese economy and steady increase in per capita wood consumption coupled with the post-World War II rundown in Japanese domestic wood supply capacity made it clear by the mid-1960s that Japan would have to take steps to secure external supplies of wood. The process of securing foreign supplies of wood by way of GTC investments began more than twenty years ago, and joint ventures became the important form of investment even before host countries began to tighten regulations on incoming FDI.[4]

---

4. Especially the Hymer-Kindleberger diversification and spreading of risk, and oligopolistic reaction explanations summarized in Buckley and Casson (1979).

Substantial American investment in Canada goes back to the 19th century. American entrepreneurs seeking new and low-cost minerals and timber spilled over the border into Canada in an age when and in places where the border was regarded as rather meaningless. The majority of American stakes in Canadian resources and industry were by individuals or small companies, but they added up: in 1909 one survey indicated that 90 percent of the timber in British Columbia was controlled by American companies or individuals.

American capital was attracted by accessible supplies of raw materials, low-cost power, and abundant water resources. Government policies on both sides of the border played a role, too. Canada's National Policy made it relatively difficult to export to Canada from the U.S. but relatively easy to cross the border and set up shop with an investment, and in 1911 the U.S. admitted Canadian newsprint duty free. Indeed, American investment in the Canadian paper and pulp industry accounted for more than half of American investment in Canadian manufacturing through 1914. See Mira Wilkins, <u>The Emergence of Multinational Enterprise</u> (Cambridge: Harvard University Press, 1970).

1970s: A Decade of Shocks

By now the "shocks" of the last decade are well known: (1) The Bretton Woods system collapsed by the middle of the decade and was replaced by nothing. (2) Two rounds of OPEC price increases raised real energy costs, shifted countries' comparative advantage, and, coupled with the Bretton Woods collapse, helped set the stage for the current semipermanent bank-debt crisis of the less developed countries (LDC). (3) Increased capital costs of minimum-economic-scale pulp mills, set against lackluster growth in traditional markets (OECD economic growth in the 1960s was much better than the 1970s, while world industrial wood consumption increased 3.7 percent per year in the 1960s, but only 1.8 percent per year in the 1970s), gave reason for caution in expansion at home or abroad. (4) The "unlimited" availability of ever-expanding new sources of lower cost timber appears to have passed away forever. (5) The willingness of both OECD and LDC governments to resort to (or threaten to resort to) sometimes disguised and sugar-coated, sometimes barefaced protectionism and local processing requirements is by now painfully clear.

A cursory examination of the impact of these "shocks" on the patterns of trade and investment suggests the following kinds of changes:

1. Lower economic growth, reduced per capita consumption of wood, higher energy costs, and increased protectionism all lead to a slowdown in trade and investment with perhaps more attention to recycling. For example, in 1978 Japan's use of waste paper in paper production amounted to 75 percent of domestic pulp production (Japan Paper Association 1980).

2. Higher energy and capital costs increase transportation costs of both the raw material (logs) and the finished product, but since wood processing is both a weight- and volume-losing and value-adding process, one might expect to see an increase in investment (foreign and/or domestic) in processing nearer the source of raw materials.[5] One can see how value increases with processing in table 1.

---

5. Takeuchi (1974) estimates that converting tropical hardwood logs to sawnwood, plywood, and veneer cuts about 40 percent of volume and weight from raw log material. Since ocean freight (in 1974) amounted to 25 to 33 percent of the

Table 1. Unit Value of Imports and Exports in 1978 (price in U.S.$ per cubic meter).

Product	Japan	U.S.	Canada
Coniferous (c)			
Sawn and Veneer Logs:			
Imports	93	42	22
Exports	265	72	51
Nonconiferous (nc)			
Sawn and Veneer Logs:			
Imports	77	50	43
Exports	287	205	70
Sawnwood (c):			
Imports	146	95	126
Exports	384	142	89
Sawnwood (nc):			
Imports	183	192	163
Exports	633	156	191
Plywood:			
Imports	422	267	271
Exports	847	378	217

Source: FAO Yearbook of Forest Products 1978.

3. Increased government intervention in the form of protectionism and local processing requirements, coupled with conditions in item 2 above, could encourage FDI--giving an extra fillip to natural shifts in comparative advantage.
4. Increased protectionism and local processing requirements can coincide with increased nationalism and an unappealing if not hostile

---

average cost, insurance, and freight price of tropical hardwood imports in Japan and Korea, and since the cost of raw log material accounts for 65 to 85 percent of the total cost of producing sawnwood plywood and veneers, any factor price increase (energy) that would increase the relative price of ocean freight would seem to encourage more local processing in the log exporting countries.

regulatory environment, however, thus discouraging FDI. The sectors of the wood industry of most interest to the local processing goals of the developing countries--sawmilling and plywood--are not judged to be particularly difficult to master; that is, the contribution of the foreign investor is not highly prized by the host government in these industries. (During the summer of 1981 the author was repeatedly told in Singapore, Malaysia, and Indonesia that foreign investment was not particularly necessary or welcome in sawmilling and plywood because the capital and management requirements were not especially difficult and the technology could easily be acquired from vendors in Scandinavia, Germany, the United Kingdom, Japan, and even in Taiwan and Korea.) In the pulp and paper sector, where the capital and technology requirements more clearly suggest FDI, foreign investors are correct to focus on the political risks associated with large immobile and vulnerable capital investments dependent on consumption of natural resources with uncertain tenure in, say, Sumatra or the Philippines.

Joint Ventures

Defining the Term

The term "joint venture" means different things to different people. A literature search today is likely to turn up more references to activity or proposed activity in or with centrally planned economies like China and Eastern Europe than articles on FDI in OECD or LDC hosts, but joint ventures were well-established themes in international business literature as early as 1961. The term "joint venture" used in connection with the People's Republic of China (PRC) or the Council for Mutual Economic Assistance (CMEA) countries today can refer to an activity that actually does involve some kind of joint ownership, or it may refer to various hybrid forms of coproduction, compensation trade, countertrade, and licensing agreements. But many ventures in the PRC and CMEA countries going under the name joint venture are in fact more properly understood as temporary devices for accomplishing some specific task, at which time most U.S. countries enter joint ventures with the expectation that they are going to last although the venture is expected to end. Some U.S. managers see joint ventures in OECD and LDC hosts as learning periods and steps toward total acquisition.

As we use the term here, a joint venture is any joint ownership relationship between one or more foreign (i.e., foreign to the host country) firms and one or more local (i.e., indigenous to the host country) firms. There are several possible categories of partners to a joint venture: private or public entities from the home country, private or public entities from the host country, private or public companies from third countries, and the general shareholding public in the host country.

From the perspective of the home-country company the joint venture may be majority owned (51 to 95 percent owned by the company from the home country), 50/50, or minority owned. (Entities more than 95 percent owned are considered to be wholly owned subsidiaries.) The possible combinations, while not exactly endless, are quite numerous.

A Quick Survey of the General State of Affairs of Joint Ventures

William H. Davidson's 1982 book on global strategic management draws heavily on the Harvard Business School's Multinational Enterprise Management study (Davidson was one of the co-authors of the 1977 edition of this MNE source book) in his discussion of 1,023 key joint ventures made by 180 of the most important U.S. MNEs. Of these 1,023 joint ventures, 620 were of the type with one U.S. partner and one host-country partner; 143 joint ventures were with other MNEs from outside the host country. These joint ventures were principally in large, capital-intensive natural-resource industries such as mining, chemical refining, iron and steel, nonferrous metals, paper, glass, electrical machinery, and motor vehicles. Davidson suggests that this type of joint venture may be a way to avoid confrontation with large foreign competitors in foreign markets. Of the 1,023 joint ventures studied, 115 had a widely dispersed foreign ownership pattern and were concentrated in OECD host countries and Mexico and India, where relatively well-developed stock markets make such an ownership pattern possible. Only 81 of the 1,023 joint ventures were with host governments as partners, and these ventures were concentrated in refining, chemicals, and tires in LDC host countries (Davidson 1982).

Why Firms Enter a Joint Venture

Joint ventures were a familiar if not common form of ownership long before the last decade's increased hostility to FDI in general and FDI by MNEs in particular. But few American executives really liked the idea of joint ventures ("If it's worth doing, it's worth doing ourselves") and some have been adamantly opposed: "General Motors holds that unified ownership for coordinated policy control of all its operations throughout the world is essential for its effective performance as a worldwide corporation" ("GM Position on United Control of Foreign Corporations," GMC, February 11, 1963). Other firms (IBM) would prefer divestment or not going abroad to sharing ownership, but some companies seem to have managed for years to cope very well with joint ventures, and even to advocate them: "I have a strong opinion that every company establishing a manufacturing facility in Mexico, or anywhere else for that matter, should organize it with local capital" (Celanese Corporation vice-president, quoted in Richard Robinson, "Management Attitudes toward Joint and Mixed Ventures Abroad," Western Business Review, February 1962).

One thing is certain. The heady days of superpowerful MNEs leaping across national boundaries in a single bound and challenging the sovereign power of governments are over. And it is strange to recall just how heady some statements were in the days when MNEs appeared to be taking over the world: "The nation-state is just about through as an economic unit" (C. P. Kindleberger, American Business Abroad, 1969); "it is the most powerful agent for... internationalization of human society.... And the social responsibilities, the social awareness of the multinational corporation is one of its best characteristics" (Center for the Study of Democratic Institutions, April 1971); "In the new global economy we will see small transnational companies run in West Africa, by London telecommuters who live in Honolulu" ("The Future of International Business," The Economist, January 22, 1972).

By the early 1970s, host governments were already demanding and getting more, not less, sovereign control over MNE access to their resources, markets, and local financing. But the benefits MNEs had to offer their increasingly critical hosts--technology, management, and access to world markets and foreign exchange--are real, and one solution to the conflict between FDI and sovereign host governments (especially, but not exclusively, LDC hosts) is the joint venture form of ownership as a condition of entry. This is not the place to discuss the investment regulations in the thirty-odd Pacific Rim countries, but we can say with some confidence that a willingness to share ownership with a local partner is an explicit and implicit condition for entry by many LDC hosts. Put more directly, many joint ventures in the Pacific Rim today, especially resource-using joint ventures, are joint ventures because they have to be.

Host governments encourage or require shared ownership because they hope to increase the flow of benefits to the local economy and perhaps to change the style of management of the venture such that local society and political economy derive more benefit from the joint venture than they would have derived from a wholly owned subsidiary. The core threat of FDI by MNEs to an LDC host is that the MNE has some special strengths and opportunities to think and act in ways that extend beyond the small and necessarily narrow vision of the host--a special ability to use resources located beyond the jurisdiction of the host, but in ways that override the interests of the host. This ability is worrisome both to governments wanting sovereign control over their territory and to local businesses that must compete with the large MNEs. MNE power is also obnoxious to host-country intellectuals who would at least challenge and change, if not destroy, the status quo represented by the host government--hence the widespread appeal of joint venture requirements as a condition of entry. There is strong evidence that explicit and implicit government policies have had a major impact on increasing the number of joint ventures (Hood and Young 1979).

Overall Trends Toward More Joint Ventures

Virtually all recent studies and interpretations of MNE data indicate an increase in the use of the joint venture form of foreign investment. Let us consider first the major joint venture trends recorded in these studies before we turn back to the question of whether or not the joint venture form is in fact in the narrowly defined best interests of the host countries that insist on having them. (See tables 2-5).

This is not the time or place to discuss the details of the movement to joint ventures in LDCs, but some "megatrends" stand out:

. Before 1951 more than 58 percent of the overseas affiliates established by the biggest 180 U.S.-based MNEs were wholly owned subsidiaries, and only 11 percent were minority owned joint ventures. By

Table 2. Ownership Patterns* of 1,276 Manufacturing Affiliates† of 391 Multinational Corporations Established in Developing Countries, by Period of Establishment, 1951-1975.

Home Country and Type of Ownership	Before 1951	1951-60	1961-65	1966-70	1971-75
Affiliates of 180 U.S.-based Corporations					
Total number	214	229	281	303	231
Wholly owned	125	102	105	140	101
Majority owned	26	49	54	54	40
Co-owned	12	18	32	34	24
Minority owned	24	43	61	65	65
Unknown	27	17	29	10	1
Affiliates of 135 European and U.K.-based Corporations					
Total number	266	244	416	694	---
Wholly owned	104	77	87	131	---
Majority owned	41	49	65	114	---
Co-owned	14	16	46	46	---
Minority owned	26	68	149	292	---
Unknown	81	34	69	111	---
Affiliates of 76 other Multinational Corporations					
Total number	73	42	159	279	---
Wholly owned	20	7	17	17	---
Majority owned	6	11	20	23	---
Co-owned	9	3	10	21	---
Minority owned	12	18	106	207	---
Unknown	26	3	6	11	---

Source: United Nations 1978.
*See note to table 4.
†See note to table 4.

Table 3. Distribution of Ownership Patterns* of 1,276 Manufacturing Affiliates† of 391 Multinational Corporations Established in Developing Countries, by Period of Establishment, 1951-1975.

Home Country and Type of Ownership	Number Established as Percentage of Total				
	Before 1951	1951-60	1961-65	1966-70	1971-75
Affiliates of 180 U.S.-based Corporations					
Total number	100.0	100.0	100.0	100.0	100.0
Wholly owned	58.4	44.5	37.4	46.2	43.7
Majority owned	12.2	21.4	19.2	17.8	17.3
Co-owned	5.6	7.9	11.4	11.2	10.4
Minority owned	11.2	18.8	21.7	21.5	28.1
Unknown	12.6	7.4	10.3	3.3	0.4
Affiliates of 135 European and U.K.-based Corporations					
Total number	100.0	100.0	100.0	100.0	---
Wholly owned	39.1	31.6	20.9	18.9	---
Majority owned	15.4	20.1	15.6	16.4	---
Co-owned	5.3	6.6	11.1	6.6	---
Minority owned	9.8	27.9	35.8	42.1	---
Unknown	30.5	13.9	16.6	16.0	---
Affiliates of 76 other Multinational Corporations					
Total number	100.0	100.0	100.0	100.0	---
Wholly owned	27.4	16.7	10.7	6.1	---
Majority owned	8.2	26.2	12.6	8.2	---
Co-owned	12.3	7.1	6.3	7.5	---
Minority owned	16.4	42.9	66.7	74.2	---
Unknown	35.6	7.1	3.8	3.9	---

Source: United Nations 1978.
*See note to table 4.
†See note to table 4.

Table 4. Ownership Patterns* of 2,997 Manufacturing Affiliates† of 391 Multinational Corporations in Developing Countries, at Date of Entry and in 1970s.

Home Country and Type of Ownership	Number of Affiliates Classified by Ownership Pattern at Entry					
	Wholly owned	Majority owned	Co-owned	Minority owned	Unknown	Total
Affiliates of 180 U.S.-based Multinational Corporations, 1975						
Total	580	203	81	237	1	1,102
Wholly owned	472	26	5	20	–	523
Majority owned	43	137	9	14	1	204
Co-owned	14	6	64	11	–	95
Minority owned	24	21	2	175	–	222
Unknown	27	13	1	17	–	58
Affiliates of 211 Multinational Corporations Based Elsewhere than in U.S., 1970						
Total	358	439	352	434	312	1,895
Wholly owned	186	81	70	43	12	392
Majority owned	70	315	34	7	2	428
Co-owned	32	24	228	14	–	298
Minority owned	50	18	16	348	9	441
Unknown	20	1	4	22	289	336

Source: United Nations 1978.
*Affiliates of which the parent firm of the system owns 95 percent or more are classified as wholly owned; over 50 percent, as majority owned; 50/50 as co-owned; 5 to under 50 percent, as minority-owned.
†The affiliates of U.S.-based corporations are those in which the U.S.-based parent of the multinational enterprise held a direct equity interest: the affiliates of corporations based in the United Kingdom, Western Europe, and Japan include those in which parent companies held equity interest indirectly through other affiliates.

Table 5. Foreign Manufacturing Subsidiaries of 391 Multinational Enterprises, Classified by Ownership Patterns.*

Country and Ownership Patterns	180 U.S.-based Enterprises		135 Europe- and U.K.-based Enterprises		61 Japan-based Enterprises		All 391 Enterprises in Sample‡	
	No.	%	No.	%	No.	%	No.	%
All Countries								
Total subsidiaries	5,727	100.0	4,661	100.0	562	100.0	9,601	100.0
Wholly owned†	3,730	65.1	2,278	48.9	34	6.0	4,907	51.1
Majority owned	1,223	21.4	1,320	28.3	74	13.2	2,177	22.7
Minority owned	723	12.6	712	15.3	431	76.7	1,537	16.0
Unknown	51	0.9	351	7.5	23	4.1	980	10.2
Industrialized Countries								
Total subsidiaries	3,603	100.0	3,207	100.0	46	100.0	6,060	100.0
Wholly owned	2,612	72.5	1,788	55.7	6	13.0	3,634	60.0
Majority owned	657	18.2	802	25.0	8	17.4	1,260	20.8
Minority owned	302	8.4	404	12.6	30	65.2	626	10.3
Unknown	32	0.9	213	6.6	2	4.3	540	8.9
Developing Countries								
Total subsidiaries	2,124	100.0	1,454	100.0	516	100.0	3,541	100.0
Wholly owned	1,118	52.6	490	33.7	28	5.4	1,273	36.0
Majority owned	566	26.6	518	35.6	66	12.8	917	25.9
Minority owned	421	19.8	308	21.2	401	77.7	911	25.7
Unknown	19	0.9	138	9.5	21	4.1	440	12.4

Source: Vernon 1977 (Harvard Multinational Enterprise Project).
*Data for U.S.-based enterprises are provisional, as of 1975; others are final, as of 1970.
†Subsidiaries of which the immediate parent in the system owns 95 percent or more are classified as wholly owned; 50 percent of more, as majority owned; 5 to 49 percent as minority owned.
‡Includes 15 multinational enterprises based elsewhere than in the United States, Europe and the United Kingdom, or Japan.

1971-75, only 44 percent of the affiliates established in LDCs were wholly owned subsidiaries, and 28 percent were minority owned joint ventures.
- European and other (mostly Japanese) MNEs were always more likely to use the joint venture form, but over the period 1951 to 1975 they moved even further away from the wholly owned subsidiary form.
- At least through 1975, U.S., European, and Japanese MNEs were much more likely to use joint ventures in LDC host countries than in industrialized host countries, but European and Japanese joint ventures tend to be much more stable (i.e., likely to remain the majority, 50/50, or minority joint venture they started out as) than U.S. joint ventures.

While the megatrends are clear, there are some interesting minor trends to discuss in the following sections, and the wisdom of the trends is not so clear.

Vernon (1977) and others have raised serious questions for more than a decade about the economic wisdom of host governments' requiring joint ventures as a condition of entry. Use of a social cost/benefit analysis approach in evaluating a joint venture clearly suggests that local participation in ownership will reduce the future outflows associated with profit remittances and might also reduce outgoing transfer payments and result in more local purchases of inputs (Hood and Young 1979), but even here the evidence is mixed. Nevertheless, a hard look at how the local host partnership capital might otherwise have been used (opportunity cost or next-best use of resources), along with some unsystematic and anecdotal evidence that the social performance of joint ventures might actually deteriorate in some host countries upon conversion from wholly owned subsidiary, does not make a compelling economic case for the joint venture form of ownership.[6]

Joint Ventures: Advantages, Disadvantages, and Strategic Considerations

The emphasis in the preceding section was on joint ventures performed at the point of a gun: "If you want in, you must take on a local partner." It is, however, important to remember that many joint ventures are voluntary marriages if not exactly "love matches." The following sections summarize the advantages and disadvantages of joint ventures and provide some insight into the strategic considerations that lead management to avoid joint ventures.

Advantages of Joint Ventures

The international business literature on joint ventures lists the following as advantages associated with joint ventures:

---

6. The Harvard Business School case studies UNICAP (A) and UNICAP (B), published in Vernon and Wells (1981), illustrate how unattractive the use of very scarce local capital in a joint venture can be to a host society from a social cost/benefit point of view.

1. Politically enhanced image in nationalistic environment; presentation of a good image sympathetic to strong local character at a time and place and in a product where local identification might be important.
2. Access to resources--especially technology, market information, distribution networks, sales forces, and knowledge of local business practices--which might not otherwise be available for ready purchase on the market.
3. Ability to spread risk and thus for the same amount of capital participate in several different and geographically diverse ventures, especially in natural resource ventures where the existence and quality of resources sought are unknown.
4. Benefit from the "experience curve" effect through the joint venture partner.
5. Access to economies of scale (lower unit costs of production) not possible without the extra production and scale provided by the partner.
6. Better position to sell to the government and qualify for government incentives and grants.
7. Where the local partner is the government or a government agency, advantage 6 may be especially true, and the local partner is well placed to act as a go-between in dealing with state agencies and serving as antennae to anticipate changes in regulations and lend stability to the rules of the game under which foreign joint ventures operate.
8. The financial advantage--access to local capital--was an advantage, especially to Italian, British, and Japanese MNEs during periods of exchange control, but was usually a secondary consideration.

Disadvantages of Joint Ventures

The disadvantages of joint ventures are usually reckoned to be as follows:
1. One whole set of disadvantages stems from the conflict of interest inherent in the situation where the MNE partner seeks unification of markets, rationalization, and integration of production on a world scale in the context of a marriage with a local partner whose vision is limited to his own domestic national market. More specifically, these local-versus-global conflicts

include such issues as structure of taxation, pricing in local markets versus world markets, sourcing and transfer pricing, reinvestment versus distribution of earnings, and localization versus standardization of product quality and marketing policy.

2. Fundamental unwillingness on the part of the foreign partner to share earnings derived from a technological innovation or advancement in which the local partner played no part in developing.

3. Diverging disclosure procedures, especially where policies on protection against foreign exchange losses may deviate from local orthodoxy.

4. Use of scarce international management time spent settling disputes between headquarters and the local partner.

5. Legal constraints, especially real or imagined U.S. constraints on trade ("The advantage of a wholly owned subsidiary vis-à-vis the U.S. Justice Department is that you cannot conspire with yourself").

6. Finally, the most encompassing disadvantage of the joint venture form is most fully developed by Franko in his classic study (1971).[7] Stated simply, this disadvantage is the loss of flexibility in a rapidly changing competitive environment. Private corporations capable of investing overseas are likely to be dynamic and evolving entities. An arrangement with a local partner might be highly appropriate today, but can quickly become inappropriate because of changes in the international competitive environment and changes in the corporate strategies necessary to respond to these environmental changes; hence one of the marriage partners in the joint venture outgrows the relationship.

---

7. In his study Franko investigates the reasons for joint venture <u>instability</u> (instability is defined as changes in joint venture status ranging from the buying out of the venture by the home-country parent to divestment) in <u>voluntary</u> joint ventures. He concludes that the most proximate cause of change in joint venture ownership stems from a change in the home-country parent corporate strategy on issues of breadth or concentration of product line, multinational standardization and specialization, and international cost cutting.

Strategic Considerations: When to Avoid a Joint Venture

If Franko's book is the classic work on stability of joint ventures, Stopford and Wells (1972) is probably (along with the Davidson book already cited) the classic work on organization and ownership of MNE subsidiaries. Using the Harvard MNE source data as the basis of further research, Stopford and Wells found that most of the 187 U.S. MNEs studied had some experience with joint ventures: only 33 had had no joint ventures and only two had had all joint ventures, so there was a great disperson of experience with the joint venture ownership form. But out of this vast variety of experience some important patterns emerged. Joint ventures tended to be more common in extractive industries, in acquisitions and diverse manufacturing and marketing activity, and less common where there was a high level of advertising and marketing expense, large research and development efforts, and centralized organizational structures.

Stopford and Wells argue that the key to understanding a firm's willingness to participate in a joint venture is an understanding of the firm's basic strategy. If that strategy dictates the need for unambiguous control, the firm will press for wholly owned subsidiaries, especially where the perceived contribution of the foreign partner is low. Four basic strategies seem to dictate the need for control, hence the avoidance of joint ventures: marketing, rationalization of production, innovation of new products, and security of raw material supply. Let us consider each of these strategies briefly.

1. Firms that have relied on strategy of product differentiation through heavy advertising, uniform global marketing programs, or other intensive marketing efforts (e.g., ethical drugs) exhibit a strong preference for wholly owned subsidiaries. Also research by Franko and Davidson as well as Stopford and Wells indicates that firms are much less likely to form a joint venture in their main product line (three-digit SIC); the wider the product line produced and marketed abroad, the more likely or more amenable the firm is to a joint venture; the narrower the product line, the less amenable the U.S. MNE to joint venture ownership.

2. Some firms, especially firms with maturing products, respond to the threat of maturity by attempting to lower production costs by moving manufacturing to lower cost production environments overseas. This strategy requires a high level of centralization of decision making in the home office and is thus incompatible with shared local ownership.

3. Other firms follow a strategy based on a

continuous flow of new innovations stemming from a high level of centralized research and development in the U.S. and see insurmountable difficulties in reaching an agreement on division of the profits from such innovations. Such firms also express concern about control of secrets and quality control over their output.

4. Stopford and Wells found in their 1972 study (based on data already several years old--data from the pre-OPEC and pre-NIEO era) that firms whose strategies depended on secure access to raw materials were willing to form joint ventures with other MNEs (e.g., Japanese and American firms in Broken Hill, Australia) in OECD countries, but were generally not willing to share ownership with local host partners, particularly in LDCs. This situation has, of course, changed dramatically in the last decade at the insistence of more assertive LDC host governments.

Japanese Joint Ventures in the Forest Products Sector

The general Japanese tendency to go abroad in the joint venture form is well established in the literature on Japanese FDI (e.g., Kojima 1978, Ozawa 1979, Sekiguchi 1979, Yoshino 1976). But it is also clear that Japanese FDI that initially used joint ventures may now be shifting away from joint ownership, thus illustrating the points about the dynamics of international competition and changing corporate strategy made by Franko, Stopford and Wells, and Davidson. Since 1970, Japanese MNEs have turned away from the joint venture form as they have moved into more complex products with increased demand for quality control and more elaborate distribution systems, and as they have moved outside the isolated East Asian markets to North America and Europe, where they encountered an increased need for tighter control. Finally, there is less need for Japanese firms to have access to local capital qua capital, since foreign exchange is no longer the scarce resource in Japan that it was in the first days of Japanese FDI.

The use of joint ventures by the Japanese forest products sector is probably as great as in any sector in Japan. The full story is a fascinating one but can be related here in only barest outline form.[8] One of the

---

8. J. Frederick Truitt, "Logs in the Asia-Pacific: A Renewable Resource Industry under Pressure to Change," in International Business Strategies (White Plains: JAI Press, forthcoming).

principal lessons from this story reinforces the ideas put forth by Franko, Stopford and Wells, and Davidson, especially the idea that joint venture survival and patterns of change depend on changes in the international competitive environment and corporate strategy.

If there is one overarching theme in the story of Japanese FDI in the wood industry, it is that joint ventures, especially joint ventures with GTCs as partners, played an extremely important but changing role in the process of securing wood for Japanese consumers.

Japan's dependence on imported energy is well known. Less well known is Japan's dependence on imports of wood: since the early 1970s, imported wood has been supplying more than 60 percent of Japan's wood needs. Japan's demand for wood soared in the 1960s (from 45 million cubic meters in 1955 to an all-time peak of 117 million cubic meters in 1973), but the share of demand filled by domestic supply decreased steadily because of the depletion of forests during the war and in the postwar period, and the late and small-scale start in reforestation that really did not get under way until the early 1960s. Imported supplies from North America, Southeast Asia, and the U.S.S.R. rapidly and efficiently replaced and supplemented domestic supply. Imports of wood went from 5 percent of total demand in 1955 to about 68 percent today.

The internal circumstances which generated Japan's appetite for wood are well known: (1) superrapid economic growth through the first oil shock; (2) rapid increases in per capita paper consumption and housing starts; and (3) construction of tidewater pulp and paper and lumber mills to take advantage of the innovations in chip (mostly foreign) manufacture and transportation.

The external circumstances which alleviated Japan's outward dependency were as follows:
1. The Columbus Day Storm of October 12, 1962, in the Pacific Northwest blew down about 27 billion board feet of trees (about 50 times what Mount St. Helens blew down), which were then suddenly available for export to a Japanese market that was suddenly able to pay for imports of wood.
2. The GTCs had been developing (or redeveloping) their capacities to supply Japan's raw material-hungry, tidewater process industries. They had the contracts, financing, and foreign expertise to do this at a time when most Japanese resource users were not particularly capable of engaging in overseas ventures.
3. Pacific Rim exporters of wood--not yet preoccupied

with the value-adding, job-creating issues of the 1980s—were happy to supply the wood in just about as raw and unprocessed a form as the Japanese importers wanted it. In 1955, 79 percent of Japan's wood imports came as logs; in 1980, 77 percent came as logs and chips, the next best thing to a log.

In the summer of 1981, I did an extensive survey of Japanese investment in the Pacific Rim to secure sources of wood. Although the survey was reasonably accurate and complete in October 1981 (surely there were gaps then when it was fresh), it is a bit out of date today. A summary of that survey, which identifies and provides some basic information on more than 45 Japanese ventures in nine Pacific Rim countries, is included here in my paper as an Appendix. Several points about these joint ventures stand out and are summarized below.

1. <u>The key institution in preparing the way for Japanese investment in wood was the GTC.</u> It is widely acknowledged that the aggressive domestic sales activities of GTCs in the early 1960s played a critical role in the large and sudden increase in Japanese wood imports from North America. Primarily interested in handling large volumes of standardized products on low margins, the GTC derives its profits from financing and handling invoicing and other paperwork associated with both international and domestic transactions. Of key importance in the GTC's success in selling North American logs and lumber was the very favorable transaction terms regarding promissory notes used for purchases by wholesalers and sawmillers. Typically a GTC was paying the foreign log seller with a line of credit or cash and accepting a five- or six-month promissory note from the Japanese sawmiller or log wholesaler. And in some areas the GTC was also extending credit to small loggers and log exporters.

GTCs of course had been supplying the large and important volume of tropical hardwood log flow (mostly from the Philippines) into Japan through the 1950s <u>and</u> handling the large export volumes of Japanese hardwood plywood in the 1960s. It was only natural, given the expertise of the GTCs, the small scale of the log consumers (in 1960 there were 24,229 Japanese sawmills in operation compared to about 8,500 in the U.S.), and the fragmented nature of the log and lumber distribution system, that the GTC would extend its external services to customers by expanding from tropical hardwood logs into temperate softwood logs as that demand emerged. The move was not always easy, but played an important role in establishing the transpacific trading and investment patterns for the product which is still Japan's second largest (after energy) consumer of foreign exchange and the one product category in which the U.S. has <u>gained</u> market share in Japan over the period 1968-70 to 1976-77).

GTCs were able to provide the connecting link between overseas supply of raw materials and small domestic consumers because of their obvious strengths: an extensive domestic and foreign network organization capable of clear and rapid communication; a great deal of foreign and international experience--they were especially strong in English-speaking personnel--and, most important, they had capital and provided access to financing which they extended to sawmillers and wholesalers desperately squeezed for short-term capital. The post-World War II rebuilding period with imported hardwood logs through the North American softwood import boom of the 1960s and 1970s was characterized by large increases in representation and expanding purchasing power of GTCs.

Initially, the GTC move into the Pacific Northwest log market was small, tentative, and indirect. At first the GTCs did not deal directly with the American forest product giants--Weyerhaeuser, Georgia-Pacific, or Crown Zellerbach--but dealt instead with export brokers, because their first purchases were small volume contracts for lower quality logs which could not be disposed of in the U.S. domestic market. As volumes increased substantially after 1962 and GTCs paid better prices than local markets, the use of brokers ceased and the trading companies began to deal directly with log exporters.

In contrast to logs, (the much smaller) import of lumber required more stable markets and a much more stable relationship between exporter, GTC, and wholesaler/consumer in Japan. Because Canada prohibits the export of logs, and because some Canadian companies (MacMillan Bloedel and Seaboard) showed an interest in exporting lumber to Japan quite early, the task of the GTC in organizing this trade was different from the task of organizing U.S. log export trade. Major adjustments and longer-term, less flexible commitments were necessary. First, the GTC had to guide and pressure Canadian producers to manufacture lumber to the different and more diverse metric Japanese sizes and more demanding quality specifications--straighter grain, fewer knots, and smoother surface. These different specifications called for existing equipment to be run more carefully at slower speeds and in some cases called for new equipment altogether. This new style of manufacture also required a domestic market and/or a chip facility to utilize the raw material not recovered as lumber. The whole concept of sawmilling is different in Japan, where recovery rates from logs are 80 percent and above, while in the U.S. and Canada average recovery is about 50 percent. Thus in North America, sawmilling and chip production are quite naturally complementary parts of the same operation. In contrast to the U.S., the GTC relationship to Canadian forest products companies was required to be more complex and more permanent--a complexity and permanence which, as we see below, was often manifested as a joint venture investment in a mill and chip operation combined.

2. *Several important constraints shaped the way Japanese investment to secure wood evolved*. As noted in the previous section, the first Japanese action to secure wood supply was not direct private investment, in the current textbook sense of the term. Rather, it was investment by GTCs in developing the personnel, contacts, data sources, and communications systems necessary to perfect their trading abilities in Southeast Asia (Philippines, Malaysia, Indonesia) and North America.

The next phase of Japanese activity *did* require investment in the more traditional meaning of the word: capital tied up in concessions, plant, and equipment, as well as personnel. However, this second phase of activity was a bit messy in its evolution and therefore difficult to describe neatly, because it happened in different stages, with different combinations of partners in different host countries. But as we look at the Appendix, "Summary of Japanese Investment in Pacific Rim to Secure Raw Materials," several constraints on the pattern of Japanese investment should be kept in mind:

- GTCs, because of their early acquired ability to work internationally, were quite often a partner in these overseas adventures. They were useful in introducing partners and reducing risks of currency fluctuations and unpredictable changes in volume requirements. (But interviews revealed a tendency to value the GTC less as time passed and international abilities became more widespread.)
- In most of the countries with the best raw material sources there have been log export restrictions (Canada, U.S.), or at least nominal local processing requirements for foreign concession holders (Indonesia, Sarawak) as a quid pro quo for log exports. Although some of these requirements were irregularly enforced until recently (Indonesia), they have helped shape the pattern of Japanese investment.

- GTCs have been more reluctant to move into lumber and plywood production abroad than chip production, because of the delicate, peculiar, and protected Japanese market in these two sectors.

    Sale of foreign lumber and plywood would work against long-standing (and sometimes exclusive) relationships between the GTCs and the 650 Japanese plywood mills that are almost wholly dependent on imported logs, and the larger and more efficient segment (15,000 of 23,000) of the Japanese sawmilling industry, which also consumes imported logs.

    Production of lumber and plywood of sufficient quality for acceptance in the Japanese market has been a serious problem. One company confided that their Japanese customers rejected 20 percent to 50 percent of Sarawak hardwood lumber. Unskilled labor, unmaintained machinery, old and incomplete equipment result in low-quality plywood which cannot meet even Philippine plywood specification standards, let alone the superdemanding Japanese standards.

- But in most places outside of Japan, lumber and chip production are usually complementary halves of the same operation, because of low (by Japanese standards) recovery rates of lumber from logs.

- Therefore, where a Japanese foreign investor is involved in lumber or plywood, there is usually a large domestic market for GTC connections to third-country markets that can absorb the unwanted (in Japan) manufactured output. (If you want the dowry you must also take the bride; where you get rid of her is up to you.)

3. <u>Some observations can be made about the pattern of Japanese joint venture investments</u>. Most Japanese investment is in areas that are richest in wood resources, and have the most strict log export regulations: Canada, Indonesia, and Malaysia. Two major sources of wood in the past--the Philippines and the U.S.--do not seem to have attracted a proportional share of Japanese direct investment. For most of the postwar period easy cutting regulations in the Philippines made investment unnecessary. The Japan-Philippines commercial treaty which would have secured such investment was not signed until 1976 and by then there simply was not that much left in the Philippines to cut. Although U.S. log export restrictions place a constraint on the total quantity of West Coast logs available for export, those that are left are nicely available through GTC trading function, hence there is no

need to invest. More permanent and secure access (i.e., land or concession purchase/investment) to large supplies of standing timber is denied to potential Japanese investors by the pattern of private landholdings, government ownership, and regulation. Thus the GTC trading solution is the most sensible way to continued access to U.S. wood resources in most cases. In some cases, however, a joint venture might provide access to raw material (e.g., Jujo-Weyerhaeuser, North Pacific Paper joint venture).

For reasons mentioned above, most investments involve a GTC. An additional reason for a GTC partner in a chip or pulp venture is found in the scale of the new venture. Sometimes the minimum economic scale of a new chip or pulp venture produces more product than the paper company partner can immediately use in Japan because of the small scale in Japan, domestic supplier commitments, or a down business cycle. In this case, the GTC's function is to dispose of the unneeded supply of chips or pulp.

Almost all investments are joint ventures involving a local partner. The local partner is necessary, particularly in recent years, to secure access to forest concession and to "legitimatize" the exploitation of a natural resource by a foreign country.

Reflecting the recognition of the need to secure new supplies of raw materials, Japanese pulp and paper companies (without GTC partnership) invested in seven trial plantations of 3,000 hectares in rapid-growth pine, <u>Agathis</u>, <u>Eucalyptus</u>, and <u>Albizzia</u> in Malaysia, Indonesia, Papua New Guinea, and the Solomon Islands. Results have been mixed. For example, Kawasaki expected to be able to harvest giant ipil-ipil from its Mindanao plantation at three years eight months, based on small-scale experiments; but on a large scale, harvest will be delayed at least several years. Furthermore, the original purpose of tropical plantation-raised chips was to supply pulp and paper factories in Japan with imported chips. Today the economics of making paper in Japan from imported chips is not as favorable as it was in the early 1970s when most trial plantations were started.

Japanese investments in the "last frontier" of Pacific Rim log supplies--Irian Jaya, Papua New Guinea, the Solomon Islands--have only just begun on a trial planting basis as noted above and Honshu Paper's Jant chip company in Papua New Guinea. Results of this chip operation have not been encouraging.

Japanese forestry investments in Indonesia--currently <u>the</u> principal source of tropical hardwood logs--have a high mortality rate. The first two investments listed under

Indonesia in the Appendix, PT Triomas and PT Zedsko, have ceased functioning since 1979. Also JETRO Jakarta reports five other Japanese forestry projects valued at U.S.$14 million have been "revoked."

References

Aird, K. L., and W. A. J. Calow. 1982. Pacific trade and investment in forest products. In: H. E. English and A. Scott, eds., Renewable resources in the Pacific, pp. 73-81. International Development Research Center, Ottawa.

Buckley, P. J., and M. Casson. 1979. The future of the multinational enterprise. Longman, London.

Davidson, W. H. 1982. Global strategic management. John Wiley, New York.

Franko, L. G. 1971. Joint venture survival in multinational corporations. Praeger, New York and London.

Hood, N., and S. Young. 1979. The economics of multinational enterprise. Longman, London.

Japan Paper Association. 1980. Pulp and paper statistics.

Kojima, K. 1978. Japanese direct foreign investment: A model of multinational business operations. Charles E. Tuttle, Tokyo.

Ozawa, T. 1979. Multinationalism, Japanese style: The political economy of outward dependency. Princeton University Press, Princeton, N.J.

Sedjo, R. A. 1982. Forest plantations, production, and trade in the Pacific Basin. In: H. E. English and A. Scott, eds., Renewable resources in the Pacific, pp. 97-102. International Development Research Center, Ottawa.

Sekiguchi, S. 1979. Japanese direct foreign investment. Allanheld, Osmun, Montclair.

Stopford, J. M., and L. I. Wells, Jr. 1972. Managing the multi-national enterprise. Basic Books, New York.

Takeuchi, K. 1974. Tropical hardwood trade in the Asia Pacific region. Johns Hopkins Press, Baltimore.

United Nations. 1978. Transnational corporations in world development: A re-examination. United Nations, New York.

Vernon, R. 1977. Storm over the multinationals: The real issues. Harvard University Press, Cambridge, Mass.

Vernon, R., and L. T. Wells, Jr. 1981. Manager in the international economy. Prentice-Hall, Englewood Cliffs, N.J.

Yoshino, M. Y. 1976. Japan's multinational enterprises. Harvard University Press, Cambridge, Mass.

APPENDIX: SUMMARY OF JAPANESE INVESTMENT IN PACIFIC RIM TO SECURE RAW MATERIALS

Date Started	Japanese Investor	Capital Ratio	Name of Local Company	Share Capital (millions) (+loans)	Number of Employees	Major Business Lines	Annual Sales or Production	Local Business Partner	Investment Objectives	Comment
AUSTRALIA										
1971	Daishowa Paper C. Itoh & Co.	62.5% 37.5%	Harris Daishowa (Aust.) Pty., Ltd.	A$6.0	153	Chip production from eucalyptus	A$23.0	None	a	
CANADA										
1969	Jujo Paper Sumitomo Forestry	12.5% 12.5%	Finaly Forest Industries	C$8.0	326	Lumber and pulp	C$28.65	Cattermole Timber, Ltd., 75%	a,b	Goal was to secure pulp, took long time (15 years) to become profitable
1969	Mitsubishi Corp. Honshu Paper	25.0% 25.0%	Crestbrook Forest Industries, Ltd.	C$30.7	1,087	Pulp, lumber, plywood	C$65.12	Local capital 49.9%	a,b,h,i,j	Deficit
1970	C.Itoh	80%	CIPA Lumber Co.	C$0.10	90	Lumbering	n.a.	Pacific Logging 20%	a	Absorbed by Q.C. Timber
1970	Mitsubishi Corp.	70%	Mayo Lumber Co.	C$2.0	315	Lumber mill	n.a.	Local capital 30%	a	Absorbed by Mayo Forest Products Ltd. 1978
1972	C.Itoh	100%	Q.C. Timber	C$11.0	238	Lumber mill and veneer	n.a.	none		
1972	Daishowa Paper Marubeni Int'l Ltd.	25% 25%	Daishowa Marubeni Int'l Ltd.	C$12.0	17	Pulp export	C$4.3	Weldwood of Canada 50%	a,b,f,h	Paying dividends
1978	Mitsubishi	40%	Mayo Forest Products Ltd.	C$8.0	100	Lumber	80,000 m³	Pacific Forest Products, Ltd. 60%	a	Reorganized 1978
1981	Daishowa	n.a.	n.a.	C$64	n.a.	Pulp	500 ton/day cap.	West Fraser Mills n.a.	a	Planned for 1981 or 1982
1981	Oji Paper Mitsui	n.a.	n.a.	US$50	n.a.	Paper	n.a.	Int'l Paper	a,b	Joint venture to increase capacity of Int'l Paper's New Brunswick printing paper mill

PACIFIC RIM JOINT VENTURES

Year	Japanese Partner	%	Venture Name	Capital	Employees	Business	Production	Local Partner	Notes	Status
INDONESIA										
1968	Mitsui & Co. and Oji Paper	42.5%	PT Triomas Forestry Development Indonesia	US$1.0	11	Forestry exploitation	n.a.	C.V. Alas 15%	a	Ceased functioning 1979—shares to Indonesia PMON
1968	Sanyo-Kokusaku Pulp	50%	PT Zedsko Indonesia	US$.4 [+1.276]	110	Forestry exploitation, Lumber & chips	US$1.02	Zeds Trading	a,b,d,h, 1	This venture has retired as of 1981
1969	Mitsubishi Corp.	80%	PT Balikapapan Forest Industries	US$8.0 [+2.7]	467	Forestry exploitation	n.a.	PT Meranti Jaya 20%	a	n.a.
1970	Sumitomo Forestry	65%	PT Kutai Timber	US$5.0 [+2.540]	905	Logs, plywood	Logs 100,000 m³ Plywood 4 mil pcs Rp8, 800 mil	PT Kaltimex Jaya	a,h,j	Paying dividend
1972	U.M.W. Timber	70%	PT U.M.W. Marabunta Timber	US$6.0	740	Forestry exploitation	n.a.	PT Marabunta Djaj	a	n.a.
1972	Toyo Menka Kaisha Ltd.	49%	PT East Kalimantan Timber Industries	US$0.8	150	Forestry exploitation	n.a.	PT Satya Djaya Raya Trading 51%	a	n.a.
1973	Konan Tsusho Kaisha	70%	PT Swoody Ltd.	US$1.25 [+0.865]	150	Forestry exploitation	n.a.	PT Kumala Mas Palembang 30%	a	n.a.
1973	Mitsui Overseas Forestry Development	49%	PT Katingen Timber	US$1.0 [+1.785]	233	Forestry exploitation and sawmill	n.a.	PT Sarvhana 51%	a	n.a.
1974	MDI (Jujo, Kawasaki, Sanyo-kokusaku & Oji)	70%	PT Chipdeco	US$2.0 [+4.3]	165	Mangrove chips	n.a.	PT Karya Kencana 30%	a	n.a.
1977	Mitsubishi Corp.	70%	PT Mangole Timber Producers	US$3.124	n.a.	Forestry exploitation	n.a.	Local capital 30%	a	n.a.
1978	Arrow M. Industrial	50%	PT Arrow M. Gobel	US$1.5 [+1.068]	156	Forestry exploitation	n.a.	Pabrik Diesel 40%, Matsushita Electric Trading 10% PT Indo Karya 30%, Shizuoka Indonesia Corp. 10%	a	n.a.
1978	C. Itoh Forestry	60%	PT Arut Bulik Timber Co.	US$1.5 [+4.50]	390	Logging and Sawmill	n.a.		a	n.a.
MALAYSIA										
1966	Mitsubishi Corp.	5%	Sentosa Plywood	M$5.5	490	Plywood	n.a.	Local capital 95%	a	Kluang Johore, Peninsular Malaysia

Date Started	Japanese Investor	Capital Ratio	Name of Local Company	Share Capital (millions) (+loans)	Number of Employees	Major Business Lines	Annual Sales or Production	Local Business Partner	Investment Objectives	Comment
1967	Daishowa Paper Shukko Shoji	60% 30%	Daishowa (M) Wood Products Sdn. Bhd.	M$5.8	9	Rubberwood & sawmill waste chips	n.a.	Tan Teiton 10%	a,h	Selangor Peninsular Malaysia
1967	C. Itoh	35%	Limbang Trading Sdn.	M$2.7	197	Felling & processing trees	n.a.	Local capital 65%	a	Limbang, Sarawak
1969	Kohjin Co. Nippi Boeki	36% 3.9%	Sarawak Wood Chip Co. Sdn. Bhd.	M$1.0	184	Mangrove chip	n.a.	Jasa Sarawak Sdn. 60%	a	Sarikei, Sarawak
1969	C. Itoh Daiken Trade	39% 10%	Mados-C. Itoh-Daiken Sdn. Bhd.	M$4.0	800	Plywood	M$15	Mados 51%	a,1	Deficit/Johore, Peninsular Malaysia
1971	Oji Paper	100%	Oji Malaysia Plantation Sdn. Bhd.	M$10.5	1	Trial planting 451 ha of pine trees	n.a.	n.a.	a,c	Johore, Peninsular Malaysia
1972	Daishowa Paper	89.5%	S.E.A. Afforestation Sdn. Bhd.	M$1.3	10	Trial plantation 133 ha of pine trees	n.a.	Local capital 10.5%	a,g,h	Johore, Peninsular Malaysia
1973	Sumitomo Corp.	30%	Ladalam Timber Complex	M$2.0	96	Sawmill	n.a.	Lahad Datu Plywood 30%, Local capital 40%	a,d	Sabah
1973	MDI (Jujo, Sanyo-kokusaku, Kohjin, Kawasaki & Oji)	51%	Jaya Chip Sdn. Bhd.	M$9.0	550	Chips	n.a.	Local capital 49%	a	Business results are good/ Sandakan, Sabah
1974	Nissan Nohrin Kogyo	50%	Nissan Majulah (Sabah) Sdn. Bhd.	M$.4	8	Timber Lumber	M$.125	Syarikat Majulah (Tawau) Sdn. Bhd.	a,e,h	Deficit/Taway, Sabah
1974	Mitsubishi	30%	Daiya Malaysia Sdn. Bhd.	M$.2	160	Lumber	n.a.	Local capital	a	Sarawak
NEW ZEALAND										
1969	Tokai Pulp	25%	Nelson Pine Forest Ltd.	NZ$1.5	17	Softwood and hardwood chip production	NZ$3.1	N.Z. Forest Products, Ltd.	a,1	Breaking even

Year	Japanese Partner	%	Local Partner	Investment	Employees	Products	Local %	Notes	Status
1971	Sanyo-kokusaku Oji Paper	20% 20%	Carter Oji kokusaku Pan Pacific Ltd.	NZ$36.0	337	Pulp and lumber	Carter Co. 60%	NZ$49.0	a,b,h	Paying divident, good investment
1980	Sumitomo Forestry					Cooperative arrangement with Odlins Ltd. Sumitomo lent Odlins NZ$1.5 million to provide Sumitomo with first option on Odlins' logs and sawn lumber. Arrangement involves technical assistance from Sumitomo to improve yields and quality to meet Japanese specifications.				
1981	C. Itoh	30%	n.a.	US$.3	n.a.	Beechwood chips	n.a.	a	Export to Oji Paper, Chuetsu Paper and Kawasaki Paper
PAPUA NEW GUINEA									
1974	Honshu Paper Co.	n.a.	Jant Pty., Ltd.	n.a.	n.a.	Hardwood chips & 1,511 hectares trial planting of Eucalyptus deglupta	n.a.	200,000 tons pa	a
n.a.	Sanyo-kokusaku	n.a.	Stettin Bay Lumber Co. Pty., Ltd.	n.a.	n.a.	257 hectare planting of Eucalyptus deglupta	n.a.	n.a.	n.a.
PHILIPPINES									
1956	Mitsubishi Corp.	25.5%	Agusan Wood Industries, Inc.	n.a.	200	Forestry development and plywood sales	Ayala Co. 43.1%, Others 11.4%	n.a.	a
1974	Mitsui & Co.	30%	International Agro-Forestry Development	n.a.	37	Forestation, logs	J.G. Sanvictores 70%	n.a.	a
SOLOMON ISLANDS									
n.a.	Japanese Overseas Afforesting Association	n.a.				JOAA is an overseas pulpwood planting organization jointly established by 11 leading paper companies and a forestry company in Japan. In the Solomon Islands it has planted 300 hectares in Terminalia brasii and Albizzia falcata on a trial basis. Initial studies indicate that an investment of 5.9 billion yen would yield about 520,000 cubic yards of chips/year.				
UNITED STATES									
n.a.	Alaska Pulp	n.a.	Alaska Lumber and Pulp Co.	n.a.	n.a.	Lumber and dissolving sulfite pulp	n.a.	530 st/day n.a.	a,b
1972	Daishowa Paper Daishowa Overseas Development	50% 50%	Daishowa America Co.	US$.006	9	Export paper, raw materials, logs	n.a.	n.a.	a,f,h	Breaking even

Date Started	Japanese Investor	Capital Ratio	Name of Local Company	Share Capital (millions) (+loans)	Number of Employees	Major Business Lines	Annual Sales or Production	Local Business Partner	Investment Objectives	Comment
1972	Mitsui & Co.	50%	Kodiak Lumber Mills, Inc.	n.a.	4	Woodchips	n.a.	Mitsui & Co. Inc. USA 50%	a	
1977	Mitsui & Co.	100%	North American Forestry Development, Inc.	US$.008	n.a.	Forest development	n.a.	none	a	
1979	Jujo Paper	10%	North Pacific Paper Corp.	¥110	n.a.	Production Newsprint	500-600 tons/day	Weyerhaeuser 90%	a,b	

Notes: Investment Reasons and Objectives
a. Procurement of raw materials
b. Easy local production due to abundant natural resources
c. Utilization of inexpensive labor and reduction of costs
d. More profitable local production due to industrial promotion and protection policies taken by governments
e. Expansion of sales to local and third-country markets
f. Date collection
g. Other purposes
r. Gaining of royalties

Product Sales
h. exports to Japan
i. local market
j. exports to third countries

Sources: Special Survey--Japan's Overseas Investments, Oriental Economist, 1978-80; Japan Paper Association, Pulp and Paper Statistics, 1981; JETRO Jakarta Center, List of Japanese Investment Projects in Indonesia 1967-1980, April 1981; JETRO Kuala Lumpur Office, Japanese Related Companies in Malaysia, April 1980; Asian Wall Street Journal Weekly; and company interviews.

BANK FINANCE AND SUPPORT FOR EXPORTERS

Robert E. Biehl

International trade has grown in scope in recent years. To meet the needs of companies that must now be especially competitive if they are to win orders in overseas markets, the truly international banks have become extremely sophisticated in their ability to provide creative financing and trade support services to their clients.

Whether a company has years of exporting experience or is just beginning, the international banking institution can be of tremendous assistance in providing market surveys for the company's product, country information and risk factors, credit-worthiness of a potential buyer, and letters of introduction.

Once the exporter has identified a buyer for his product, international payment terms and degree of risk must be considered. The most advantageous term for the exporter is cash in advance of shipment, and although this is not an uncommon occurrence, there must be tremendous faith on the part of the buyer. The smallest degree of risk for both exporter and buyer is obtained by using letters of credit or documentary collections (either sight or time). Sales made on an open account run the highest degree of risk for the exporter, depending of course on the credibility of the buyer.

The letter of credit is an irrevocable letter of instructions issued to an exporter by a bank at the request of its customer, the buyer. In its narrowest sense, it is a specialized and technical instrument used to finance a shipment of goods from one party to another. The letter of credit is often recommended by international bankers because of the following advantages:

1. The seller relies on the credit extended by a bank rather than credit extended by the merchant.
2. The seller is less apprehensive that payment for his goods might be delayed or otherwise jeopardized by political acts or foreign exchange problems in the buyer's country. Holding a bank's commitment, the seller "feels safe."

3. The existence of a commercial credit in the exporter's favor may provide the basis whereby the exporter can obtain financing to purchase or manufacture the goods prior to shipping under the credit. Usance (time) credits allow the exporter to place himself in funds under banker's acceptance financing shortly after shipment, and sometimes prior to export, despite having granted credit terms to the buyer.
4. Letters of credit may be confirmed by the advising bank wherein payment is conditionally guaranteed to the exporter.

Often an exporter is comfortable enough with his buyer and the buyer's country not to require a letter of credit, but not comfortable enough to sell on open account. This is usually based on information provided by his international bank. Under this circumstance, the seller will present all shipping documents to his international bank to forward on a documentary collection basis to the buyer's bank. The instruction to the collecting bank is either to release documents to the buyer against payment or, if the exporter has granted credit terms to the buyer, against acceptance.

Most exports from the United States are quoted and paid for in U.S. dollars, irrespective of the payment mechanism, and in this instance the overseas buyer assumes the foreign exchange risk. There are, however, some countries which have stringent foreign exchange regulations that prevent the local buyer from purchasing merchandise or services in any currency other than the currency of that country. The exporter, if he wishes to deal with those countries, will have to assume the exchange risk.

In every exporting transaction there is a risk that exchange rates will fluctuate between the date a price is agreed upon with the overseas buyer and the date payment is received from him.

Foreign exchange rates vary mainly as a result of supply and demand for a particular currency. Many factors contribute to these fluctuations. Among these factors are inflation, interest rates, political events, and economic indicators. Foreign exchange rates are quoted either as "spot" or "forward."

Spot is the rate of exchange at which foreign currency is bought or sold now, for delivery now, and is used for day-to-day dealings in currencies.

The forward rate is quoted now for the purchase or sale of a stated amount of foreign currency at a specified time in the future, no matter how the spot rate might change in the intervening period.

Most international banks encourage exporters who will receive foreign currency on their sales to "hedge" the foreign exchange risk by entering into a forward foreign exchange contract. The forward contract fixes the U.S. dollar amount to be received on a fixed date in the future in return for foreign currency, even if the rate of exchange changes in the meantime. The exporter will not gain if there is improvement in the exchange rate during that time, but neither will he lose if the rate moves against him.

If the exporter is not sure of the exact date on which the foreign currency will be available, he can enter into an "option" forward contract, allowing him to deliver the currency on any day between two specified future dates.

The great advantage of forward contracts lies in the fact that the exact U.S. dollar value of the export contract can be known, even though payment by the buyer is to be made in foreign currency at a future time. Banks will enter into contracts for most of the world's major currencies for all periods of up to one year ahead. Contracts in the main trading currencies can be obtained for longer periods.

It may be necessary to offer credit terms to the overseas buyer in order to secure a contract, and to find financing for the credit period. Working capital may also be required in order to acquire and/or manufacture the product in the first place. the following are some of the alternative types of financing available:

   1. The international bank may establish a credit line in the exporter's favor, with recourse, based on the usual types of collateral and security and the exporter's ability alone to repay any advance. This type of credit facility is particularly suitable for financing the manufacture of goods prior to shipment.

      At the time goods are shipped, if no prior financing has been arranged and the credit-worthiness of the buyer has been thoroughly established by the bank, a bills discounted or advance against bills for collection facility may be established. The normal procedure is for the exporter to present to the bank a bill of exchange (draft), either for sight or time payment with normal shipping documents. The bank then sends them to a bank overseas for collection from the buyer. At the time of presentation, the bank will either discount the bill and pay the exporter with recourse to the buyer, or advance a loan to the exporter, charging interest until payment is received from the collecting bank.

2. The banker's acceptance is one of the more viable and flexible financing alternatives available to the exporter. A banker's acceptance generally arises out of a letter of credit wherein the exporter (beneficiary) has extended credit terms of up to 180 days to the buyer before payment is due. At the time of negotiation of documents under a letter of credit, the exporter may request that the draft presented be accepted and discounted by the paying bank. The exporter, although granting terms to the buyer, can obtain immediate funds at a rate generally lower than prime. Additionally, the exporter may receive banker's acceptance financing for readily marketable goods in storage, provided that the draft or bill of exchange is secured at the time of acceptance by a warehouse receipt or other document conveying or securing title.

In addition to those facilities briefly described, the international bank participates in various government-sponsored programs, such as the Federal Credit Insurance Association and the Export-Import Bank, to further support and assist the exporter.

APPENDIX

Table 1. International Services Market Profile.

	Import	Export
Manufacturers	XX	XX
Processors	XX	XX
Growers		XX
Trading companies	XX	XX
Wholesalers	XX	
Retailers	XX	
Service industries		XX

- There is no typical importer, exporter, or small multinational.
- Common users of international services are the following:

Table 2. Preexport Services.

- General introduction to export procedures, especially financing
- Country information--market opportunities and risk factors
- Market-matching--trade opportunities
- Bank introductions ⟶ Trading partners introductions
- Credit information
- Preexport financing

Table 3. International Payment Terms (in descending degrees of risk to the importer and ascending degrees of risk to the exporter).

Method	Time of Payment	Goods Available to Importer	Risk to Exporter	Risk to Importer
Cash in advance	before shipment	after payment	none	completely relies on exporter to ship goods as ordered
Letter of credit	when shipment is made	after payment	very little or none, depending on credit terms	is assured shipment has been made but relies on exporter to ship goods as described in documents
Sight draft (D/P)	on presentation of draft to importer	after payment	if draft is not paid, must dispose of goods, side risks: FX availability	same as above
Time draft (D/A)	on maturity of draft	after "acceptance"	relies on buyer to pay draft when due	must "accept" draft prior to taking delivery goods, but does not pay until after he has taken possession
Open account	after delivery	before payment	relies on buyer to pay invoice as agreed	none

BANK FINANCE AND SUPPORT   489

Table 4. Exporter Concerns.

Payment Terms	Degree of Risk	Advantages
Cash in advance	↑	"Gravy"
Letter of credit		
. confirmed	R	Payment is conditionally
. advised	I	guaranteed
. sight		
. time	S	Possible basis for financing
Documentary collections	K	
. sight (D/P)		May induce buyer to place an
. time (D/A)		order or to increase
. clean		size of orders
Open account	↓	Serves as a mechanism for financing weaker customers or exporter's subsidiaries

Table 5. Exporting Under a Confirmed Time Letter of Credit.

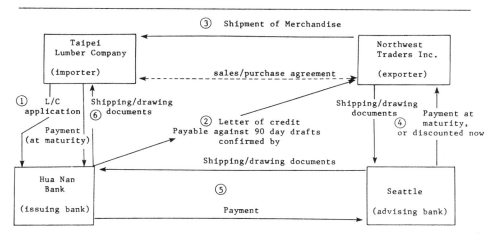

Table 6. The Foreign Exchange Concern of Exporters.

Issue:	What is the U.S. dollar value of present or future receivables which are denominated in foreign currencies?

Example:	Northwest Trade Inc. has shipped an order of Chanterelle mushrooms to Germany on 90-day terms under a time letter of credit.
	    The full face value of the L/C is Deutschemarks (DM) 225,000.
	        How can Northwest Trade assure itself of earning at least $100,000 when it receives the funds at maturity of the draft?

Strategy:	Northwest Trade Inc. can enter into a 90-day forward contract to sell DM. Based on today's market for spot and forward DM, Northwest Trade can agree to exchange DM 225,000 for $100,350.
Cost:	Northwest Trade's cost of "insurance" can be determined after completion of the transaction when the forward contract rate can be compared to the spot rate on January 24, 1982.

Table 7. Banker's Acceptance.

A time draft drawn on a bank and "accepted" by that bank--
to pay a specified amount of funds to a designated payee on a specified date

Discounting of the BA ⟶ BA financing

Advantages of BA Financing

Borrower

. Lower-than-prime cost

. Availability of funding in a "tight money" situation

. Fixed-rate financing (over the term of a transaction)

Bank

. Means of increasing and holding market share

. Overcomes liquidity limitations

. Good investment for bank's portfolio

. Allows lending over normal limits

Table 8. Banker's Acceptances: Requirements for Eligibility.

"Eligibility" — Exemption from reserves

Common Requirements

1. Short term (180 days) financing only
2. Not to be used for "working capital"
3. Amount and maturity of acceptance must be supported by a commercial transaction meeting one of the following:

Transactions Requirements

1. The importation or exportation of goods, including shipments between other countries. Preexport--supported by a sales contract at acceptance.
2. The domestic shipment of goods--when the accepting bank is directly secured by the commercial transaction at the time of acceptance.
3. The domestic storage of readily marketable goods--when the accepting bank is directly secured at the time of the acceptance by a bonded warehouse receipt.

Table 9. Banker's Acceptance Formulas.

. Net proceeds formula

$$\text{draft par value} - \left[\left[\frac{\text{draft par value} \times \text{"all-in rate"}}{360 \text{ days}} \right] \text{\# days in term} \right]$$

. Effective cost/yield formula (bond equivalent rate)

$$\left[\frac{\left[\frac{\text{discounted \$ amount}}{\text{net proceeds \$ amount}} \right]}{\text{\# days in term}} \right] 360 \text{ days}$$

Table 10. Acceptance Financing of an Export Under a Time Letter of Credit (exporter pays discount).

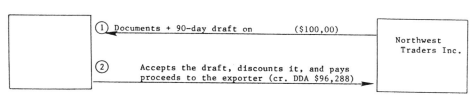

Later, at maturity, receives $100,000 from the issuing bank

Table 11. Acceptance Financing of an Export Under a Time Letter of Credit (importer pays discount).

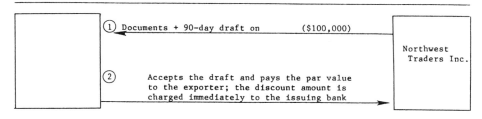

Later, at maturity, receives $100,000 from the issuing bank

Table 12. Acceptance Financing of the Domestic Storage of Goods.

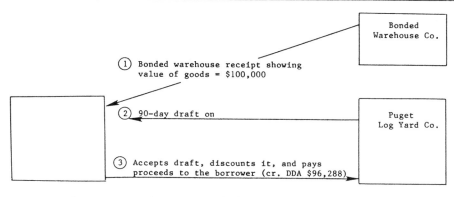

Later, at maturity, receives $100,000 from the borrower

Table 13. Major FCIA Programs.

Agricultural	Short Term Buyer Credit	Short Term Supplier Credit	Medium Term
Insures U.S. banks against risks of dealing with L/C's issued for up to 360 days by foreign banks	Insures U.S. banks against risks of lending directly to foreign buyers of U.S. goods or services	Insures U.S. banks against risks of purchasing a reasonable spread of risk of an exporter's foreign receivable portfolio	Insures U.S. banks against risks of lending directly to foreign buyers of U.S.-made capital goods
98% commercial 100% political	90% commercial 100% political	90% commercial 100% political	90% commercial 100% political

Table 14. Export-Import Bank Programs.

	Direct Loans	Bank Guarantees
Purpose:	To support sales of high value U.S.-made capital goods	To protect U.S. banks against commercial and political risks inherent in export loans
Qualifications:	$5MM minimum contract value	10% exporter's commercial risk retention
	15% down payment	15% down payment
Extent of EXIM involvement:	EXIM funds up to 65% of contract value	Commercial banks fund the entire 85% financed portion of a contract
	Other sources fund at least 35% of contract (including the down payment)	EXIM Bank also guarantees loans in the main currencies
Term:	Typically 8-10 years, may be longer to meet foreign competition	Medium term (181 days to 5 years)
Rate:	EXIM funding for: rich country: 12.15% intermediate: 12.00% poor: 11.00% Plus the commercial bank's maximum spread of 1% pt.	Market rate for commercial loans (prime or labor-based) or at a fixed rate, supported by EXIM discount loan
Fees:	2% front-end fee	Guarantee fee payable once at the time of issuance, varies with type of buyer and term of loan

Table 15. EXIM Bank Programs.

	Small Manufacturers Discount Loan Program ($25MM annual gross sales)	Medium-term Credit Program
Purpose:	To provide U.S. banks with a fixed-rate source of funds for medium-term export loans which will support U.S. exporters' sales of U.S. goods	To provide U.S. banks with a fixed-rate source of funds for medium-term export loans which will enable a U.S. exporter of U.S. goods to meet foreign subsidized, officially supported export credit competition (copies of actual foreign offers required)
Qualifications:	$2.5MM maximum contract value 15% down payment	Contract value not expected to exceed $5MM 15% down payment
Extent of EXIM involvement:	EXIM commits to fund up to 85% of contract value	EXIM will fund up to 65% of contract value
Term:	Medium term (366 days to 5 years)	Medium term (366 days to 5 years)
Rate:	EXIM funding for: rich country: 12.15% intermediate: 12.00% poor: 11.00% Plus the commercial bank's maximum spread of 1% pt.	EXIM funding to meet the foreign competition
Fees:	EXIM front-end commitment fee based on term of financing (5 years--.55%)	To meet the foreign competition
Other:	Can be combined with: a) Bank guarantee program b) FCIA medium-term coverage	Can be combined with: a) Bank guarantee program b) FCIA medium-term coverage

GRADES AND SPECIFICATIONS IN WORLD TRADE OF WOOD PRODUCTS

Robert L. Ethington

Introduction

In any country that has an internal trade in forest products, people with various degrees of skill and interest are involved in commerce. They represent the private sector and several levels of government; and their areas of expertise include supply, merchandising, transportation, manufacturing, regulation, law, and labor. Each person involved in the commerce of wood products interacts with others, and the keystone to any interaction is the definition of the product--or, in some instances, the lack of a definition.

Even the simplest of products needs definition to foster trade, but simple products are not always simple to define. Most of us would consider a log a relatively easy product to define; a log is a segment of tree stem nearly round in cross section. But log descriptions, or rules, have always been and continue to be a source of confusion in trade because there is no universally accepted rule, and in the commerce of logs a specific rule must be cited for good communications between buyer and seller.

The problems in maintaining a free flow of trade increase as trade in forest products becomes international. The number of people involved is now doubled or tripled as counterparts in other nations are dealt with, and there are many more differences between them than just language. As you have heard throughout this symposium, laws are different, traditions of use are different, environments in which products are used are probably different, and cultures are different. Almost certain to be different are the product descriptions used in the commerce of other nations because of development under entirely different circumstances.

Product definitions have always been different, but it seems to me that within the United States as well as between the United States and countries it trades with, product definitions are moving inexorably to a common point. That's the good news. The bad news is that while much has been accomplished in the last 100 years, there is more work to be done. I doubt there is an end point--a place where no more standardization is needed--because nothing is perfect. And the numbers of products with at least some subtle differences increase as resources become more scarce and technology accelerates.

The purpose of this discussion is to look at the experiences we have had in the trade of forest products and to discover any guiding principles of product standardization that may help to simplify and accelerate international trade. There is usually someone involved in international trade who, in frustration, will lecture that no standardization would be most helpful; but I don't think that is a commonly accepted notion. Although a standard product description might impede trade in a specific case, it is hard to imagine anything but marketing chaos if descriptions did not exist, just as it is difficult to imagine a nation without laws. That potential for chaos is the driving force that moves us toward increasing standardization. Standards are used to ameliorate disputes in commerce, but more important for the entrepreneur, standards can be used to attract business.

The Nature of Standards

It is not easy to generalize, because standards do not fall into neat classes. That is not surprising, since they are developed by different people for a variety of purposes. I like to think of them as falling into a spectrum that ranges from entirely descriptive to entirely inferential.

An entirely descriptive standard would define a product only by criteria that are either obvious or easily measured. For example, one could imagine a standard for woodblock flooring that would specify oak, give the dry dimensions of the block, the direction of grain, and that's all. An entirely inferential standard would imply how the product will perform. This kind of standard allows the manufacturer much greater latitude in changing raw material into a product but carries with it a need to "certify" conformance because performance is harder to measure than things like size and appearance. It is often called a performance standard. Using the woodblock flooring example again, a performance standard might specify that the floor will wear satisfactorily for twenty years as long as certain criteria for installation, maintenance, and use are followed.

There are probably no entirely descriptive or entirely

inferential standards, but they tend one way or the other. Even if the descriptive standard for woodblock floors were used, the floor might get wet, swell, and fail. The owner's wrath would then lead the manufacturer to include some guidelines for installation and use, and thus offer some inference about how the product can be expected to perform.

Most products are more complex, however, and harder to describe. An all-inclusive list of wood products standards is not humanly possible, but the kinds of things often found are size, appearance, moisture content, species, additives, adhesives, and recommended strength and stiffness for use in design. The more inferential standards contain criteria for performance such as acceptable exposure to weather and acceptable spans for typical loads.

A product standard may appear to stand alone; that is, the product standard may seem to be all that is needed to describe, specify, and use the product effectively. Yet frequently the standard is intimately linked to widely accepted but not well-described uses; or to well-described but highly technical uses. The practice usually develops first; the standard comes later and is "calibrated" to the practice. Changing a standard, therefore, often requires changing the use. Imagine the complications in trade when a buyer bumps into this problem, especially if he is dealing with a foreign source of supply, and he knows little about the standards and practices of the exporting country! An excellent example of this dilemma can be seen in standards for lumber for structural use. In most nations, there is a history of traditional building practices that have been combined with imperfect engineering principles and evolving grading procedures. Two documents or more usually result from this: (1) a description of grades of lumber (the standard), and (2) a description of the design and construction practices (a specification or, in some cases, a tradition) that will lead to acceptable performance when the standard grades of lumber are used. The documents and traditions are likely to develop in parallel fashion in various countries, but the details will be different. When commerce begins between two countries, only the standard--the product description document--may be involved in the trade agreement. The final buyer may end up using lumber manufactured to a foreign standard for his domestic design specification. The results can be disappointing or disastrous.

Even in a more enlightened trade, where buyer and seller know at the outset that all of these subtleties exist, the implications may not be understood, especially if the buyer is not the final user. Or the buyer and seller may feel helpless to do anything about the problem because standards tend to become institutionalized. The standards are woven through much of the fabric of a country--its trades, its codes, its culture, and the preferences of its people. It is, therefore, possible to see that, while buyers and sellers know there will be standard and specification barriers when foreign products are introduced to new markets, they do not have the power to bring about change. So who is responsible?

Organizations for Standardization

Just as it was difficult to generalize about the nature of standards, so is it difficult to describe standardizing processes for all nations. The prevailing system for standardizing wood products and specifying practices for use is highly developed and complex in the United States, as it is in a number of other countries.

Within any given nation, the government is usually involved with the private sector in the development of standards. Sometimes the government is involved simply because it is a major purchaser; for example, the U.S. Department of Defense develops and publishes its own standards for many wood products. In other instances the government is involved as a representative of the consuming public. The Federal Housing Administration has been involved in the development of standards for wood building materials that may be used in homes with federally backed mortgages; and at a more local level, building code authorities also have a stake in standards. Government sometimes is involved in standards development in order to foster trade for the economic good of the constituency, be it country, state, region, or municipality. For years, as a result of legislation, the U.S. Department of Commerce has worked with the private sector to promulgate product standards: The American Lumber Standard PS20-70 is an example. And finally, government may participate in standards on behalf of a special interest group; for example, the U.S. Department of Labor may represent the building trades when wood products are being discussed. Sometimes standards emerge with no apparent government involvement at all, although in wood products this would be a rare exception.

It is interesting to compare the U.S. with other countries in the development of standards. Wood products standards have developed in Canada as voluntary standards, much like those in the United States. It is my impression,

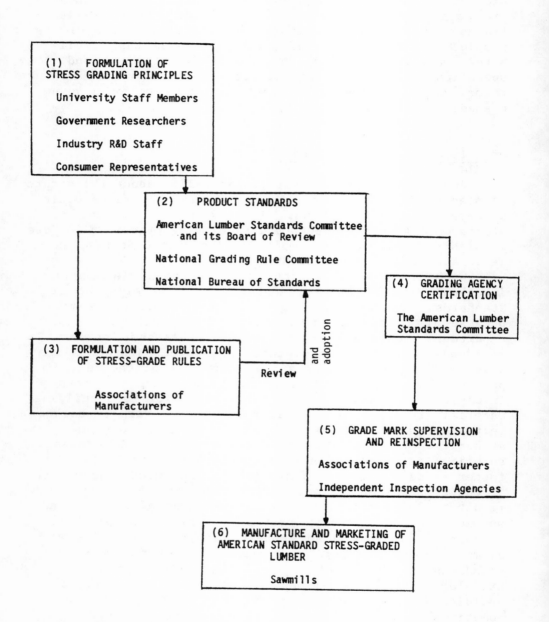

FIGURE 1. Responsibilities for stress grading in North America.

however, that in many other countries the national government plays a larger, usually more regulative, role. Japan, for example, has a vigorous economy based on international trade and a well-recognized marriage between public and private sectors that is evident in its product standardizing processes. The government of Great Britain, where wood has usually been imported, has mandatory codes of practice for consumer protection and satisfaction. I am confident that careful study of any two nations would reveal some differences that would be important to understand if international trade were to be encouraged.

Probably in any country, standards are created by committees. But in the U.S. it is impossible to turn to a single organization as a focal point for wood products standards. I have previously mentioned the U.S. Department of Commerce, which calls committees together to develop and publish certain product standards (lumber and plywood standards are significant examples). The American National Standards Institute (ANSI) and the American Society for Testing and Materials (ASTM) are two private corporations that play similar roles for a significant number of other standards. And there are many others.

Reading a standard critically and understanding it is a difficult task for a couple of reasons. It is the character of standards to describe, or to limit, and that usually takes place in a less than self-explanatory way. Each standard is supported by literature (usually not referenced within the standard) that may include published research reports, unpublished reports, minutes of meetings, letters, and even recollections of committee members. In addition, a single product is frequently associated with not one but several standards; an example is stress-graded lumber in North America. Figure 1, taken from the Wood Handbook, shows a circuitous system of responsibilities for development, manufacture, and merchandising of stress-graded lumber. Although it is not apparent from the figure, there are four standards associated with formulation of stress-grading principles (box 1 in the figure) and two standards with box 2. If grade rule books are themselves standards or specifications, as they would be from the perspective of a buyer or seller, then there are a number of those associated with boxes 3, 4, 5, and 6. The point is that users of standards are frequently perplexed about some restriction within the standard, but tracking down the reason for the restriction, or the process for bringing about change, may not be an easy task.

Not surprisingly, there are also standing and ad hoc committees that deal with standardizing the same generic product, but across national boundaries. They function under government, private, or combined auspices. An example where this has been done successfully is the national grading rule for stress-graded lumber, developed by the American Lumber Standards Committee, a quasi-public committee of the U.S. Department of Commerce (figure 1, box 4). The national grading rule is in fact international because the standard and the quality-control process overseen by the committee are applied equally in the U.S. and Canada, and the described product moves freely across the international border.

There are many international bodies dealing with standardization of forest products. It is important to mention the International Standards Organization (ISO), U.S.-Japan Forest Products Committee, the Timber Committee of the Economic Commission for Europe, and related organizations such as the International Council for Building Research Studies and Documentation (CIB). They each play a significant role in international standardization and this will have implications for trade in wood products by the U.S. They represent only a small sample of myriad organizations throughout the world, many not known to me.

West European countries have historically not had standards for stress-graded lumber, or at least have not had standards that are so universally applied as those in North America. When the Common Market increased trade between countries, there was undoubtedly a considerable need felt by participating countries for interchangeable products. About ten years ago, ad hoc meetings of experts on grading of coniferous softwoods began to take place. From those beginnings a proposed standard has evolved for stress grading of coniferous sawn timber. It is being promoted by the Timber Committee of the Economic Commission for Europe (ECE), an arm of UNESCO. The goal of ECE is to elevate it eventually to an ISO standard, which would give it greater stature and, probably, wider application. The grades are different from those in the North American National Grading Rule.

Simultaneously with the development of the ECE draft standard, a committee of CIB turned its attention to development of standardized specifications for use of wood in engineered structures. The resulting design code provides design methods compatible with the proposed ECE grades.

At about the same time that standardization within Western Europe was taking place, the need for housing built up in Japan, and lumber and plywood manufacturers in North America began to see new market opportunities for their products in Asia. As that market was investigated, it became apparent that wood products standards and building practices in Japan loomed as barriers to the use of lumber and plywood from North America. Many meetings followed, along with visits by delegations from one country to another, and a log of negotiation. This was necessary to encourage producers from North America to manufacture their products to Japanese standards, and to encourage the use of those products and North American building practices in Japan. Both goals are being reached, and the U.S.-Japan Forest Products Committee continues to provide the means for communication; it does not function as a standardizing body.

Where Do You Start, If You Want to Sell Internationally

In the interest of reasonable brevity, I have focused on standards and on standardizing organizations. They are related to regulation, to culture and tradition, to language and definition, and to the infrastructures for commerce and product use unique to each country. The more we know about these things in relation to a country where we wish to trade, and the more clever we are about adapting, the more successful that trade will probably be.

As you embark on a new international wood products venture, you can almost depend on the foreign markets having a different concept than you have of what the product should be. Their standards will be different from yours. You will ask for their standard, and it will be in a different language. You will get it translated and find it is hooked to another standard, and the cycle will repeat. Or you will find that some parts of it seem unnecessary for good product performance, but no one seems to know why those parts exist or how to change them. Or there is no desire to change them. Or the committee isn't meeting this year.

Frustration is ensured. Patience, persistence, and flexibility are essential. You are likely to need the help of some technical experts even to understand the issues. You, too, will probably feel at times that no standardization would be a refreshing change. Standards are a vital communication link, however, and if we did not have them, we would surely invent them.

INFORMATION NEEDS FOR FOREST PRODUCTS TRADE STRATEGIES

Juan E. Sève

Introduction

In a review of recent United States literature on international trade in forest products several statements can be found referring to this country's potential as a net exporter. This is, perhaps, the main strategic question regarding forest products trade in America today. Statements on this topic range from claims that the United States, and particularly its southern region, can become the world's "woodbasket" to projections of an indefinite continuation of current U.S. trade patterns as a net forest products importer.

Between these two extreme views one encounters more cautious statements such as, "The U.S. has great potential for increased forest products and, specifically, wood products exports" (NFPA International Trade Comm. 1982), or, "America's advantages contribute to making it highly competitive in world wood markets. It is thought by many to be the world's lowest cost producer of many softwood and some hardwood products" (ibid.). More modest yet is the statement, "With more intensive management and improved utilization, enough timber can be provided to meet foreseeable domestic and export demands" (USDA Forest Service 1982).

In addition to hypothetical statements of this kind, this question has also been addressed as part of the Forest Industries Council timber supply goal in 1979: "...the nation's commercial forest land should be managed to...build the potential for an international net trade surplus of forest products" (API/NFPA Task Group 1981).

Although the eventual capability of the U.S. to become a net forest products exporter has been argued in various contexts and with varying degrees of intensity, the focus of these arguments is almost invariably on potential.

Despite this vastly argued potential, however, the inescapable fact remains that the U.S. is, and has been for several decades, a net importer of forest products mainly in the areas of softwood lumber, newsprint, and woodpulp. As a result, a broad gap exists between potential and actual performance.

The potential for forest products exports is often discussed on the basis of perceived comparative advantages. Among those mentioned are: (1) resource availability, growth, and quality; (2) possession of the largest forest products industry in the world; (3) excellent infrastructure and port facilities; and (4) competitive costs (NFPA International Trade Comm. 1982). The reality of the situation, however, is much more complex than this.

In order better to assess the U.S. potential as an exporter of forest products, a few crucial questions need to be asked:
1. Is the "net exporter" status a realistic goal?
2. If this is a realistic goal, what is the strategy to turn the situation from net importer to net exporter?
3. In what specific product groups can the U.S. maintain or attain comparative advantages?
4. What kind of information is necessary in order to determine such product groups?
5. How do world forest products trade flows get restructured as the U.S. progresses toward its "net exporter" status?

Perhaps one of the reasons why this country continues its net importer status despite its perceived potential is that none of these questions have been properly addressed. Otherwise, there would not be so much disagreement and speculation regarding the future role of the U.S. in the world forest economy.

So much speculation and disagreement most likely stem from a lack of understanding of what the comparative advantages really are in the U.S. forest products sector. According to the theory of international trade, the principle of comparative advantage stipulates that: (1) a given country will tend to export those goods which it is relatively more efficient at producing and import those which it produces relatively less efficiently; and (2) trade flows and the degree of specialization of each country in the production of specific commodities will also depend upon demand conditions for each product in each country (Takayama 1972).

In other words, the precise determination of a country's trade potential with respect to others requires knowledge of specific supply and demand conditions of all goods considered in all countries considered. As a result, comparative advantages are essentially a matter of <u>calculation</u> rather than <u>speculation</u>.

The need to calculate comparative advantages turns the central questions of U.S. forest products export potential into an extremely complex and multidimensional problem. Obviously, the more complex the problem, the harder it is to define adequately. Perhaps one viable approach is to separate the central question into a number of specific issues.

Review of Major International Trade Issues

From the standpoint of comparative advantage analysis, the main strategic question of U.S. trade potential can be subdivided into a number of issues. For better focus, they can be grouped into seven major categories: (1) foreign supply of fiber and products; (2) foreign demand for specific products; (3) cost and supply structure; (4) new markets; (5) trade policies of other countries; (6) shifts in current markets; and (7) trade barriers.

These categories do not respond to any particular theory. They are simply tools to organize a large number of interactive issues which will affect, in one way or another, the United States' potential in forest products trade.

Neither the enumeration nor the discussion below is exhaustive. The purpose here is simply to illustrate the vastness and complexity of problems of international trade strategy and to set the stage for the coming discussion on analytical approaches.

Issues on Foreign Supply of Fiber and Products

This group of issues focuses on the fiber supply potential of foreign countries. This country has been faced for several years with the largely debated Canadian subsidization of stumpage prices. How long can these subsidies be maintained, and how will they affect Canada's future ability to increase its exports of lumber, newsprint, and pulp? Additionally, what other countries are subsidizing their forest products industries, and what are the possible trade effects of such subsidies?

Another issue of this type which has been discussed for several years is the possibility of a major Soviet expansion of softwood supply. Most American opinions maintain that Soviet locational and infrastructural problems are virtually insurmountable. Nevertheless, there does not yet appear to be a clear perception of what kind of effort it would take to transform the U.S.S.R.'s enormous softwood inventory into increased economic supply.

More recently, the issue of fast-growing plantations has emerged. Most of these plantations are being established in tropical or temperate Southern Hemisphere countries. Do these plantations constitute a net increment of world fiber supply? Are they merely compensating for the deterioration of tropical and subtropical forests? Although these issues have been addressed (Sedjo 1980; Laarman 1980), the industrial wood potential of newly established plantation forests is not at all clear.

Other issues in this category include the trade implications of recent structural changes in Southeast Asian forest industries such as the dramatic Indonesian plywood capacity expansion and the associated log export ban.

Issues on Foreign Demand for Specific Products

Whereas, on the supply side, trade with other countries can affect U.S. resources as well as products, on the demand side, trade issues are mainly product-related.

An issue of this type which is currently the subject of considerable discussion is the potential for increasing platform construction in Europe, Japan, and other parts of the world. If this potential develops, what would be the effects on the lumber and plywood industries of the United States, Canada, and Japan?

In less-developed countries there are also various relevant questions, such as: What are the prospects for increases in demand for long-fiber paper and board products? Which will be the sources of long-fiber supply? How will increasing needs for fuel in less-developed countries affect fiber demand and the trading potential of such regions?

Finally, another issue of current importance relates to the effect of electronic communications on the future of paper demand.

Issues on Cost and Supply Structure

These types of trade issues are perhaps the most difficult to address; however, they are certainly among the most important. This is because whoever speaks of comparative advantages speaks of production efficiencies, which are directly dependent upon production costs and resource supplies.

Discussions on U.S. forest products trade potential normally focus on forest resource considerations to justify such potential. Timber, however, is only one input in the cost structure of forest products manufacturing. The costs of nonwood inputs and their evolution are elements that can directly affect the comparative advantages of a country. In addition, the age and technological level of a country's capacity for a given product are directly connected with production costs, and so is the degree of integration between resources, pulp production, and solid-wood products manufacturing.

Closely related to the production cost issue is the question of availability of sites for new facilities. While in the case of solid-wood products this is not a major problem, world expansion of pulp production may have definite limits, considering that pulp mill sites are determined by specific conditions of wood, water, labor, energy, and transport.

Continuing with cost aspects, different parts of the world are endowed with different production factors. As a result, differences in capital/labor relationships and economies of scale need to be considered.

Regarding resource cost and supply, the ownership structure of forest resources has had considerable impact on U.S. wood costs. This impact has been largely related to federal timber supply policies. How is ownership structure expected to affect raw material costs in the future? Are there any changes anticipated in federal policies? Considering the importance of raw materials in the cost structure of some forest products, these issues are highly relevant to the country's future comparative advantages. In addition, since the structure of resource ownership varies from region to region in the U.S., it is important to consider these regional differences and their impact on long-term production costs.

Issues Regarding New Markets

Over the past decade, new markets have emerged for U.S. forest products. One of these is the Middle East. This market was created essentially as a result of the high transfer of wealth to these countries through the high oil prices of the seventies. What are the long-term prospects for these markets? Are there any structural changes anticipated in the world oil economy that would affect market growth? How competitive is the U.S. in these markets?

Another market that has recently surfaced is the People's Republic of China, with its sizable imports of raw materials and various forest products. China's enormous population reflects attractive market potential; however, considering China's current level of economic activity, how large a market can really be expected? How will it evolve? What changes can be expected in China's trade policy?

A third market with considerable potential is Southeast Asia, particularly the Association of Southeast Asian Nations (ASEAN), which includes Indonesia, Malaysia, the Philippines, Singapore, and Thailand. This area has a population of 260 million. Most of these countries experienced annual economic growth rates higher than 7 percent during the 1970s and are expected to continue this trend in the 1980s and 1990s. While most of these countries are endowed with important forest resources, softwoods are in short supply. As a result, the ASEAN region could present interesting marketing opportunities for long-fiber-based products, mainly paper and board.

Issues on Trade Policies of Other Countries

As previously mentioned, the Forest Industries Council has established the goal of building the potential for an international net trade surplus of forest products (API/NFPA Task Group 1981). The United States, however, is not the only country aiming at export expansion. Australia, for example, while being at present a net importer of forest products, is attempting to become a net exporter by the turn of the century (Batten 1981). Additionally, countries such as Brazil, Chile, and New Zealand, currently net exporters, are already trying to increase their exports.

Another type of policy being developed in some countries is that of long-term import substitution. Examples of these policies are the aggressive plantation programs started by Japan and France after World War II and more recently by Colombia. These countries are currently net importers of forest products and will probably remain so for several years; however, they are committed to reducing their dependence on foreign timber in the future.

Issues on Shifts in Current Markets

Perhaps the best example of shifting markets is Japan, the United States' largest single overseas forest products market. Several changes appear to be taking place.

Japanese housing, which had been rapidly expanding over the 1960s and early 1970s, is now expected to remain at current levels or to decline over the next two decades (Ueda and Darr 1980). Changes are also taking place in Japan's foreign sources of supply. One of these is the Indonesian plan to phase out log exports by 1985 (Japan Lumber Journal, various issues). Since Indonesia is one of the most important sources of supply for the Japanese plywood industry, the impact of such change on U.S. potential exports of logs or plywood could be favorable; however, the United States would probably have to face competition from New Zealand and the Soviet Union in these new markets.

Shifting markets can also be found at home. While U.S. housing starts (including mobile homes) are expected to recover and reach the two million mark by the late 1980s, demographic pressures are expected to subside, perhaps leaving the U.S. in search of new markets for its forest products.

Issues on Trade Barriers

Among the issues reviewed in this discussion, tariffs and quotas have probably been more thoroughly analyzed than all others. Some examples are the work of Adams and Haynes on U.S.-Canadian lumber trade restrictions (1980), Radcliffe's paper on the effect of multilateral trade negotiations on U.S. forest products trade (1980), and an econometric simulation of a log export ban by Haynes, Darr, and Adams (1980).

Except for Radcliffe's paper, these studies deal with either Canadian lumber exports to the U.S. or log exports to the Far East. While these two issues are covered very thoroughly, several basic questions remain. Some of these deal with the broad gamut of forest products other than softwood lumber and logs, and with parts of the world other than Canada or the Far East as they may affect U.S. trade in the future. Radcliffe's paper attempts to address the trade barrier question with respect to several products on an aggregate commodity basis. However, his geographic scope continues to be concentrated on Japan and Canada with the addition of the European Economic Community.

Many other relevant trade issues could probably be raised in addition to those just discussed. The point here is not to examine every possible trade question, but to give an idea of the extreme complexity involved in analyzing the U.S. forest products trade potential from an objective standpoint.

Many of the points discussed in this section relate to basic international economic structure; others are related to policy options. The importance of most of these issues for the future of U.S. forest products trade suggests the need for new systems of analysis and information. These should be built upon specific supply and demand relationships and should permit the analysis of policy alternatives.

The Need for New Tools of Analysis

While the U.S. has a long history of forest products trade, research on this subject is a fairly recent phenomenon. Among the organizations involved in this research, the Forest Service and Resources for the Future are probably the most important. Of these two, the Forest Service has recently been given the specific mission of analyzing "present and anticipated uses, demand for, and supply of renewable resources, with consideration of the international resource situation, and an emphasis of pertinent supply and demand and price relationship trends" (Darr 1981). This mission was prescribed by the Resources Planning Act of 1974 (RPA) as part of a Renewable Resources Assessment to be conducted every five years by the Secretary of Agriculture. The first Resources Planning Act Assessment containing a thorough update of the U.S. timber situation was conducted by the Forest Service in 1979. The Foreign Trade Analysis Work Unit of this agency was charged with developing projections of long-term trade in timber products as part of this Assessment. The final results were published in December 1982 (USDA Forest Service 1982). This work constitutes the most comprehensive long-term projection of U.S. forest products trade to date.

The RPA international trade projections covered a period of 50 years and focused on imports and exports of lumber, veneer plywood, pulp, paper and board, pulpwood, and logs. With the further breakdown of lumber, veneer/plywood, and logs into softwoods and hardwoods, there are nine product groups in the final results. In general, the projections show that U.S. forest products trade will continue along the same lines of the recent past; that is, imports should continue to outweigh exports by wide margins. In addition, while imports continue to increase in total volume terms, exports are projected to stagnate at present levels. Current trade patterns remain essentially unchanged, with imports being dominated by lumber, pulp, and paper (mainly newsprint) from Canada, while exports continue to concentrate on logs, lumber, paper, and board, mainly to Japan and Western Europe.

These trade projections were presented and discussed at an international trade workshop in 1980 (Darr 1980). On this occasion, both the results and the methodology were the subject of severe criticism. Perhaps the most thorough discussion on the Forest Service's approach was presented by Zivnuska at that same workshop. Zivnuska's critique focused on the fact that the Forest Service's export projections are a complement to domestic forecasts of wood supply and demand based on a "gap" model. Briefly, this type of model attempts to determine the difference between the potential production (supply) and consumption (demand) of a given product over a period of time at specified prices. As a result, supply and demand are forecast independent of each other and there is usually a gap between the two over the projection period.

In reality, of course, this gap cannot occur because prices will move to bring supply and demand into equilibrium. While the 1979 RPA Assessment contained supply and demand projections based on both gap and equilibrium models, however, the trade projections were linked only to the gap scenario. Since the prices of the "gap" scenario are considerably lower than equilibrium prices, gap projections of domestic consumption appear to be highly exaggerated. One main methodological question is, therefore, the credibility of trade forecasts linked to unrealistic projections of domestic consumption.

Aside from forecast credibility, the approach can also be criticized from the standpoint of policy analysis. As mentioned by Darr, "For the purposes of responding to the terms of the Resources Planning Act, we need projections of how U.S. trade patterns might change in the future in absence of any changes in policies and we need projections of trade patterns following any changes in policies. The need for these projections implies that we know or must somehow consider specific types of information about U.S. and other markets for timber products" (1981).

The fact is that the Forest Service analysis is built by stacking assumptions upon other assumptions, without sensitivity to equilibrium prices. Additionally, the analysis derives trade flows from a projected supply and demand gap within the United States while little attention is paid to economic forces in other parts of the world. As discussed in the first two sections of this report, the network of international trade in forest products is highly intricate, and the current approach does not provide an economic framework to analyze either major structural issues or the impact of policy changes. In other words, the international trade information requirements of the Resource Planning Act are still far from being satisfied.

Concerns about these methodological weaknesses have been voiced and discussed several times over the past couple of years. Most discussions have taken place within industry groups (Forest Industries Advisory Council 1982; API/NFPA Task Group 1982), but some views on the subject have also emanated from the Forest Service itself (Darr 1981; Haynes and Adams 1981). The thrust of these discussions is that projections of international demand and supply should be based on evaluations of economic competitiveness and production costs and be sensitive to economic policy changes.

In any event, the analytical tool which has been used to date does not address itself to any of the crucial strategic questions mentioned at the beginning of this paper, and largely ignores virtually all the issues discussed in the second section. As a result, this method cannot even begin to assess long-term comparative advantages, and unless there are drastic improvements in the system, the U.S. will not be able to identify specifically its trade potential or develop consistent trade strategies.

Characteristics of an Adequate Approach

The introductory section of this paper concluded that the precise determination of a country's trade potential required knowledge of specific supply and demand conditions of all goods considered in all countries considered. The second section, by briefly reviewing a number of important trade issues, illustrated the intricacies and complexities involved in identifying the necessary supply and demand conditions. The third section added that the analysis of trade potential required a level of thoroughness and flexibility that is lacking in the currently used framework.

Some people may think that the task of constructing a model or procedure to analyze trade potential in such a rigorous context could be an insurmountable task. They may have a point. The fact is that to date no global model exists to handle interactions between nations in forest products trade (Lönnstedt 1982). But global models do exist in fields such as energy (American Academy of Arts and Sciences 1982), and agriculture (Abkin 1981). On the other hand, a few countries have models that explicitly address the interaction of supply and demand in several forest product markets within a competitive equilibrium framework at the national level. One of these countries is the United States, and its national forest economic model is the Timber Assessment Market Model (TAMM).

Although the geographic scope of the TAMM model is essentially limited to the United States, a review of its structure may be useful to illustrate the potential of adapting its methodology to trade problems. TAMM's emphasis is not on international trade, but the model segregates the country into regions which trade forest products among themselves.

Technically speaking, TAMM is a spatial price equilibrium model. Models of this type are generally constituted by three or more regions trading homogeneous products. For each product in each region, supply functions relating local production to prices and demand functions relating local use to prices are estimated. Additionally, the regions are separated by transportation costs per unit of each product. Given these supply and demand functions and transportation costs, the model can solve for competitive equilibrium prices, volumes produced and consumed in each region, and volumes traded among regions (Adams and Haynes 1980b). In other words, prices and volumes of the products considered are internally determined by the model instead of assumed. Additionally, production, consumption, and prices are determined simultaneously and consistently.

The TAMM model was used to develop the equilibrium scenarios of the 1979 RPA Assessment referred to in the second section of this paper. Although this system still has some weaknesses, it represents considerable improvement over the gap approach. One of these experiments is the consistency of price and volume projections. Upon examining the 1979 Assessment, one discovers that projected price levels will not permit the roundwood consumption levels projected with the gap method. TAMM, on the other hand, overcomes this inconsistency by solving simultaneously for prices, production, and interregional trade. Obviously, as illustrated in table 1, the projection differences between TAMM and the gap model are substantial in both volumes and prices.

Another major improvement brought about by the TAMM model is the economic linkage it provides between the final product sector (lumber, plywood, pulp) and the stumpage sector. The simultaneous solution provided by the model permits the evaluation of impacts originated at the market end, the resource end, or the intervening production processes. This feature has allowed the testing of policy changes and their impacts on the whole U.S. forest economy.

Table 1. Comparison of TAMM and Gap Analyses—Softwood Prices and Volumes 1979 RPA Assessment.

		Softwood Stumpage Deflated Price Index (1967=100)		Softwood Roundwood Use (MMCCF)	
		Southern Pine Region	Douglas-fir Region	South	West
1976	TAMM	138.9	164.2	41.8	45.7
	Gap	138.9	164.2	41.8	45.7
1990	TAMM	230.6	275.0	56.0	49.3
	Gap	132.9	119.0	61.2	53.7
2000	TAMM	281.6	228.2	61.5	47.6
	Gap	139.1	132.8	71.3	53.8

Source: USDA Forest Service 1982, pp. 151, 205, and 207.

Policy simulations which have been analyzed with TAMM include the intensification in private forest management (Adams and Haynes 1980b), the effects of a tariff on U.S. imports of Canadian lumber (Adams and Haynes 1980a), and the potential impact of a ban on log exports from the U.S. to the Far East (Haynes, Darr, and Adams 1980).

The TAMM results and their desirable characteristics of consistency between prices and volumes, supply and demand, and resources and products have been well received by government and industry users; however, like any other analytical procedure, TAMM has some limitations. Among those worth mentioning are the exclusion of hardwoods from the model solution, the lack of an age-class characterization of inventory and growth projections, the exogenous treatment of the pulp and paper sector, and the fact that international trade is not included in the model except for Canadian lumber exports to the U.S. All these limitations are now being corrected.

For purposes of this paper, the relevant weakness of TAMM is, of course, the lack of an international trade sector. In the model's current version, U.S. trade flows are determined on a judgmental basis. Ongoing improvements include an effort to integrate Canada fully into TAMM so that not only lumber but all other forest products can have internally determined trade flows. Another plan is to add

Japan as a demand region for softwood logs and lumber. Finally, trade relationships for softwood lumber and plywood shipments to Europe are also expected to be included in the future (Darr 1981; Haynes and Adams 1981).

These improvements, however, continue to fall short of the international analysis requirements of the Resources Planning Act of 1974. In fact, even with these new features, TAMM would permit only a limited analysis of "how trade patterns might change in the future in absence of any changes in policies and...following any changes in policies" (Darr 1981). While Canada, Japan, and Western Europe are currently the major trading partners of the United States in forest products, the TAMM improvements cover only a small number of products and ignore several regions of the world which could become either customers or competitors in the future. As a result, the new version of TAMM would still provide insufficient information for the development of long-term forest products trade strategies.

Thus far it can safely be concluded that the supply-and-demand equilibrium approach can be successfully used for analyzing and projecting the production, consumption, trade, and price dimensions of the forest sector of a large economy such as the United States. As of this date, however, there is no analytical framework capable of handling these same dimensions in the broader context of long-term comparative advantages of the United States with respect to other countries.

Since the determination of comparative advantages requires knowledge of supply and demand conditions for several products in several countries, an ideal solution would be to build models similar to TAMM for various world regions and to link them in a world equilibrium framework. This linkage framework would interact with the various country or region models by providing them with broad information on production, consumption, trade flows, and world prices of wood products. The region or country models, in turn, would furnish the linking framework with resource constraints, local supply, and demand structures and policy changes. The final output would be a world equilibrium of supply, demand, and prices for various products and various countries. Among the results would be the comparative advantages of each country or region with respect to others in various types of forest products. With these results, a country would be able to identify its competitive opportunities, threats, strengths, and weaknesses, and outline strategies for future trade.

An effort of this kind, large as it may seem, can build upon several elements which already exist. In addition to the United States, several countries such as Australia, Austria, Finland, Italy, Japan, Sweden, and New Zealand currently have forest sector models of varying levels of detail. Other countries such as West Germany, Hungary, Norway, and the U.S.S.R. are in the process of building them (Adams, Kallio, and Seppälä 1982). Moreover, even though these country forest sector models are not yet linked by a global framework, experience of global models already exists, as previously mentioned, in the agriculture and energy sectors.

Based upon these experiences, a global trade model for the forest-based sector is currently being constructed by the International Institute for Applied Systems Analysis.

Ongoing Work at the International Institute for Applied Systems Analysis: The Forest Sector Project

The International Institute for Applied Systems Analysis (IIASA) is an applied research organization founded in 1972 by the governments of the United States and the Soviet Union. Its offices are located in Laxenburg, Austria. At present, the Institute is supported by 17 countries in North America, Asia, Eastern Europe, and Western Europe (American Academy of Arts and Sciences 1982). The main purpose of IIASA is to conduct interdisciplinary and multinational research efforts on global problems in a nonideological and nonpolitical environment. In its decade of existence, the Institute has had significant research achievements in varied areas such as energy, agriculture, demography, optimization systems, and ecology. In general, these efforts have resulted in practical applications in several countries. Research at IIASA is conducted on the basis of multidisciplinary teams, involving scientists of various specialties, from several countries, representing universities, government agencies, and private concerns. This form of international cooperation has enabled scientists to establish overseas contacts in their fields and to develop a more global perspective of problems in their home countries.

In 1978, it was suggested that IIASA should look at problems related to the forest sector. After a few preliminary efforts in 1979 and 1980, IIASA's current Forest Sector Project began in September 1981. The two major objectives of this project are:
1. To help participating countries in the development of detailed national models for the forest sector (Task A).

2. To develop a global model for the analysis of world trade in forest products, based on regional component models and a linkage system (Task B) (Adams, Kallio, and Seppälä 1982).

IIASA's role in Task A will be to develop a generic or prototype model for the forest economy of a country. This model will include supply and demand for industrial products, roundwood, and stumpage. Biological elements such as inventories, growth, drain, and mortality will also be included, as well as institutional factors like forest practices regulations. This type of framework will permit the determination of production, consumption, and prices of pulp products, solid-wood products, raw material requirements, and stumpage. The prototype model will serve as a blueprint for the detailed national models which will be built by participating countries (Adams, Kallio, and Seppälä 1982; Lönnstedt 1982). The United States already has TAMM, which is a detailed national model, therefore this task is not of primary importance for this country. However, since these national models will eventually be linked through the global trade model, the U.S. needs to ensure that its national model is compatible with those of other countries. Additionally, the experience the U.S. has acquired by building TAMM can be very valuable in helping other countries to build or adjust their own models.

Task B has the mission of building a global trade system, which will be accomplished in various stages. The first of these will be the development of a linkage mechanism involving basic assumptions on the form and nature of the behavior of world markets. A second stage will consist of constructing the regional components of the Global Trade Model (GTM). These will contain the basic elements of the forest economy of a country or region, but will be more aggregate than the detailed country models. IIASA believes that the linking of highly detailed country models in a system to explain global trade patterns appears to have little chance of success (1982). Therefore, it was decided to build the Global Trade Model on the basis of more simplified regional components, although the Global Trade Model and its regional components will be kept compatible with the detailed national models. After completion of all the regional components, these will be calibrated, tested, and assembled to form the Global Model.

For purposes of constructing the Global Trade Model, IIASA is initially proposing the product categories and country groupings shown in table 2. Although both country

Table 2. Proposed Product Categories and Country Groupings.

Product Categories	Country Groupings
1. Coniferous logs	1. Africa
2. Nonconiferous logs	2. Canada
3. Pulpwood	3. United States
4. Fuelwood	4. Brazil
5. Coniferous lumber	5. Latin America excluding Brazil
6. Nonconiferous lumber	
7. Panels	6. China
8. Pulp	7. Japan
9. Newsprint	8. ASEAN
10. Other printing and writing papers	9. Asia excluding China, Japan, and ASEAN
11. Other paper and board	10. Finland
	11. Sweden
	12. Western Europe excluding Finland and Sweden
	13. Hungary
	14. Eastern Europe excluding Hungary
	15. Australia
	16. Oceania excluding Australia
	17. U.S.S.R.

Source: Adams, Kallio, and Seppälä 1982.

groupings and product categories may be debatable, IIASA has considered these as a minimum for the assembly of the data base required for model building and testing (IIASA 1982).

Once the Global Trade Model is built and tested, the next stage will be to develop joint simulations in which the detailed country models and the Global Trade Model will interact. Since the Global Trade Model will be much more aggregate than the detailed country models, disaggregation procedures will need to be built on a country-by-country basis. The expected result of this state is the simulation of scenarios depicting the future development of world trade and its interaction with the forest-based sector of specific countries. These simulations will also permit the analysis of policy options emanating from individual countries and their effects on world trade as well as domestic impacts on resources, production, consumption, and prices.

No doubt the IIASA Forest Sector Project is ambitious, but if successful, the Project can make invaluable contributions to the information needs for U.S. trade strategies. The interaction of TAMM with IIASA's Global Trade Model will permit the identification of the United States' trade potential in terms of specific comparative advantages. In addition, this combination of models will largely satisfy the international requirements of the Resources Planning Act of 1974.

Work on the IIASA Forest Sector Project to date has produced a number of results, including:
- The development of a network of scientists in several participating countries
- Technical studies about demand, supply, and trade modeling for forest products
- Preparation of theoretical structures for country prototype models and the Global Trade Model

Work is now progressing in the assembly of the Global Trade Model's data base, on the Global Model's linkage mechanism and the construction of its regional components. Plans are for IIASA to complete the Forest Sector Project by mid-1985.

As of last September, the collaboration network included the countries and international organizations listed in table 3. More recently, IIASA has been seeking additional cooperation from countries in South America and Southeast Asia (Kallio and Seppälä, pers. comm., November 29, 1982). Australia, Finland, West Germany, Hungary, Japan, Sweden, and the United States seem to be among the most active participants.

United States participation in the project to date has been largely the role of the Forest Service. Other American organizations which have collaborated include Data Resources Incorporated, Oregon State University, Resources for the Future, University of Washington, University of Wisconsin, and Weyerhaeuser Company. In addition, the National Forest Products Association (NFPA) and the American Paper Institute (API), as well as several forest products companies, have kept abreast of IIASA's progress without yet making a direct commitment to the effort.

The contributions of United States participants have been in the form of financial support, visiting scientists, and direct collaboration in the writing of IIASA papers. The Forest Service is now beginning the construction of the United States' regional component of the IIASA Trade Model, which should be completed in the second half of this year (Haynes, pers. comm. 1982).

Table 3. Countries and International Organizations Participating in the IIASA Forest Sector Project.

Countries	International Organizations
Australia	FAO
Austria	FAO/ECE Timber Section
Canada	UNIDO
Finland	IUFRO
France	
Germany (FRG)	
Hungary	
Italy	
Japan	
Netherlands	
New Zealand	
Norway	
Sweden	
United Kingdom	
United States	
U.S.S.R.	

Source: Adams, Kallio, and Seppälä 1982.

While direct participation of private industry has been limited to this date, API and NFPA are trying to find ways to improve this situation. Forest products companies could significantly contribute to the project in developing realistic supply-and-demand relationships for specific solid-wood and paper products. On the other hand, industry participants will benefit from direct access to international data, foreign contacts, and new techniques of trade analysis.

Although it is too early to tell whether the IIASA Forest Sector Project will succeed, it is the only international study that integrates forest products markets and resources in a supply-and-demand context and on a worldwide basis. Additionally, the models are being built for the purpose of examining policy alternatives. Considering the increasingly complex and important role of the United States in world trade of forest products, it behooves this country to support the project in every way it can to ensure its success. It must be remembered that the effort is already in progress and that several countries are participating. As work progresses, participating countries will gain major informational advantages regarding world trade of forest products, and therefore the United States should commit its full cooperation.

References

Abkin, M. H. 1981. Concepts behind IIASA's food and agriculture model. Paper presented at the North American Conference on Forest Sector Models, Williamsburg, Virginia, December 1981.

Adams, D. M., and R. W. Haynes. 1980a. U.S.-Canadian lumber trade: The effect of restrictions. Proceedings of a Workshop on Issues in U.S. International Forest Products Trade. Resources for the Future, Washington, D.C.

Adams, D. M., and R. W. Haynes. 1980b. The 1980 softwood timber assessment market model: Structure, projections and policy simulations. Forest Science Monograph 22. Society of American Foresters, Washington, D.C.

Adams, D. M., M. Kallio, and R. Seppälä. 1982. Structural change in the forest sector: The Forest Sector Project. International Institute for Applied Systems Analysis, Laxenburg, Austria.

American Academy of Arts and Sciences. 1982. Strengthening the international interdisciplinary activities of the American Academy of Arts and Sciences: Support for the International Institute for Applied Systems Analysis (IIASA). Cambridge, Mass. November.

API/NFPA, International Trade Research Task Group. 1981. Background Paper. July 13.

API/NFPA, Task Group on World Supply/Demand Trends. 1982. Research recommendations. Presented at a meeting on January 5, 1982.

Batten, D. F. 1981. Towards an interdependent system of Models for Australian forest sector analysis. Paper presented at the North American Conference on Forest Sector Models, Williamsburg, Virginia, December.

Darr, D. R. 1981. Methods of analysis of bilateral and multilateral trade among major forest products trading partners. Paper presented at the North American Conference on Forest Sector Models, Williamsburg, Virginia, December.

Darr, D. R. 1980. U.S. exports and imports of some major forest products--The next fifty years. Proceedings of a Workshop on Issues in U.S. International Forest Products Trade. Resources for the Future, Washington, D.C.

Forest Industries Advisory Council. 1982. Strategies for growth. Proceedings of the 1982 FIAC. American Forest Institute, Washington, D.C.

Haynes, R. W., and D. M. Adams. 1981. Research on TAMM and other elements of the U.S. timber assessment system. Paper presented at the North American Conference on Forest Sector Models, Williamsburg, Virginia, December.

Haynes, R. W., D. R. Darr, and D. M. Adams. 1980. U.S.-Japanese log trade--Effect of a ban. Proceedings of a Workshop on Issues in U.S. International Forest Products Trade. Resources for the Future, Washington, D.C.

IIASA Forest Sector Project, Demand, Supply and Trade Group. 1982. Considerations in future development of the IIASA forest sector project: Model structure, product demand models, product category, definition geographical aggregation and data availability. International Institute for Applied Systems Analysis, Laxenburg, Austria, WP-82-108.

Laarman, J. G. 1980. Discussion on R. A. Sedjo's paper on world forest plantations. Proceedings of a Workshop on Issues in U.S. International Forest Products Trade. Resources for the Future, Washington, D.C.

Lönnstedt, L. 1982. Mathematical formulation of the forest sector prototype model--The first step. International Institute for Applied Systems Analysis, Laxenburg, Austria.

NFPA, International Trade Committee. 1982. Increased wood products exports: A bonus for the industry and nation. National Forest Products Association, Washington, D.C.

Radcliffe, S. J. 1980. U.S. forest products trade and the multilateral trade negotiations. Proceedings of a Workshop on Issues in U.S. International Forest Products Trade. Resources for the Future, Washington, D.C.

Sedjo, R. A. 1980. World forest plantations--What are the implications for U.S. forest products trade? Proceedings of a Workshop on Issues in U.S. International Forest Products Trade. Resources for the Future, Washington, D.C.

Takayama, A. 1972. International trade. Holt, Rinehart and Winston, New York.

Ueda, M., and D. R. Darr. 1980. The outlook for housing in Japan to year 2000. USDA Forest Service, Research Paper PNW-276, August.

USDA, Forest Service. 1982. An analysis of the timber situation in the United States, 1952-2030. Forest Resource Report No. 23, December.

Zivnuska, J. A. 1980. Discussion on D. R. Darr's paper on U.S. exports and imports of some major forest products--The next fifty years. Proceedings of a Workshop on Issues in U.S. International Forest Products Trade. Resources for the Future, Washington, D.C.

INTERNATIONAL TRADE STRATEGIES

George E. Taylor

The economy of the Pacific Northwest depends on international trade for the employment of one-fifth of its labor force and the maximum production of its major industries. This condition is irreversible, unless we are willing to accept a lower standard of living. At stake are well over 300,000 jobs and, at the very least, the future prosperity of our agricultural, forest products, and aerospace industries, all of which are internationally competitive.

The international trade of the Pacific Northwest depends directly on the successful functioning of the open international trading system as we have known it since World War II. At the heart of that system is the institution known as the General Agreement on Tariffs and Trade (GATT), set up in 1947 to arrange for a lowering of the obstacles to trade. The successes of GATT made possible the spectacular increase in world trade of the last three or four decades. GATT is the main barrier in the way of trade wars and economic chaos because it provides for the arbitration of international trading disputes and for general acceptance of most favored nation treatment. As long as GATT's eight-odd members observe the spirit and the letter of its agreements, the system can function--much as it needs repair.

GATT is essential, but it is itself dependent on the voluntary cooperation of the industrial democracies. The international trading system is more than a series of economic arrangements. It is part and parcel of political and security arrangements such as NATO, the Organization for Economic Cooperation and Development, the European Community, and bilateral security alliances between the members of the Atlantic-Pacific system. While the alliances depend on the healthy functioning of the economic base of international trade, that trade is influenced by political and security problems that have a life of their own. To endanger trading relationships, therefore, is to endanger the security of the industrial democracies. And to endanger security arrangements is to endanger trade.

The international trading system has created an unprecedented degree of economic integration among the industrial democracies. This has come about largely through the success of multinational corporations in speeding the mobility of capital and production facilities, and partly through the international institutions and agreements that support and monitor international finance, such as the World Bank, the International Monetary Fund, and the former Bretton Woods agreement. The growth of international economic penetration makes it almost impossible for any important country to withdraw and survive. The prosperity of the part depends on that of the whole.

The international trading system is in crisis for a variety of reasons. One of the main ones is the global banking crisis. Various countries have accumulated more foreign debt than their economies can support. Many of our banks, on the other hand, have more foreign loans at risk than they have capital and other reserves. Efforts are being made to at least keep interest payments coming. The threat of renunciation of debts by countries such as Mexico, Argentina, and Brazil is met with warnings that all their assets abroad would be seized as payment for defaulted loans. There are well-known scenarios describing how the Clearing House Interbank Payment System (CHIPS), the electronic nerve center of the $2.5 trillion Eurodollar market, can be brought to a standstill.

What will happen? The answer to the debt problem, it is argued, is not to lend debt-ridden countries more money at higher interest rates. An alternative has been suggested that might lead to the gradual recovery of debtor nations. This is to reduce or withhold interest payments for a period of time on the $300 billion commercial loans. The banks created the situation and should therefore--it is proposed--pay most of the cost. By cutting interest rates in half, they would not ruin themselves and might give the debtors a chance to avoid total default. There is every reason to believe that the future of international trade need not be disastrous if, as Karin Lissakers put it, the parties who helped create the crisis will bear part of the cost of resolving it.

The international trading system is also in crisis because of protectionist pressures and the growing avoidance of GATT regulations by important countries, including the United States. Recession has hit the industrial democracies, and unemployment is universal. The OECD forecasts that by the end of 1983 there will be 32 million unemployed in its 24 member countries. The combination of slow growth and increasing competition, of continuing structural and technological changes, creates a situation in which governments must choose between facilitating economic adjustment and protecting industries and workers, thus undermining the international trading system as well as national economies.

The pressure on governments to save jobs and industries threatened by structural changes is powerful indeed, especially in Europe. That is why the proportion of world trade subject to nontariff barriers has steadily increased. It is estimated that at least half of world trade is subject to such barriers. As it is much easier to blame others for domestic problems, the search for scapegoats is inevitable. Japan, for example, is called on to take the blame for the decreasing productivity of major American industries.

Our future prosperity depends, therefore, on avoiding the temptation to go in for protectionist measures. Put positively, the U.S. needs to use all its influence to reaffirm support of an open trading system and to act to insist on observance of the rules of GATT. The stand of the administration seems to encourage the view that it will not be stampeded into protectionist measures. But it could be pushed if public support for protectionism becomes stronger.

It is well to keep in mind that the trading position of the U.S., taken as a whole, including current accounts, is quite sound. The problems arise from structural changes such as the increase in trade in services (now 40 percent of U.S. exports) and the decline of manufacturing. Protectionism would prevent structural adjustments at the expense of increasing unemployment and international trade wars. These, no one can afford.

International trade is not a matter of choice. For the second time in American history we have to trade to live. There is no such thing as fully protectionist or free trade economies. The question is one of emphasis and direction. International trade is the fuel, the lubricant of economic growth. And growth solves more of our problems than protected stagnation does. The future, therefore, is as bright or as gloomy as our collective attitude and national will. The American colonies survived the first fifty years because they developed an international trade that was both worldwide and interdependent. They started from scratch at a time when the international trading system was dominated by great trading companies, backed by powerful navies. The U.S. today does not start from scratch, but the decisive element is still one of national will based on understanding of the economic and political realities of our day.

Contributors

Takashi Akutsu, Director-General, Japan Trade Center (JETRO), San Francisco
Richard L. Atkins, Vice President, Marketing, Burlington Northern International Services, Inc., Portland, Oregon
James S. Bethel, Professor, College of Forest Resources, University of Washington, Seattle
Robert E. Biehl, Assistant Vice President and Manager, International Banking, Barclays Bank International Ltd., Seattle
Charles W. Bingham, Executive Vice President, Weyerhaeuser Company, Tacoma
Brian J. Boyle, Commissioner of Public Lands, State of Washington, Olympia
Susan J. Branham, Forest Service, U.S. Department of Agriculture, Washington, D.C.
David G. Briggs, Assistant Professor, College of Forest Resources, University of Washington
Richard V.L. Cooper, Partner, Coopers & Lybrand, Washington, D.C.
John B. Crowell, Jr., Assistant Secretary of Agriculture, U.S. Department of Agriculture, Washington, D.C.
Charles E. Doan, Director, Trade and Industrial Development, Port of Tacoma
Robert L. Ethington, Director, Pacific Northwest Forest and Range Experiment Station, Forest Service, U.S. Department of Agriculture, Portland, Oregon
M. T. Fast, Director, International Operations, American Plywood Association, Tacoma
Marco A. Flores-Rodas, Assistant Director-General, Forestry, Food and Agriculture Organization of the United Nations, Rome
William D. Hagenstein, Consultant, Portland, Oregon
David Haley, Associate Professor, Faculty of Forestry, University of British Columbia, Vancouver
John O. Haley, Professor, School of Law, University of Washington
Vernon L. Harness, Director, Forest Products Staff, Foreign Agricultural Service, U.S. Department of Agriculture, Washington, D.C.

CONTRIBUTORS

Robert M. Ingram III, Vice President and Manager, Forest Products Commercial Banking Center, Rainier National Bank, Seattle
John Larsen, Vice President, Weyerhaeuser Company, Tacoma
Vivien F. Lee, Attorney at Law, Seattle
Bronson J. Lewis, Executive Vice President, American Plywood Association, Tacoma
Bruce R. Lippke, Manager, Marketing and Economic Research, Weyerhaeuser Company, Tacoma
Irene W. Meister, Vice President, International, American Paper Institute, New York
Timothy J. Peck, Chief, Timber Section, FAO/ECE Agriculture and Timber Division, Geneva
Pedro M. Picornell, Senior Vice President, Paper Industries Corporation of the Philippines, Manila
Stephen B. Preston, Associate Dean, School of Natural Resources, University of Michigan, Ann Arbor
H. A. Roberts, Executive Vice President, Western Wood Products Association, Portland
Gerard F. Schreuder, Professor, College of Forest Resources, University of Washington
Roger A. Sedjo, Director, Forest Economics and Policy Program, Resources for the Future, Washington, D.C.
Juan E. Sève, Planning Coordinator, Fiber Resources, Container Corporation of America, Chicago
Shizuo Shigesawa, President, Japan Plywood Inspection Corporation, Tokyo
Melvin E. Sims, General Sales Manager, Foreign Agricultural Service, U.S. Department of Agriculture, Washington, D.C.
Eugene W. Smith, Senior Industrial Development Officer, ITC-DREE Regional Office (P.O. Box 49178, Bentall Postal Station, Vancouver V7X-1K8)
W. Ramsay Smith, Associate Professor, College of Forest Resources, University of Washington
John Spellman, Governor, State of Washington, Olympia
Charles L. Spence, President, Pacific Lumber and Shipping Company, Seattle
Kenji Takeuchi, Senior Economist, Commodities and Export Projections Division, Economic Analysis and Projections Department, Economics and Research Staff, The World Bank, Washington, D.C.
George E. Taylor, President, Washington Council on International Trade, Seattle
David B. Thorud, Dean, College of Forest Resources, University of Washington

J. Frederick Truitt, Associate Professor, School of Business Administration, University of Washington

Yoshio Utsuki, Assistant Director, Forest Products Division, Forestry Agency, Ministry of Agriculture, Forestry and Fisheries, Tokyo

John Wadsworth, President, BIS Marketing Research Ltd., London

John Ward, Director, International Trade, National Forest Products Association, Washington, D.C.

Ross S. Whaley, Director of Forest Resources Economics Research, Forest Service, U.S. Department of Agriculture, Washington, D.C.

James G. Yoho, Professor, Department of Forestry and Natural Resources, Purdue University, West Lafayette, Indiana

Index

Abaca pulp, 267
Adhesives, 211, 213, 338
Africa: potential for expansion, 14, 15, 58, 439; fuel-gathering and grazing, 12; future area in plantations, 13; growing stock, 14; wood supplier, 27; arable land, 34; size of closed forest, 34; and fabricated housing, 106; and U.S. plywood trade, 108-9; pulp supplier, 271; hardwood production, 440; as an export trading company market, 384; consumption of pulp, 113; linerboard, 115; paper and paperboard, 117; industrial wood, 333-36; hardwood, 442. See also individual areas and countries of Africa
Agathis, 475
Agricultural Act of 1981, 160
Agricultural Trade Expansion Act of 1978, 160
Agriculture: trade patterns, 29-32; management transition compared with forestry, 46-47; agricultural lands being replaced by forests, 52; surpluses and exports, 90; U.S. exports, 143, 144, 147. See also Food production
Alaska, 235, 261, 271
Alaska Pulp Company, 260, 481
Albizzia, 264, 267, 475, 581
Amazon area, 439, 443
American Lumber Standards Committee, 498, 500
American National Standards Institute (ANSI), 499
American Paper Institute (API), 120, 518, 519
American Plywood Association (APA), 21; Performance Standards, 101-2; export activities, 102, 126-27; cooperation with FAS, 102, 126; plywood approval in West Germany, 107; and housing in Chile, 109; activities in Latin America, 127
American Society for Testing and Materials (ASTM), 499
Amling, Frederick, 159
Angola, 439
Antiboycott laws, 119, 163
Antibribery enforcement, 166

Antimonopoly and Fair Trade Law (Japan), 151
Antitrust action: and export trading companies, 127-28, 158, 164, 430; and Japan, 149-56; Zenith case, 155, 156; and shipping legislation, 171-72
Argentina, 35, 37, 173, 524
Arrow M. Industrial, 479
Asahi Chemical Industry Company, 260, 261
Asia: trade with, 6-7; future area in plantations, 13; growing stock, 14; potential for expansion, 15, 122; wood supplier, 27, 29; arable land for agriculture, 34; size of closed forest, 34; and immigration to U.S., 37; growing paper and pulp market, 80, 112, 113, 114, 117, 118; lumber standards, 88; trade restrictions, 123; trade with British Columbia, 245; joint ventures, 258; competition for Asian markets, 271; share of world trade, 341; trade with Europe, 358; hardwood vs. softwood, 413-14; financing methods, 428, 430; trade with Japan, 434; hardwood production, 440; prices, 443. Consumption: linerboard, 112, 115; fiberboard, 293; industrial wood, 333-36; hardwood, 442. See also individual areas and countries of Asia
Asian Wall Street Journal, 260
Aspen, 51
Association of Southeast Asian Nations (ASEAN), 112, 114, 119, 262-68, 507
Australia: trade with Weyerhaeuser, 7; "other woodland," 12; U.S. trade, 96, 97, 122, 279; tariffs, 238; joint ventures, 257, 469; currency devaluation, 271, 447; among largest importers and exporters, 306, 307; trade with Europe, 356; trade with Pacific Northwest, 370; plantations, 451; Japanese investment in, 478; net importer, 507; forest sector model, 515. Consumption: hardwood, 289, 442; newsprint, 294; logs, 434. See also Pacific Rim countries
Austria, 257, 354; softwood lumber

production, 295; particleboard production, 299; among largest importers and exporters, 303, 305; consumption trends, 412; forest sector model, 515
Austrian Paper and Packaging Company, 257
Automotive industry: panel products, 104, 105, 107, 213; protectionism, 186, 188

Bahamas, 103
Bank for International Settlements, 36
Bank of China, 259-60
Banks and banking: debt crisis, 35-37, 455, 524; credit programs, 126, 127, 146-47; export trading companies, 127-28, 377, 380, 393, 397, 399, 430; financing small- and medium size firms, 425-31; finance and support for exporters, 483-93. See also Export-Import Bank; finance and investment
Bark, 123
Belgium: U.S. plywood imports, 102; use of panel products, 105; particleboard consumption and production, 291, 299; among largest importers and exporters, 303-7; percentage of imports, 354; current difficulties, 447
Best Conventional Technology, 132
Best Practicable Treatment, 132
BIS Marketing Research Ltd., 410, 413
Bleached paperboard, 112, 113, 116, 118
Block, Secretary of Agriculture John R., 144
Bloedel-Donovan Lumber Mills, 281
Boeing Company, 111-12
Boggs, Lindy, 172, 173
Boise Cascade Corporation, 257
Bolcom-Canal Lumber Company, 277
Bonker, Don, 69, 169
Boren, David L., 170, 171
Brazil: future forest area, 13; indebtedness of, 35, 37, 524; exporter of pulp, 58, 116, 356; U.S. shipping agreement, 173; joint ventures, 257, 260; among largest importers and exporters, 304-6; trade with Europe, 345, 356; Amazon area, 439, 443; efforts to expand, 507. Consumption: softwood, 288, 289; hardwood, 288; plywood, 290, 291; fiberboard, 292; pulp, 292; printing and writing papers, 294; paper and paperboard, 295. Production: softwood, 295; hardwood, 297, 433, 440; plywood, 297; fiberboard, 298, 299; pulp, 300
Bretton Woods system, 455, 524
Breweries, 104

British Columbia: major exporter, 53, 241-42; future supplies, 55, 59, 250-53; ocean transport facilities, 134; markets, 242, 243, 244, 246; volume of timber harvested, 242, 251-52; forest products production, 242-43; interior vs. coast, 243-44, 246, 250; mills, 244, 250; role of provincial government, 246-54; dimension lumber, 243, 247; export controls, 248-49; licensing, 249; government incentives, 250; growth rates, 253; competitor with U.S., 271, 279; foreign ownership in, 453. Production: lumber, 242, 243-44; woodpulp, 243, 245; newsprint, 243, 245; plywood, 243, 246
British Columbia Ministry of Forests Acts, 247
Broken Hill, Australia, 469
Bryant Lumber Company, 277
Bulk cargo, 172-73
Burlington Northern International Services, 256
Burma, 433, 439
Business Accounting and Foreign Trade Simplification Act, 165, 166

Cabinet Counsel on Commerce and Trade, 171
California, 232-33, 256
Cameroon, 433, 439, 443
Canada: exporter of forest products, 3, 27, 29, 53, 191, 232, 240-41, 357; in relation to U.S., 3, 67, 85, 88, 100, 104, 192, 235, 271, 357; U.S. competition, 122, 134, 271; trade with Weyerhaeuser, 7; supplier for U.S., 27, 29, 91, 242, 243, 244, 245; supplier of pulp to U.S., 111, 242, 245; U.S. plywood imports, 102; comparison with U.S. plywood, 104, 177; market for U.S. paper and paperboard, 117; tariffs on U.S. molding and millwork, 158; imports from U.S., 187-88; and Smoot-Hawley Act, 188; and U.S. Congress, 189, 190; and U.S. protectionism, 185-93; trade with Washington State, 270; and Department of Commerce decision, 271; U.S. investment in, 453-54; supplier for Europe, 27, 98, 242, 244, 245, 246, 350, 358; trade with Japan, 65, 66, 67, 208, 209, 242, 244, 247, 472; competition for Japanese market, 68, 85, 86, 88, 98, 209; Japanese investment in, 478; and Southeast Asia, 263; trade with China, 421; "other woodland," 12; protectionism toward, 22; wood consumption, 24, 334-35; rising demand for wood products, 26; per capita forest area, 62; controls on

INDEX   533

log exports, 66, 88-89, 248-49, 472; log imports, 434; tariff disadvantage in pulp and paper, 116, 118; market for lumber and plywood, 121; ocean transport facilities, 134; provincial landownership, 241; joint ventures, 257, 258, 260; among largest importers and exporters, 303-7; restrictions on partly processed products, 312; trade in paper and paperboard, 350; trends in wood prices, 363; future shortages, 359; value added in processing, 456; standards, 497, 500; subsidization of stumpage prices, 504; inclusion in TAMM study, 513. Production: softwood, 295; plywood, 296, 297; particleboard, 299; fiberboard, 299; pulp, 300; newsprint, 300; printing and writing papers, 301; paper and paperboard, 301. Consumption: softwood, 288, 359; hardwood, 442; plywood, 290; particleboard, 290, 291; fiberboard, 292; pulp, 292; newsprint, 294; printing and writing papers, 294; paper and paperboard, 295. See also British Columbia
Canadian Forestry Service, 359
Canadian Softwood Plywood (CSP), 246
Cargo handling facilities, 133-34. See also Bulk cargo
Caribbean, 109, 127, 433
Cariboo Pulp and Paper, 257
Cartels: in Japan, 149-56; and U.S. action, 155; as trade barriers, 238
Cedar: Japanese, 68; ban on export of logs, 69-70, 88; world's largest mill, 277
Celanese Corporation, 459
Cenibra, 260
Central Africa, 433, 434, 435, 439
Central African Republic, 433, 443
Chafee, John H., 166
Champion International Corporation, 257, 277
Chile: housing symposium in, 21, 109; plantations, 52, 315, 451; exports of logs and processed wood products, 58; potential for exports, 59, 254, 507; competition for Japanese market, 68, 69; housing needs, 109; joint venture, 256; trade with Northwest, 279; among largest exporters, 303, 315; trade with China, 421
China (PRC): opportunities for trade with, 6, 89, 99, 119, 122, 239, 266, 416-24, 448, 507; U.S. quotas on textiles from, 6, 22; trade with Weyerhaeuser, 7; normalization of relations with, 30; U.S. shipping agreement, 173; trade with Pacific Northwest, 270, 279-80, 281, 285; trade with U.S., 99, 122, 419, 420; exports to Japan, 208; land in plantations, 13; housing needs, 99; need for grading standards, 99; special economic zones, 259, 424; joint ventures, 259-60, 457; sawmills, 280; importer of logs, 285; among largest importers and exporters, 304, 306, 307; forest area, 417; shipping delays, 419; shortage of foreign exchange, 419; sources of imports, 419-21; conservation efforts, 421-23, 451; decentralized log buying, 427. Consumption: linerboard, 112, 115, 239, 419, 448; plywood, 266, 448; softwood and hardwood, 288; fiberboard, 292; pulp, 292, 419; newsprint, 294; printing and writing papers, 294; paper and paperboard, 295, 337; firewood, 416, 419. Production: softwood, 295; hardwood, 297, 440; plywood, 297; fiberboard, 299; pulp, 300; newsprint, 300; printing and writing papers, 301; paper and paperboard, 301; timber, 419; logs (by province), 416-17, (by year), 418. See also Pacific Rim
Chinese Timber Corporation, 418
Clearing House Interbank Payment Systems (CHIPS), 524
Cline, William, 190
CLS dimension lumber, 247
CoCom (Coordinating Committee for Multilateral Export Controls), 150
Colombia, 21, 109, 507
Columbus Day Storm (1962), 470
Committee on Renewable Resources for Industrial Materials (CORRIM), 330-32
Commodity Credit Corporation (CCC), 126, 144, 146; GSM-102 Export Credit Guarantee Program, 126, 144, 145, 146-47
Common Market. See European Economic Community
Competitive Shipping and Ship Building Act of 1983, 172
Congo, 433, 439
Congress, U.S.: position on trade, 189; and Canada, 189-90; hearings on log export question, 284
Conservation. See Forest management; Sustained yield
Control Data Corporation, 387
Cooperative Overseas Market Development Program (COMDP), 247
Corporation and Labour Union Returns Acts, 453
Council for Mutual Economic Assistance (CMEA, or Comecon), 357, 457
Council of Forest Industries of British Columbia (COFI), 68, 247

Countertrade, 387, 388
Cox, Thomas R., 278
Credit: programs, 126, 127, 143-48; blended credit program, 144; Commodity Credit Corporation, 126, 144, 145, 146; credit guarantees, 146, 485-86; usance, 484. See also Banks and banking; Finance and investment
Crown Zellerbach, 256, 472
Currency exchange. See Exchange rates
Customs Service, 169
Cypress (Japanese), 68, 210
Czechoslovakia, 303

Daiei, 155
Daiken Trade, 480
Daishowa Paper Manufacturing Company, 257, 260, 478, 480, 481
Danforth, John C., 168
Data Resources Incorporated, 518
Davidson, William H., 458, 468
Defiance Lumber Company, 281
Dempsey Lumber Company, 281
Dendrothermal energy, 268
Denmark, 102, 105-6, 303-6
Department of Agriculture, U.S.: role in U.S. exports, 121, 125, 143, 145, 147, 148, 231-39; and Egyptian wheat deal, 144, 145; and government organization, 165; management regulations, 233; renewable resources studies, 509-11. See also Foreign Agricultural Service; Forest Service; Renewable Resources Assessment
Department of Commerce, U.S., 21; and standards, 101, 497, 499, 500; International Trade Administration and model home in Chile, 109; and export trading companies, 128, 164; and Canadian lumber imports, 271
Department of Commerce and Economic Development, Washington State, 21, 109
Department of Defense, U.S., 497
Department of Justice, U.S., 166, 467
Department of Labor, U.S., 497
Department of Trade and Commerce (proposed), 165
Department of Transportation, U.S., 173
Department of the Treasury, U.S., 36, 161, 162, 171
Depressed Industries Law, 118
Developing countries (less developed countries, Third World): effects of recession on, 10; food production, 11; fuelwood, 12, 16, 24, 33, 432; forests in, 12-13; growing stock in, 14; wood consumption, 24, 334-36; rising demand for forest products, 26; food importer, 30; population growth, 31, 32-33; resource inequities, 33, 34; supply of non-fuel minerals, 33; bank indebtedness of, 35, 37, 455; stronger voice in political world, 38; prospects for paper and paperboard, 117, 118, 119; nontariff barriers, 120; P.L. 480 aid, 145, 146; preferred tariff rates, 167; competition for Asian markets, 271; and reforestation, 273; financial and technological agreements, 310; poor quality control, 314; efforts to increase exports of products, 407; hardwood exports, 432; softwood exports, 432; hardwood production, 433; joint ventures, 458, 460-65
Domestic content legislation, 167
Domestic International Sales Corporations (DISCs), 161-62, 170-71
Dominican Republic, 103, 109
Douglas, David, 276
Douglas-fir, 243, 246, 248, 276-78, 309
Douglas-fir Exploitation and Export Company, 282
Dow Chemical Company, 260, 261
DuPont, Washington, 133-34
Dutch Timber Research Institute (TNO), 106

Earthquakes, 107
East Indies, 53, 55-56, 59
Economic Commission for Europe (ECE); study of Europe, 352; Timber Committee, 500; and IIASA, 519
Economic Recovery Act of 1981, 163
Economics of forest resources, 44-59. See also Finance and investment
Ecuador, 21, 109
Edge Act corporations, 127
Egypt, 37, 96, 144, 145
Employment, 313-14; future unemployment, 406, 524
Energy considerations, 330-33
England: timber frame construction for housing, 97, 104, 124; joint ventures, 257, 466; different from Europe, 402-3; long-time importer, 402-3; current difficulties, 447; standards, 499. See also United Kingdom
Environmental issues: influence on world trade in wood, 129-36; water quality standards, 132-33; Weyerhaeuser experience, 133-34; exporting pollution, 362
Environmental Protection Agency, 131
Equatorial Guinea, 433, 439
Eucalyptus deglupta, 264, 267, 475, 481
Eureka, California, 256
Europe, Eastern: as an importer from the U.S.S.R., 29; as a food

INDEX   535

importer, 30; indebtedness of, 35, 37; imports from Western Europe, 357
Europe, Western: and wood shortages, 5, 6, 14, 26, 27; anti-Japanese protectionism, 6; trade with Weyerhaeuser, 7; closed forest, 13, 34; growing stock, 14; wood consumption, 23, 24, 25, 27, 96, 334-36, 349-53, 410-12; sources of supply, 353-54; softwood major import, 27; and immigration to U.S., 38; trade with U.S., 96, 97, 122, 123, 236, 350; North America, 340, 345, 350, 358-59; Canada, 345, 350; British Columbia, 242, 244, 245; Scandinavia, 53, 58; U.S.S.R., 27, 57, 340, 345, 357-58; Latin America, 345; Brazil, 345; tropics, 340, 356-57, 434, 442; Africa, 340, 435; Middle East, 340; Southeast Asia, 263, 345; Japan, 469; import-export comparisons, 82, 83, 340-49, 404; competition in pulp and paper trade, 116; prospects for paper and paperboard, 117, 118, 119; trade barriers, 123; antitrust law in, 150, 151; glued-laminated lumber industry, 210; joint ventures, 258, 461-64; place in international trade, 340-49; exports of forest products, 343, 344; supplier of paper to North America, 345; major importer, 345; rate of growth of GDP, 349, 352; housing construction, 349, 351-52, 413; recession in, 351; investments, 351; study of future prospects, 351; recycling waste paper, 353; European forest resources, 355; productivity, 355; competition from overseas suppliers, 355; as an export trading company market, 383; distinction between England and Europe, 402-3; trends in markets, 410-15; future consumption, 412; financing exports to, 430; standards, 500; inclusion in TAMM study, 514. Consumption: plywood, 103-4, 343, 344, 345, 409, 410-12; pulp, 112, 113, 271, 343, 344, 345, 358, 409, 410-12; panel products, 410-12; linerboard, 112, 115, 118; particleboard, 291, 351, 410-12; fiberboard, 293, 410-12; paper and paperboard, 114, 117, 118, 343, 344, 345, 358; hardwood, 442. See also European Economic Community; Scandinavia; individual countries
Europe, Southern: hardwood consumption, 442
European Community, 523
European Economic Community (EEC, Common Market): and forest products trade, 3, 242; duty quota on plywood, 103, 123, 158, 177, 238, 246; trade barriers to U.S., 123, 508; and oak lumber, 126; export subsidies, 145; U.S. support of, 158; U.S. access to market, 165; tariffs, 177, 238; and British Columbia exports, 242, 244; paper industry in, 345; growth of intra-European trade, 349; imports from Nordic countries, 349; hardwood production, 440; standards, 500
"European Timber Trends and Prospects, 1950 to 2000" (ETTS III), 342, 352, 359; (ETTS IV), 353
Evans, Governor of Idaho John V., 425
Exchange rates (currency), 65; floating, 5; decline for the dollar, 30; Soviet adjustments, 69; dollar overvalued, 82, 139, 271, 406, 447, 448; Japanese currency changes, 82; Japanese-U.S., 85; Canadian advantage, 85, 86; need for disciplined monetary system, 116; devaluations in Australia and New Zealand, 271, 447; future instability, 406; in Europe, 447; fluctuations, 484; spot and forward rates, 484; forward contract, 485, 489; "option" forward, 485. See also Finance and investment; Prices
Export Administration Act of 1979 (EAA), 163, 169
Export and Import Transactions Law (Japan), 150, 151
Export Credit Guarantee Program. See Commodity Credit Corporation
Export-Import Bank, 165, 170, 486, 492, 493
Export management companies (EMCs), 394
Export Sales Corporations, 170
Export Trading Company Act of 1982, 21, 127-28, 157, 163-64, 256, 261, 430; structuring an ETC, 375-401; and Japan, 139; and Washington State, 430

Fair Trade Commission (Japan), 150, 151, 152, 156
Far East. See Asia; individual countries
Federal Credit Insurance Association (FCIA), 486, 492, 493
Federal Housing Administration, 497
Federal Maritime Commission, 172
Federal Pipe and Tank Company, 277
Federal Reserve Board, 164
Federal Trade Commission, 154
Federal trust land, 270
Federation of Timber-cooperative Societies in Japan, 216
Ferry-Baker Lumber Company, 281

Fiberboard: Japanese market, 209-10; medium density fiberboard (MDF), 210, 215, 293, 351; Japanese industry, 213-14; world consumption, 292; properties of, 291; density, 291, 293; world production, 298, 299; top ten countries, 303; largest importers and exporters, 305; European exports, 343; European imports, 346-47, 410-12

Fibreco, 248-49

Finance and investment: mounting indebtedness, 10, 30, 35-38; decline in exchange rates for dollar, 30; rise of multinational corporations, 36; effects of market restrictions, 89; joint ventures, 258, 259-60, 451-82; risk, 314, 488, 489, 492; in Europe, 351; export broker, 428, 472; letter of credit, 483, 484, 486, 488, 489, 491, 492; sight letter of credit, 428, 429, 430, 449, 485, 488; documentary collection, 428, 429, 430, 483, 484, 489; banker's acceptance, 428, 429, 484, 486, 490, 491; foreign direct investment (FDI), 453; forward rate of exchange, 484-85, 489; credit guarantees, 485-86; payment terms compared, 488; insurance, 492, 493. See also Banks and banking; Exchange rates; Export Trading Company Act

Fine Hardwoods-American Walnut Association (FH-AWA), 126

Finland: use of plywood, 108; environmental issues, 133; pulp consumption, 292; softwood lumber production, 295; plywood production, 297; pulp production, 300; newsprint production, 300; printing and writing paper production, 301; among largest importers and exporters, 303-7; forest sector model, 515

Flonibra, 260

FMPA Stuttgart, 107

Food and Agriculture Organization (FAO): objectives of, 17; forecasts for pulp and paper, 114, 359; study of regional development, 119; forecasts for world lumber consumption, 121-22, 359; study of Europe, 352; and IIASA, 519

Food production: and forests, 11; U.S. trade boom, 29; in Third World, 32, 34

Foreign Agricultural Service (FAS), USDA: assistance in U.S. trade, 80, 237; and plywood exports, 102; described, 125-26; credit for exports to Jamaica, 147; and export responsibilities, 160-61; and government reorganization, 165

Foreign Corrupt Practices Act of 1977 (FCPA), 120, 161, 166

Foreign Corrupt Practices Clarification Act, 166

Foreign direct investment (FDI), 453

Foreign Exchange and Foreign Trade Control Law (Japan), 150

Foreign Investment Review, 453

Foreign Investment Review Agency (FIRA), 454

Foreign Market Development program, 126

Foreign trade. See Trade, international

Foreign Trade Analysis Work Unit, 509

Forest and Range Resource Analysis (British Columbia), 250

Forest Industries, 280

Forest Industries Council, 502, 507

Forest management, 16-18; compared with agriculture, 47; transition of old-growth forests to plantations, 47-51; time profiles based on demand, 49-50; effects of technology on, 52; USDA regulations, 233; in South, 234; in Washington State, 272-73; conservation, 10-11; reforestation, 272-73, 451; exploitation and conservation, 362; changing world practices, 364; conservation in China, 421-23, 451. See also Plantations; Sustained yield

Forest Practices Act, 273

Forest products. See Wood products

Forest resources: need for financial investment, 9-10; conservation, 10-11; in industrialized countries, 10; nature of, 11; and food production, 11; size of, 12-16; "other woodland," 12-13; "closed forest," 12-13; plantations, 13; growing stock of closed forest, 13-14; in developed countries, 14; in developing countries, 14, 33; output by the year 2000, 15, 16; management, 16-18; technological progress, 15-16, 51-52; sociopolitical considerations, 16-18; resource inequities, 32-34; economic aspects, 44-59; transition of old-growth forest to plantations, 47-51; effects of technology on, 51-52; major exporting regions, 53-58; trade in raw materials, 81-93; world lumber consumption, 95-96; national forests, 233; private ownership of land, 234; industrial ownership, 234-35; British Columbia, 242-44; reforestation, 272-73. See also Forest management; Wood consumption; Wood products

Forest Sector Project, 515-19

Forest Service, British Columbia, 251
Forest Service, U.S.: trade projections, 54, 509, 510; on U.S. shortages, 90-93, 359; timber assessments, 90; role in U.S. wood trade, 232; government-owned land, 232-33; studies on landownership, 234; research initiatives, 237, 509; nondeclining even flow, 54, 251; report on southern forestry industry, 273; participation in IIASA, 518; Resources Planning Act Assessment, 509-11, 512, 513
Foster, Richard J., 36
France: and protectionism, 6, 238; U.S. plywood imports, 103; use of panel products, 107; paper industry problems, 257; among largest importers and exporters, 303-7; current difficulties, 447; plantations, 507. Consumption: softwood, 288, 409; hardwood, 288; plywood, 290; particleboard, 290, 291; pulp, 292; newsprint, 294; printing and writing papers, 294; paper and paperboard, 295; consumption trends, 410, 411. Production: hardwood lumber, 297; plywood, 297; particleboard, 299; pulp, 300; printing and writing paper, 301; paper and paperboard, 301
Franko, Lawrence G., 452, 467, 468
French Pacific Islands, 103
Frenzel, Bill, 168, 170, 171
Fuelwood, 12-13; shortages, 16, 33; world consumption of, 24, 25, 334; decline in use of in Nordic countries, 24; growing need for, 26; ipil-ipil, 264; in China, 416, 419
Furniture: use of MDF, 210; use of particleboard, 213; and developing countries, 407
Furuya Company, 282

Gabon, 433, 439
Garn, Jake, 169
General Agreement on Tariffs and Trade (GATT): establishment of, 184, 185, 523; export subsidies code, 145; U.S. support of, 158, 525; and taxes on exports, 162, 170, 171; illegal trade practices, 168, 524; multilateral trade negotiations, 64-65, 168, 238, 360, 508; dispute settlement, 191; and plywood tariffs, 246; Ministerial Meeting, 361; importance of, 523; Kennedy round, 158, 360; Tokyo round, 139-40, 158, 177, 179, 181, 246, 360
General Electric Company, 387
Generalized System of Preferences (GSP), 167
General Motors Corporation, 36, 459

General trading companies (GTCs), 454, 470, 471-75
Georgia-Pacific Corporation, 258, 472
Germany, East, 291, 292, 303, 305, 307
Germany, West (FRG): U.S. trade with, 96, 121, 122; U.S. plywood imports, 102; use of panel products, 107; antitrust laws in, 150, 152; trade with Japan, 213; supplier of technology, 213, 457; joint venture, 257; among top ten consumers, 293, 296; among largest importers and exporters, 303-7; consumption trends, 410, 411, 412; current difficulties, 447; forest sector model, 515. Consumption: softwood, 288, 409; hardwood, 288; plywood, 290; particleboard, 290, 291; fiberboard, 292; pulp, 292; newsprint, 294; printing and writing papers, 294; paper and paperboard, 295. Production: softwood lumber, 295; particleboard, 299; pulp, 300; newsprint, 300; printing and writing papers, 301; paper and paperboard, 301
Ghana, 433, 439
Gibbons, Sam, 170, 171
Global Trade Model (GTM), 516-18
Glued-laminated lumber industry in Japan, 210-11
Gorton, Slade, 171
Gould Lumber Company, 277
Governors' Conference on Lumber Export, 425
Grading standards: willingness to adjust to foreign markets, 96, 124, 236, 314, 317-18; Japanese, 98; need for in China. See also Metric system; Quality control; Standards; Stress grading
Gray Report, 453
Grays Harbor Commercial Company, 281
Great Lakes region, 51, 58
Greece, 447
Gross domestic product (GDP), 349, 352
GSM-5 program, 146
GSM-102 program, 144, 145, 146-67
GTE International, 260
Guangdong Province, 259, 417

Hardwood: prospects for U.S. exports, 122; world consumption, 288, 289, 440-42; world production, 296, 297, 436; top ten countries, 303; largest importers and exporters, 304; European exports, 343; European imports, 345, 346-47, 356; future supply and demand, 359, 436-40; European trends, 410, 411; softwood as substitute, 413-14; tropical trade, 432-46; tropical hardwood "quasi-nonrenewable," 434; trade

outlook, 442–43; price outlook, 443–46
Harvard Business School, Multinational Enterprise Management study, 458, 465
Hatfield, Mark, 172
Hawaii, 276, 278
Heinz, John, 166, 169
Hemlock, Western, 269
Hinoki (Japanese cypress), 68, 210
Holland. See Netherlands
Honda, 154
Hong Kong, 6, 112, 266, 304, 427, 434
Honshu Paper Company, 475, 478, 481
Hoover, Herbert, 188
Housing: slowdown in building, 5, 407; symposia in Latin America, 21, 109, 127; prospects in U.S., 26, 89, 95, 407, 508; construction in Japan, 27, 63–64, 67–68, 98, 195, 196, 316, 508; prospects in Japan, 407; reliance on U.S. market, 78; effect on international market, 89–90; construction in England, 97, 104; needs in China, 99; in Belgium, 105; fabricators in Denmark, 106; "zero energy" house, 106; prefabricated housing in West Germany, 107; and earthquakes in Italy, 107; needs in Latin America, 109; growth of timber frame construction, 122, 407; wood as substitute for masonry, 79–80, 97, 127; glued-laminated lumber, 210–11; building codes, 238, 360, 361, 497; promotion of plywood, 309; in Europe, 349, 351–52, 407, 413; standards, 497. See also Timber frame construction
Hudson's Bay Company, 276, 370
Hungary, 303, 515
Hu Yaobang, 422
Hydroelectric energy, 268

Idaho sawmill owners, 425
Immigration, changing pattern of, 37, 38–39
India, 32, 281; hardwood consumption, 288; printing and writing paper consumption and production, 294, 301; among largest importers, 306; joint ventures, 458
Indonesia, 262; supplier of timber, 29, 264, 311; major exporter, 53, 55, 56; trade with Japan, 65, 208, 209; Japanese trial plantations in, 475; Japanese investments in, 475–76, 479; U.S. shipping agreement, 173; joint ventures, 258; plywood industry, 265, 505; pulp and paper, 267; restrictions on log exports, 88, 436, 437–38, 473, 505, 508; among largest exporters, 304; hardwood production, 297, 433, 440;

export of products, 407; teak production, 434; foreign investment, 457. See also Irian Jaya
Industry Structural Council (Japan), 118
Inman-Poulsen Lumber Company, 281, 282
Institut für Bautechnik, 107
Insulating board, 213, 214, 293
International Business Machines (IBM), 459
International Council for Building Research Studies and Documentation (CIB), 500
International Harvester, 260
International Institute for Applied Systems Analysis (IIASA), Forest Sector Project, 515–19
International Longshoremen's and Warehousemen's Union (ILWU), 367
International Monetary Fund (IMF), 35, 36, 185, 524
International Paper, 478
International Sales Corporation (ISC), 170
International Sales and Service Corporation (ISSC), 170
International Standards Organization (ISO), 500
International Trade Administration (ITA), U.S. Department of Commerce, 21, 109, 165, 169
International Trade Fair (Santiago, Chile), 109
International Union of Forestry Research Organizations (IUFRO), 519
Ipil-ipil, 264, 475
Iran, 30
Ireland, 103
Irian Jaya, 56, 438, 439, 475
Israel, 37
Italy: trade with U.S., 96, 103, 122; use of panel products, 107; among largest importers and exporters, 303–7; current difficulties, 447; joint ventures, 466; forest sector model, 515. Consumption: softwood, 288, 289, 409; hardwood, 288; particleboard, 290, 291; printing and writing papers, 294; paper and paperboard, 295. Production: particleboard, 299; printing and writing papers, 301; paper and paperboard, 301
Itoh and Company, C., 478, 479, 480, 481
Ivory Coast, 421, 433, 439

Jamaica, 144, 147
Japan: growth rate of, 5; decline in domestic production, 26, 85, 87, 194–95, 201; domestic economy, 65, 139; forested area, 62; plantations, 62, 69, 507; trial plantations, 475;

INDEX 539

Depressed Industries Law, 118; former aid recipient, 145, 146, 147; future of domestic forest lands, 200, 451; reforestation, 451, 470; housing construction, 27, 63-64, 67-68, 98, 195, 196, 316, 320, 407, 448, 500-501, 508; housing starts (1965-82), 77; platform frame construction, 98, 195, 407, 448; nonwooden houses, 67; use of cyprus, 67-68, 210; use of cedar, 68, 88; management practices, 153-55; number of mills, 85, 87, 98-99, 199, 202, 203-4, 260, 471; sawmill industry, 199-201, 260, 472-73; tidewater pulp and paper and lumber mills, 470; price of lauan logs, 443; lumber grading, 98, 99, 196, 236; standards, 68, 88, 195-96, 206, 448, 472, 473, 474, 499, 501; coniferous market, 68-69; plywood market, 70, 108, 122, 123, 126, 197-99, 202, 474; pulp and paper trade, 116, 118, 141-42, 475; prospects for paper and paperboard, 64, 117, 118, 119, 141-42; distribution of timber products, 196-99; particleboard and fiberboard, 209-10, 211-15; hardboard, 213-15; glued-laminated lumber industry, 210-11; recycling waste paper, 455; value added in processing, 456; supplier of technology, 457; forest sector model, 515

Consumption: need for wood, 14, 26, 27, 308; demand-supply projections, 63-64, 77, 200; demand-supply of industrial timber (1955-82), 75; pulp and chip demand, 64, 112, 141, 245, 249, 285; wood consumption, 23, 25, 27, 63-64, 87, 176, 194, 195, 288, 290, 334-36; imports of raw materials, 62, 81, 285, 342, 345; wood imports as share of total consumption, 176; softwood, 288, 409; hardwood, 288, 434, 442; particleboard, 290, 291; fiberboard, 292; plywood, 290; linerboard, 112, 114, 115, 118, 141; pulp, 292; newsprint, 294; printing and writing papers, 294; paper and paperboard, 295; among top ten consumers, 296; among largest importers and exporters, 303, 304, 306, 307; imports of major forest products (1978-82), 73; timber imports (1960-82), 76

Production: softwood, 295; hardwood, 297; fiberboard, 299; pulp, 300, 475; newsprint, 300; printing and writing papers, 301; paper and paperboard, 301; plywood, 179-80, 202-11, 297, 508

Trade: volume of trade in forest products, 3, 27, 62, 65, 87, 122, 137; opportunities for trade with, 6, 98, 122, 137-42, 236, 448; protectionism against, 6, 65, 138, 139, 149, 153, 156, 175; tariffs, 64, 71, 108, 118, 123, 126, 139-40, 149, 158-59, 175-82, 238; and U.S. tariffs, 69-70; trade balances, 138, 149; efforts to open the market, 139-40; and antitrust action, 149-52, 155-56; trade legislation, 150-52; protectionist policies of, 149-56; price fixing, 152-53; value of yen, 175, 448; shipping costs, 371-74, 455-56; joint ventures, 258, 259, 260, 261, 463, 464, 466, 469; 469-76, 478-82; currency changes, 82, 85; market restrictions, 88, 118-19, 138-40, 149-56, 284, 285; Japan Trade Center (JETRO), 137, 139; invisible trade, 137; trading companies, 164, 378, 454, 471-75; industrial associations (addresses and telephone numbers), 216; availability of capital, 469; Columbus Day Storm, 470; share of world trade, 341, 342; trade in agricultural products, 137, 140

Trading partners: U.S.-Japanese trade (tables), 73-74, 84; imports from U.S., 27, 65, 66, 69, 70, 71, 84, 121, 137, 200, 207, 208, 472; supplier of plywood to U.S., 29, 103, 108, 123, 126, 176, 180, 202, 206, 207; U.S. access to market, 165; trade with U.S West Coast, 137, 236, 270, 279-80, 282, 284, 474-75; trade discussions with U.S., 22, 284; trade with Weyerhaeuser, 7; Malaysia, 65, 208, 474; Indonesia, 65, 208, 209, 508; Canada, 65, 66-67, 207, 208, 244, 245, 247, 249, 472; North America, 201, 207, 208, 469, 471; Southeast Asia, 201, 263, 264, 266, 434, 473; investment in North America and Southeast Asia, 454; U.S.S.R., 29, 53, 57, 65, 201; New Zealand, 201; South Korea, 208; China, 208, 419, 421; Singapore, 208; Philippines, 208, 471, 474; investment in Pacific Rim region, 471-76, 478-82; West Germany, 213; Europe, 356, 357, 358, 469; tropics, 434, 442

Japan Brazil Paper and Pulp Resources Development Company, 260
Japan Fiberboard and Particleboard Manufacturers' Association, 212, 214, 215, 216
Japan Housing and Wood Technology Center, 216
Japan Laminated Timber Manufacturers'

Association, 211, 216
Japan Lumber Importer's Association, 216
Japan Plywood Distributors' Association, 216
Japan Plywood Inspection Corporation, 216
Japan Plywood Manufacturers' Association (JPMA), 202, 216
Japan Pre-finished Plywood Manufacturers' Association, 216
Japan Timber Products Storage Corporation, 216
Japan 2 x 4 Home Builders Association, 98
Japan Trade Center (JETRO), 137, 139
Japan Wood-flooring Manufacturer's Association, 216
Japan Wood Preservers' Association, 216
Japan Wood Processing Machinery Manufacturers' Association, 216
Japanese Agricultural Standard, 195-96, 206, 210; for structural plywood, 217-30
Japanese Import-Export Bank, 258
Japanese Overseas Afforesting Association, 481
JETRO Jakarta, 476
Joint ventures, 256-61, 310, 377, 395; Pacific Rim, 451-82; term defined, 457-58; advantages of, 465-66; disadvantages of, 466-67; when to avoid, 468; Japanese, 258, 259, 260, 261, 463, 466, 469-76, 478-82
Jones Act, 158
Jones, Walter, 171
Jujo Paper Company, 258, 260, 478, 479, 480, 482
Jujo-Weyerhaeuser, 475

Kalimantan, 56
Kampuchea, 433
Kawasaki, 475, 479, 480, 481
Kennedy round (GATT tariff negotiations), 158, 360
Kodiak Lumber Mills, Inc., 482
Kohjin Company, 480
Komatsu Ltd., 260
Konan Tsusho Kaisha, 479
Korea, South: trade with, 6, 122, 448; supplier of hardwood plywood, 29, 202; indebtedness of, 37; pulp consumption, 112; former aid recipient, 147; U.S. shipping agreement, 173; factories, 206; exports to Japan, 208; joint ventures, 261; and Southeast Asia, 263, 264, 266, 434; trade with Washington State, 270; importer of logs, 285, 434; plywood consumption, 290, 291; plywood production, 297; among largest importers and exporters, 304, 306; shipping costs, 373; export of products, 407; reforestation, 451; development of, 452; freight costs, 456; supplier of technology, 457
Kraft linerboard. See Linerboard

Laja, Chile, 256
Lake States region, 51, 58
Landis, Richard G., 80
Laos, 433, 439
Larch, 57
Latin America: opportunities for trade with, 7, 95, 123; trade with Weyerhaeuser, 7; fuel-gathering and grazing in, 12; size of closed forest, 13, 34; future of plantations, 13, 47; growing stock in, 14; potential for expansion, 14, 15, 58; as a market for plywood, 21, 109; programs to promote trade with, 21, 109; rising demand for wood products, 26, 109; as a wood supplier, 27; arable land, 34; and immigration to U.S., 37; emerging region, 53, 58; domestic use of production, 59; import-export comparison, 83; as a wood consumer, 95, 333-36; housing needs, 109; tariff barriers, 109, 123, 127; traditional construction materials, 124, 127; trade with British Columbia, 245; joint ventures, 258; as a pulp supplier, 271; trade with Northwest, 279; trade with Europe, 345; early trade with Pacific Northwest, 370; as an export trading company market, 383-84; trade with China, 421; hardwood supplier, 433, 434, 435; Amazon area, 439; consumption of hardwood, 442; pulp, 113; linerboard, 115; paper and paperboard, 117
Lauan, 70, 202, 206, 207, 267; prices, 443-45
Law on Temporary Measures for Structural Adjustment of Specific Industries (Japan), 142
Leeward/Windward Islands, 103
Legislation, U.S. trade, 157-73; maritime, 158, 171-72; and protectionism, 159; and 1980 government reorganization, 160; Foreign Corrupt Practices Act, 161; DISCs, 161-62; and taxes, 161-63; Export Administration Act, 163; domestic content legislation, 167; pending, 164-74; reciprocity, 168, 190. See also Export Trading Company Act of 1982; Webb-Pomerene Act
Less developed countries. See Developing countries

Levitt, Theodore, 415
Liberia, 421, 433, 439
Linerboard (kraft linerboard): in Japan, 7, 112, 114, 115, 141; as an international commodity, 85; U.S. exports of, 112-15; future trade in, 116, 118; British Columbia production, 245; joint ventures, 258; North American exports to Europe, 349, 350; shipment methods, 366; Chinese imports, 419
Liner Code, 173
Lissakers, Karin, 524
Long, Russell, 172
Longview, Washington, 258
Lumber. See Sawnwood
Luxembourg: U.S. plywood imports, 102; use of panel products, 105; particleboard consumption, 291; particleboard production, 299; among largest importers and exporters, 303-7

MacMillan Bloedel, 472
Malaysia, 262, 448; supplier of timber, 29, 264, 434; major supplier, 53, 55, 56; per capita forest area, 62; exports to Japan, 65, 208; hardwood production, 297, 440; among largest importers and exporters, 304; export of products, 407; hardwood supplier, 433; restriction of log exports, 436, 437; price of Meranti, 443, 444; plantations, 451; foreign investment in, 457; Japanese trial plantations in, 475; Japanese investments in, 479-80. See also Sabah; Sarawak
Maltese Cross label, 277
Management. See Forest management
Maple Flooring Manufacturers Association (MFMA), 126
Marché Tropicaux et Méditerranée, 443, 444
Maritime legislation, 158, 171-72
Marketing for foreign markets, 319-20
Market price mechanism, 311-12
Marubeni Corporation, 257, 478
Mass transit, 189
Matsushita, 154
McLaughlin, John, 276
MDI, 479, 480
Mears, John, 370
Medium and Small Enterprise Promotion Law (Japan), 151
Medium density fiberboard (MDF), 210, 215, 293, 351
Meranti sawnwood, 443, 444
Metric system: modification of mills for production, 7, 317, 472; Far Eastern lumber standards, 68, 88; U.S. lack of interest in producing in accord with, 314; as a trade barrier, 360; financing mill alteration, 429. See also Standards
Mexico: indebtedness of, 35, 36, 37, 524; and immigration to U.S., 37; U.S. plywood imports, 103; forestry conditions, 308; joint ventures, 458
Middle East: growing import demands, 7, 266, 342; trade with Weyerhaeuser, 7; wood consumption, 96; U.S. trade with, 96, 97, 108-9, 122-23, 404; and fabricated housing, 106; trade with Europe, 340; as an export trading company market, 384; and Southeast Asia, 434; and Pacific Rim, 452; as a new market, 506; consumption of softwood, 409; pulp, 113; linerboard, 115; paper and paperboard, 117
Mills: modified for metric-size production, 7, 317, 429; standards in Japan, 68; decline in number of Japanese, 85, 87, 98-99, 199, 200, 260; U.S. small mills' requests for subsidies and protection, 85; foreign grading specifications, 96, 317; U.S. environmental standards, 132; in British Columbia, 243, 244, 250; pulp mill joint venture, 256; Japanese interests in Pacific Northwest, 260; in Southeast Asia, 265; assistance for small mills, 274, 425-31; Chinese sawmills, 280; adjusting to foreign markets, 317-18; quality control, inspection, sampling, 320-27; location of, 365; hardwood vs. softwood, 413-14; recovery rates, 473
Mindanao, 475
Minerals, nonfuel, 33
Ministry of Agriculture, Forestry and Fisheries (MAFF), Japan, 26, 196
Ministry of Foreign Trade (PRC), 423
Ministry of Forestry (PRC), 418
Ministry of Forests (British Columbia), 247, 250, 252
Ministry of Industry and Small Business Development (British Columbia), 247
Ministry of International Trade and Industry (MITI), Japan, 118, 214; and trade legislation, 150; procartel policies, 150-51; and automobile and electronics industries, 154-55
Mitsubishi Corporation, 478, 479, 480, 481
Mitsui and Company, 260, 282, 478, 479, 481, 482
Mitsui Overseas Forestry Development, 479
Monetary system. See Exchange rates
Morrison Mill Company, 281
Morse, Wayne, 284

542 INDEX

Motor Mill, 277
Moynihan, Daniel Patrick, 165
Mozambique, 439
Multilateral trade negotiations (MTN), 64-65, 168, 238, 360, 508. See also General Agreement on Tariffs and Trade (GATT)
Multinational corporations, 36, 310, 379, 387; multinational enterprise (MNE) studies, 458-76; threat to host country, 460; in developing countries, 458-65; in all countries, 464; success of, 524. See also Joint ventures

Nakasone, Japanese Prime Minister Yasuhiro, 140
Nanaimo, British Columbia, 134
National Academy of Sciences, 330
National Bureau of Standards, 498
National Forest Products Association (NFPA), 121, 122, 126, 161, 518, 519
National forests, 233
National Grading Rule Committee, 498
National Lumber and Manufacturing Company, 281
National Lumber Exporters Association (NLEA), 126-27
National Research Council, 330
Near East, 12, 24
Netherlands: U.S. plywood imports, 102; use of panel products, 106-7; platform frame construction, 247; joint venture, 256; plywood consumption, 290, 291; newsprint consumption, 294; among largest importers and exporters, 304-7; percentage of imports, 354; softwood imports, 409; consumption trends (1962-85), 410, 411; current difficulties, 447
New Brunswick International Paper Company, 260
Newly industrialized countries (NICs), 452
Newsprint: demand in Japan, 141; as an international commodity, 85; tariffs on, 192; British Columbia production, 243, 245; joint ventures, 258; world consumption, 293, 294; world production, 298, 300; top ten countries, 303; largest importers and exporters, 306; European exports, 343; European imports, 346-47, 349, 350; North American exports, 349, 350; duty-free imports from Canada, 454
New Zealand: plantations, 52, 58, 451; competition for Asian markets, 68, 69, 271, 508; U.S. trade with, 96; potential for exports, 254, 507; joint ventures, 257; currency devaluation, 271; among largest exporters, 306; trade with Europe, 356; log imports, 434; hardwood consumption, 442; Japanese investment in, 480-81; forest sector model, 515. See also Pacific Rim countries
Nichimen Company, 258
Nigeria: hardwood consumption, 288, 289; hardwood production, 297, 433; net importer, 439
Nippi Boeki, 480
Nippon Electric Company, 260
Nissan Nohrin Kogyo, 480
Nondeclining even flow, 54, 251
Nordic countries. See Scandinavia
NORPAC, 258
North Africa: and fabricated housing, 106; U.S. plywood trade, 108-9; trade with Europe, 340; growing import demands, 342; trade with China, 421
North America: exportable surpluses of softwood, 3; closed forest in, 13, 34; extent of growing stock, 14; production by the year 2000, 14-15; wood consumption, 24, 25, 95, 334-35; trade with Japan, 27, 66, 82, 84, 141, 200, 469, 471; major exporter, 53, 54-55, 58, 59; import-export comparison, 82, 83, 341; U.S.-Canadian competition, 85, 86, 88; prospects for paper and paperboard, 117, 119, 141, 349; softwood panels, 206, 209; glued-laminated lumber industry, 210; use of insulating board, 293; plywood production, 296; trade with Europe, 27, 98, 340, 345, 347, 348, 349, 350, 357, 358-59; share of world trade, 340, 341, 342; U.S.-Canadian trade, 342; trade with Southeast Asia, 345, 434; total imports, 345; paper and paperboard exports to Europe, 349, 350; recovery from recession, 351; competition with Europe, 355; trends in wood prices, 363; hardwood consumption, 434, 442; hardwood production, 440; foreign investment, 453-54; sawmill recovery rates. See also Canada; United States
North American Forestry Development, Inc., 482
North American National Grading Rule, 500
North Atlantic Treaty Organization (NATO), 523
Northeastern Lumber Manufacturers Association (NeLMA), 126
North Pacific Paper Corporation, 475, 482
Norway: use of plywood, 108; environmental issues, 133; newsprint

production, 300; among largest importers and exporters, 306, 307; forest sector model, 515

Oak, 122, 123, 126, 238
Oceania: opportunities for trade with, 7, 96; emerging region, 53, 58; wood consumption, 96, 97, 334-35; trade with British Columbia, 245; hardwood production, 433, 440; hardwood consumption, 442. See also Australia; New Zealand
Odlins Limited, 481
Office of Strategic Trade, 169
Office of Trade Ombudsman (Japan), 140
Oil crisis, 30, 46, 139
Oil nations: effect on world economy, 35, 405; OPEC, 46, 138, 447, 455
Oji Paper Company, 260, 478, 479, 480, 481
Operating characteristic (OC) curve, 323-25
Oregon: port facilities, 7; government-owned land, 232-33; export of railroad ties, 281; volume of lumber exported, 283-84; restriction on log exports, 285; export history, 370. See also Pacific Northwest
Oregon State University, 518
Organization for Economic Cooperation and Development (OECD), 452, 455, 523, 524; Positive Adjustment Policy, 142; joint ventures, 458
Organization of Petroleum Exporting Countries (OPEC). See Oil nations
Ottinger, Richard L., 167
Overseas Private Investment Corporation (OPIC), 165

Pacific Basin. See Pacific Rim countries
Pacific Basin Economic Council, 80
Pacific Northwest: importance of trade, 1, 523; attitude toward exports, 4; Weyerhaeuser Company, 7; ports, 7; favorable for trade, 19; major exporting region, 53; future supplies, 54-55, 59; ban on export of cedar logs, 69-70, 88; growth of exports, 79; importance of Pacific Rim market, 80; trade with Japan, 137; national forests, 233; competition with British Columbia, 246; Japanese interests in sawmills, 260; transportation costs, 271, 285, 371-74; history, 276-86, 370; volume of lumber exported, 283-84, 285; restrictions on partly processed products, 312; shipping facilities, 366, 367; trade with Australia, 447; Columbus Day Storm (1962), 470; and general trading companies, 472. See also Oregon; Washington State
Pacific Rim countries (Pacific Basin): opportunities for trade with, 6, 8, 80, 95-96, 98-99, 265, 286; and Weyerhaeuser trade, 7; growing supplier of hardwoods, 27; imports from Soviet Union, 53, 57; area defined, 80; import-export comparison, 82, 83; wood consumption, 95-96; trade with Europe, 358; as an export trading company market, 383; joint ventures, 451-82; reforestation, 451; plantations, 451; Japanese investments in, 470, 471-76, 478-82
Pacific R. J. Reynolds Industries, 80
Panama, 103
Panel products: world consumption, 25, 122, 334-35; U.S. exports, 79, 209; opportunities for trade, 101-10, 407; performance standards, 101-2; U.S.-Japanese standards, 180; composition board in Southeast Asia, 267; production problems for foreign markets, 316-27; imported by Saudi Arabia, 336; technological advances, 337-38; European exports and imports, 340, 341, 343, 344, 346-48; European consumption, 350, 352, 353, 407, 410-12
Paper and paper products: needs in Japan, 64, 141-42; Far East market, 80; preparation for world markets, 88; U.S. trade in, 111-20; increase in tonnage, 114; markets for U.S., 114-15; tariff imbalance, 116; future prospects, 114-20; technological changes, 141; joint ventures, 258; sanitary products, 258; world consumption, 293, 295, 334-35, 336, 337; world production, 298, 301, 302; top ten countries, 303; largest importers and exporters, 307, 340, 341; energy used to produce, 332-33; substitutes for, 332-33, 353; consumption in China, 337; European exports and imports, 340, 341, 343, 344, 345, 346-48, 349, 350, 356, 358; changes in European paper industry, 345, 352; North American exports, 349, 350; European consumption, 350, 352, 353; change in product use, 352; recycled paper, 353, 455; shipment methods, 366; shipping costs, 373-74; Pacific Rim production, 451; foreign investment risks, 457. See also Bleached paperboard; Linerboard; Newsprint; Printing and writing papers; Pulp
Paper Industries Corporation of the Philippines (PICOP), 266-67, 268
Papua New Guinea, 53, 55, 56, 262,

264, 266, 433, 438-39, 443; Japanese investment in, 475, 481
Parks, Richard, 89
Particleboard, 101; use of in Europe, 24, 107, 351; tariffs, 177, 181; standards, 178; Japanese market, 209-10; industry in Japan, 211-14; world consumption, 290-91; world production, 298, 299; top ten countries, 303; largest importers and exporters, 305; European exports, 343; European imports, 346-47, 410-12
Peccei, Aureho, 36
Pell, Claiborne, 172
Peninsular Malaysia. See Malaysia
People's Republic of China. See China (PRC)
Perry, Commodore Matthew, 279
Persian Gulf, 7
Peru, 21, 109, 279, 280
Philippine mahogany, 27
Philippines, 262; supplier of plywood to U.S., 29; major supplier, 53, 55, 56; U.S. shipping agreement, 173; plywood production, 202, 265, 266, 474; exports to Japan, 208, 471; supplier of logs, 264, 434; restriction of log exports, 436, 437; pulp, 267; plywood consumption, 291; plywood production, 297; among largest exporters and importers, 304; export of products, 407; hardwood supplier, 433, 440; plantations, 451, 475; risk to foreign investment in, 457; Japan-Philippines commercial treaty, 474; Japanese investment in, 481
Phoenix Shingle Company, 277
Pine, radiata, 69
Plantations: estimated extent of, 13, 58-59, 329, 505; potential for, 15; in Latin America, 47, 52; transition from old-growth forest to, 47-51; replacing croplands and pasture lands, 52; in Japan, 62; Japanese investments in, 475; in Southeast Asia, 264, 266, 268; European imports from tropics, 356; reliability of, 356-57, 505; in Africa, 433; tropical species, 433-34; in Pacific Rim, 451, 475
Plasterboard, 407
Platform frame construction. See Timber frame construction
Plywood: program to assist producers, 21; U.S. imports of, 29, 103, 108, 123, 126, 176, 180, 202, 206, 207; lauan used in Japan, 70, 202, 206, 207; U.S. exports, 79, 102-3, 123, 176, 209; opportunities for trade, 101-10, 448; standards for industry, 101-2, 123, 126, 176, 178, 180,
217-30; and EEC, 103, 123; tariffs, 123, 177, 180-81, 238, 246; Japanese production, 179, 197-99, 202-11; softwood structural panels from North America, 206, 209; Japanese Agricultural Standard for Structural Plywood, 217-30; British Columbia production, 243, 246; joint ventures, 259; in Southeast Asia, 265-67; market in China, 266; world consumption, 289-91; world production, 296, 297-98; top ten countries, 303; largest importers and exporters, 304; promotion campaign, 309; value added, 312, 456; European exports, 343; European imports, 345, 346-47; British consumption, 403; European consumption, 409, 410-12; hardwood vs. softwood, 413-14; U.S. lowest-cost producers, 414; tropical hardwood, 434; Indonesian production, 438
Poland: indebtedness of, 37; supplier to U.K., 97; particleboard consumption, 290, 291; fiberboard consumption, 292; softwood lumber consumption, 289; fiberboard production, 299; among largest importers and exporters, 303, 305
Pope and Talbot, 278
Population: growth, 31, 32-33; immigration, 37, 38-39; worldwide wood consumption, 334
Port Blakely Mill Company, 279, 281
Ports: facilities for exports, 7, 366-68; user fee legislation, 172; labor rates, 367; port charges, 367-69. See also Shipping
Portugal, 303, 306
Positive Adjustment Policy (PAP), 142
Prices: real price of natural resources, 45-46; trends over time, 47-49, 362-63, 443-46; and cartels, 152; influence on consumption, 309-10; the market mechanism, 311-12; value added, 312-13, 455-56; for foreign markets, 318-19; European stability, 355; increase for softwood, 359; increase for hardwood, 441, 443-46; cost of nonwood inputs, 506; TAMM price equilibrium model, 512-15, 516, 518
Printing and writing papers: indirect exports, 111, 112; U.S. exports of, 114; world consumption, 293, 294; world production, 298, 301, 302; top ten countries, 303; largest importers and exporters, 307; substitutes for, 332; European exports, 343; European imports, 346-47
Production problems, 316-27

Protectionism: tariff and nontariff barriers, 4-5, 8, 120, 123-24, 139-40, 149-50, 190, 192, 238, 360-64, 525; in France, 6, 238; anti-Japanese, 6, 138, 139, 149, 153, 156, 175, 179, 525; in Japan, 65, 139-40, 149-56; quotas on Chinese textiles, 6, 22; not in U.S. interest, 8; widespread tendency toward, 10, 22, 185-89, 456, 524; and pulp and paper trade, 116; current U.S. trend toward, 119, 159; trade restrictions, 310-15, 360-64; OECD and LDC, 455. See also Tariffs and quotas
Public Law 480: aid to Japan, 145, 146; aid to Europe, 146; aid to developing countries, 145, 146
Puget Mill Company, 280, 281
Puget Sound Lumber Company, 281
Pulp (pulpwood, woodpulp): future production, 15; use in Europe, 24, 112, 271; Scandinavian producers, 27; technological changes, 51-52, 141; needs in Japan, 64, 141-42, 285; preparation for world markets, 85; U.S. trade in, 111, 112, 113, 116, 118; U.S. environmental standards, 132; British Columbia production, 243, 245, 250; joint venture, 256; in Southeast Asia, 267-68; world consumption, 25, 292, 293, 334-35, 336; world production, 298, 300; top ten countries, 303; largest importers and exporters, 306; substitutes for, 333; technological advances, 338; European exports and imports, 340, 341, 343-45, 346-48, 350, 356, 358; shipment methods, 366; shipping costs, 373-74; Pacific Rim production, 451; U.S. investment in Canada, 453, 454; foreign investment risks, 457; problem of available mill sites, 506
Pulpwood Harvesting Area Agreements (British Columbia), 250

Quality control: in developing countries, 314; for foreign markets, 320-27; and joint ventures, 469; in U.S. and Canada, 500. See also Standards
Quotas. See Tariffs and quotas

Radiata pine, 69, 254
Railroads: use of Northwest wood in construction of, 279, 281; use for shipment, 366
Rainier National Bank, 425
Reagan administration, 138, 139, 144, 145, 146, 148, 171; and port modernization, 172; stand on open trade, 191, 525; weak dollar, 406
Reciprocal Trade and Investment Act, 168
Reciprocity legislation, 168, 190
Redcedar logs, 69-70, 88, 163, 248
Regulations, U.S.: environmental, 130; EPA, 131; and other countries, 131-32; permits and approvals, 133-34. See also Legislation, U.S.
Renewable Resources Assessment, 509-11, 512, 513
Renton, William, 279
Resource Management Cost Account (Washington State), 272
Resources for the Future, 509, 518
Resources Planning Act of 1974 (RPA), 509-11, 512, 513, 514, 518
Rickard, Wes, 89
Rio Doce, 260
Risk: in processing considerations, 314; in export trading companies, 391, 400-401; in foreign investment, 457, 466
Romania, 297, 299, 304, 305
Roth, William V., 165
Roundwood: opportunities for exporting, 81-93; worldwide consumption, 334-35; European exports and imports, 340, 341, 343, 344, 346-48; tropical hardwood, 432; Pacific Rim production, 431

Sabah, 262, 264, 267, 437
Sampling methods for quality control, 321-25
Sanitary paper products, 258
Sanyo-kokusaku Pulp, 479, 480, 481
Sapelli logs, 443-45
Sarawak, 262, 264, 437, 439, 473, 474
Saudi Arabia: U.S. trade with, 96, 103; among largest importers, 304, 308; imports of panels, 336
Sawmilling: foreign investment not sought, 457; Japanese mills, 199-201, 471, 474; rate of recovery, 473. See also Mills
Sawn and veneer logs: world production, 440; prices, 444-46; unit value of imports and exports, 456
Sawnwood: Canadian lumber production, 243-44; Southeast Asia, 265; softwood lumber consumption, 288, 289; hardwood lumber consumption, 288, 289; softwood lumber production, 295, 296; hardwood lumber production, 295, 296; largest importers and exporters, 303, 304, 308; distribution mechanism in Japan, 197-98; world consumption, 334, 335; European exports and imports, 341, 343, 344, 346-47, 348; North American imports, 345;

European consumption, 350; shipping charges, 369, 372-73; U.S. consumption, 403-4, 405; world imports, 409, 412; U.S. exports to China, 420; unit value of imports and exports, 456. See also Hardwood; Softwood
Scandinavia: sustainable harvest, 5, 6, 122; declining use of fuelwood, 24; wood shortages, 24, 27; wood chips from U.S. for pulp, 27; major supplier, 53, 58, 59, 308, 354; supplier to U.K., 97; U.S. plywood imports, 102, 105-6, 108; tariff advantage in pulp and paper trade, 116, 118; U.S. competition, 122, 132-33, 235; water quality issues, 132-33; supplier to EEC countries, 349; supplier of technology, 457. See also Denmark; Norway; Sweden
Scotland, 104
Scott Paper Company, 257-58
Seaboard, 472
Seattle Cedar Lumber Manufacturing Company, 277
Shanghai, 280
Sheathing, 101, 104, 105, 106
Shipping: legislation, 172-73; to foreign markets, 320, 448, 455-56; history, 370; cargo ships, 370-71; costs for logs, 371-72, 455-56; costs for lumber, 372-73, 455-56; costs for pulp and paper, 373-74; distance from foreign markets, 404; shortage of Chinese ships, 419. See also Ports; Transportation
Shipping Act of 1916, 171, 377
Shukko Shoji, 480
Siberia, 57, 358; logs for Japan, 68
Simpson, Governor, 276
Simpson Timber Company, 256
Singapore, 262; exports to Japan, 208; plywood industry, 262; among largest importers and exporters, 304; export of products, 407; importer of logs, 434; development of, 452; foreign investment, 457
Smith, Carl, 410
Smoot-Hawley Act, 188
Softwood: world consumption, 288, 289, 409, 441; world production, 295, 296, 359, 440; top ten countries, 303; largest importers and exporters, 303, 409; European exports, 343; European imports, 345, 346-47; future prices, 359; British consumption, 403; U.S. consumption and exports, 403-4; European trends, 410-12; substitute for hardwood, 413-14, 471
Solomon Islands, 433, 475, 481
Sony, 154, 155
South Africa, 58, 306, 356, 442

South America. See Latin America
Southeast Asia: supplier to Japan, 27, 179; supplier of plywood for U.S., 29; major exporter, 53, 55-56; log resources decreasing, 70, 436-40; U.S. exports to, 112, 114; opportunity for regional development, 119; joint ventures, 258-59; view of wood market, 262-68; main markets, 263; trade in logs, 264, 436; plantations, 264; lumber, 265; plywood, 265-67; composition board, 267; pulp and paper, 267-68; share of world trade, 342; supplier to Europe, 345, 356; trade with China, 421; market prospects for hardwoods, 432-46; Japanese investment in, 454; market potential, 507. See also Association of Southeast Asian Nations (ASEAN); Pacific Rim countries; individual countries
Southern Forest Products Association (SFPA), 126; activities in Latin America, 127
Southern Plywood Manufacturers Association, 309
South Korea. See Korea, South
South Sea logs, 209, 435
Soviet Union. See Union of Soviet Socialist Republics (U.S.S.R.)
Spain: particleboard consumption and production, 290, 291, 299; fiberboard production, 299; among largest exporters, 305
Special economic zones (PRC), 259, 424
Standards: U.S. Product Standards for plywood, 101; APA Performance Standards, 101-2; British, 104, 124; Nordic, 106; Japanese, 108, 123, 126, 176, 178, 217-30, 236, 472; U.S. and foreign, 124, 131-32, 236; environmental, 129-36, 362; European, 236; U.S. accommodation of metric system, 314, 317; adjusting to foreign markets, 317-18, 494-501; quality control, inspection, sampling, 320-27; as trade barriers, 360; nature of, 495-97; development of, 497-501. See also Grading standards; Metric system
Stevens, Ted, 172
Stimson Mill Company, 277, 281
St. Paul and Tacoma Lumber Company, 281
St. Regis Paper Company, 257
Stopford and Wells, 468, 469
Stress grading, 407, 498, 499, 500
Sturd-I-Floor, 101, 104, 106
Sugi (Japanese cedar), 68
Sumatra, 258, 457
Sumitomo Corporation, 480
Sumitomo Forestry, 478, 479, 481
Summer Iron Works, 280

Sustained yield, 251-52, 355, 362-63, 364, 436
Suzuki and Company, 282
Suzuki, Japanese Prime Minister Zenko, 140
Sweden: time profile reflecting levels of demand for timber, 49-50; U.S. plywood imports, 103, 108; competitive devaluations, 116; environmental issues, 133; joint ventures, 257; largest importers and exporters, 303, 305, 306, 307; consumption trends, 412; forest sector model, 515; softwood lumber consumption, 289; pulp consumption, 292. Production: softwood lumber, 295; fiberboard, 299; pulp, 300; newsprint, 300; paper and paperboard, 301; particleboard, 299
Switzerland, 305, 412

Tacoma Mill Company, 279
Taiwan: trade with, 6, 448; supplier of plywood, 29, 202; former aid recipient, 147; and Southeast Asia, 263, 264, 266, 434; pulp consumption, 112; export of products, 407; import of logs, 434; development of, 452; supplier of technology, 457
Takata and Company, 282
Tariffs and quotas: in Japan, 64, 123, 126, 139-40, 149, 159, 175-82, 238; in U.S., 64-65, 177, 246, 271; on U.S. log exports, 69; on plywood, 103, 123, 126, 158, 177, 180-81, 238, 246; Scandinavian advantage, 116, 118; in Europe, 123, 177, 238, 361; in Latin America, 109, 123-24, 127; and U.S. trade laws, 157-58, 167; in Canada, 158, 246; preferential rates for developing countries, 167; U.S. concessions, 168; president's negotiating authority, 168; on particleboard, 181; Smoot-Hawley Act, 188; overcoming barriers, 238, 360, 508-9; U.S.-Canadian relations, 271, 508; temporary, 315. See also General Agreement on Tariffs and Trade (GATT); Protectionism; Textile trade restrictions
Tax Equality and Fiscal Responsibility Act of 1982, 162
Tax problems, 161-63
Tax Reform Act of 1977, 162
Technology: and economics of forest resources, 46, 51-52, 58; and environmental standards, 132; cargo handling, 133; plywood, 206; adjusting to foreign markets, 316-27; computer-controlled bucking station, 318; and cost of wood, 329; genetic manipulation, 329; in relation to consumption, 333-38; computer use in mills, 337; and panel production, 337-38; and fiber industry, 338; easy access to, 457
Teijin Ltd., 260-61
Terrorism, international, 38
Terzawa and Company, 282
Textile trade restrictions, 6, 22
Thailand, 262, 304, 308, 315, 433, 439
Third World. See Developing countries
Timber Assessment Market Model (TAMM), 512-15, 516, 518
Timber frame construction, 97, 104, 105, 122, 124, 316, 320, 407, 505; platform frame construction, 98, 195, 247
Timberman, The, 280
"Timber-mining," 362, 364
Timber resources. See Forest resources
Time magazine, 36
Tokai Pulp, 480
Tokyo round of tariff negotiations (GATT), 139-40, 158, 177, 179, 181, 246, 360
Toray Industries, Inc., 260
Toyo Menka Kaisha Ltd., 479
Toyota, 36
Trade Act of 1974, 158, 168
Trade, international: importance of, 1-2, 78; history of forest products trade, 2-3, 276-86; U.S. preeminence, 4; growing competition, 5, 6; floating exchange rates, 5; slowing of world growth, 5; opportunities, 6-7, 78-80, 81-93, 94-100; export trading company legislation, 21, 127-28; U.S.-Japanese trade discussions, 22; volume of trade in forest products, 27-29; agricultural indicators, 29-32; resource inequities, 32-34; international lending boom, 35-38; changing political order, 38-39; political interdependence and justice in allocation of world's resources, 40, 41; Japanese timber market, 62-77, 194-230; foreign markets as dumping ground, 78-79, 152; marketing methods, 79; wood as substitute for masonry, 79-80, 97, 127; importance of Far East and Pacific Rim, 80, 98-99, 122, 451-82; raw material exports, 81-93; U.S.-Canadian competition, 85, 86, 100, 122; lumber standards, 88; market restrictions, 88-90, 103, 123-24, 149-56; western softwood opportunities, 94-100; world consumption of wood, 23-26, 95-96, 121, 333-37, 442, 455; grading and sizing norms, 96, 98, 99, 494-501; panel products, 101-10, 122; U.S.

plywood exports, 102-3; pulp and paper, 111-20; indirect exports, 111, 112; prospects for pulp and paper, 114-20; solid-wood products, 121-28; trade problems, 123-24, 237-39, 316-27; environmental issues, 129-36; U.S.-Japanese trade, 137-42, 149-56, 175-82; U.S. agricultural exports, 143-44; credit programs, 143-48; antitrust action, 149-56; U.S. trade laws, 157-73; Canadian perspective, 183-93, 240-54; role of U.S. Department of Agriculture, 231-39; joint ventures, 256-61, 451-82; Southeast Asia, 262-68; Washington State, 269-75, 276-86; product mix and flow of wood materials, 287-315; wood consumption, 287-96, 303; wood production, 295-302, 303; import-export statistics, 303-8; trade restrictions, 310-15, 360-64, 508-9; value added, 312-13; employment, 313-14; production problems, 316-27; contract negotiations, 325-27; renewable vs. nonrenewable resource products, 328-39; consumption trends, 333-38; European viewpoint, 340-64, 402-15; transportation methods and costs, 365-74; structuring an export trading company, 375-401; People's Republic of China, 416-24; financing small- and medium-size firms, 425-31; tropical hardwoods from Southeast Asia, 432-46; bank finance and support, 483-93; trade strategies, 502-20; supply-demand projections, 509-20

Trade Policy Staff Committee, 167
Trade Representative, Office of the U.S. (USTR), 165, 167; establishment of, 160
Trading companies, 164; role in Japanese plywood trade, 197-99. See also Export Trading Company Act of 1982; General trading companies (GTCs)
Transportation: to foreign markets, 370, 448; methods and costs, 365-74, 435, 455. See also Shipping
Treaty of Rome, 152
Trinidad, 103, 109
Tropical regions: drop in revenue from trade, 10; annual depletion of, 13; compared with temperate forests, 14; potential of, 14-16, 26, 436-40; trade with Europe, 340, 356-57; competition, 355; trends in wood prices, 363; domestic use, 356; conservation, 356, 436-40; trade in hardwood, 432-46; pattern of trade, 434-35; market structure, 435-36;
demand outlook, 440-42
Truck hauling, 366
Trust Pacific Islands, 103

U.M.W. Timber, 4/9
Union of Soviet Socialist Republics (U.S.S.R.): softwood exporter, 3, 29, 97, 232, 504; major supplier, 27, 53, 56-57, 59; "other woodland," 12; closed forest, 13-14, 34; growing stock, 14; future production, 15, 26, 59, 235; wood consumption, 23-24, 25, 95, 334-35; food importer, 30; effect of political actions on world economy, 30; indebtedness of, 37; exports to Japan, 65; competition for Japanese market, 68-69, 508; as a market for U.S. lumber, 95; supplier to U.K., 97; domestic needs, 97; U.S. competition, 122, 271; U.S. grain exports, 163; pipeline project, 163; lumber shipment in 1923, 282; among largest importers and exporters, 303-7; trade with Europe, 27, 57, 97, 345, 347, 348, 357; share of world trade, 341, 342; competition with Europe, 355; trends in wood prices, 363; trade with China, 419; forest sector model, 515; IIASA, 515. Consumption: softwood and hardwood, 288; plywood, 290; particleboard, 290, 291; fiberboard, 292, 293; pulp, 292; newsprint, 294; printing and writing papers, 294; paper and paperboard, 295; among top ten consumers, 293, 296. Production: softwood, 295; hardwood, 297; plywood, 297; particleboard, 299; fiberboard, 299; pulp, 300; newsprint, 300; printing and writing papers, 301; paper and paperboard, 301; among top ten producers, 302
Union pour la Commerce d'Importation de Panneau, 103
United Arab Emirates, 96
United Kingdom: importer from Scandinavia, 55, 97; timber frame construction, 97, 104, 124, 247, 407; supply sources, 97; U.S. plywood imports, 102, 104; use of panel products, 104-5; shipment facilities, 105; trade with U.S., 122, 281; trade with Japan, 202; trade with British Columbia, 246; among largest importers and exporters, 303-7; percentage of imports, 354; supplier of technology, 457; joint ventures in developing countries, 461-64; newsprint production, 300; paper and paperboard production, 301, 345. Consumption: softwood, 96, 246,

288, 409; plywood, 290; particleboard, 290, 291; fiberboard, 292; newsprint, 294; paper and paperboard, 295, 345; consumption trends, 410-12
United Nations, 38
United Nations Conference on Trade and Development (UNCTAD), 173
United Nations Educational, Scientific, and Cultural Organization (UNESCO), 500
United Nations Industrial Development Organization (UNIDO), 519
United States: volume of trade in wood products, 3, 4, 20, 27, 28, 29, 67, 79, 102-3, 121-23, 184, 232, 235, 357; net importer of forest products, 3, 20, 503; compared with Canada, 3, 67, 85, 86, 100, 242; competition with Canada, 85, 86, 88-89, 98, 104, 122, 235, 246, 251, 448; importer from Canada, 111, 192, 243, 245; Canadian-U.S. relations, 183-93; investment in Canada, 453-54; importance of soil, 5; potential of, 3-5, 235; competition with Europe for wood, 26; demand for wood products, 26, 90-91; import-export figures, 27-29; boom in agriculture, 29-32, 90; agricultural trade, 137, 143, 144, 147; immigration to, 37, 38-39; political influence on trade, 39, 131; perspective of role in the commercial world, 41; agricultural lands being replaced by forest land, 52; western softwood lumber opportunities, 94-100, 403-4; growth of western lumber exports, 79; domestic markets in Northeast, Central states, and South, 94-95, 192; the South as major supplier, 58, 59, 192, 246, 271; government-owned land in South, 232; private ownership in South, 234; South's failure to reforest, 273; South's shipping facilities, 404; Great Lakes region, 51, 58; per capita forest area, 62; export restrictions on logs, 69-70, 179, 284, 474; foreign markets as dumping ground, 78-79; importance of domestic housing, 26, 78, 89, 95, 508; marketing methods, 79, 124; opportunities for raw materials exports, 81-93; domestic vs. foreign markets, 81-82, 89-90, 95, 100, 116, 124, 237, 239, 318, 357, 403, 404; dollar overvalued, 82, 138, 271; market restrictions, 88-90, 123-24, 310-15; supply and demand studies, 90, 93; trade in panel products, 101-10, 122; exports of softwood, 102-3, 104, 403-4, 409; trade in pulp and paper, 111-20, 141, 245, 350; prospects for pulp and paper, 114-20, 141-42; tariff disadvantage in pulp and paper, 116, 118; obstacles to expansion of trade, 120; trade in solid-wood products, 85, 86, 121-28; major competitors, 122, 271; environmental issues, 129-36; judicial system, 134-35; credit programs, 143-48; antitrust activity, 150, 153, 154, 155-56, 158; federal trade laws, 157-73; protectionism in, 159, 179, 185-93; tax problems, 161-63, 170-71; shipping agreements, 173; tariff rates, 177, 246; importance of trade, 184, 235; Department of Agriculture, 231-39; government-owned land, 232-33; private and industrial land, 234-35; overcoming trade barriers, 237-39; joint ventures, 256-61; Northwest history, 276-86; among largest importers and exporters, 303-7, 309; restrictions on partly processed products, 312; producing to metric and other export specifications, 314; trends in wood prices, 363; lack of knowledge of world trade, 403; European view of, 402-15; imports of veneer, 435; value added in processing, 456; joint ventures in developing countries, 458-65; number of sawmills, 471; bank finance and support, 483-93; development of standards, 497-501; IIASA, 515

  Consumption: 24, 27, 94-95, 334-35, 336, 405; softwood, 288, 403-4, 409; hardwood, 288; plywood, 290; particleboard, 290, 291; fiberboard, 292; pulp, 292; newsprint, 294; printing and writing papers, 294; paper and paperboard, 295; among top ten consumers, 293, 296; consumer preference, 309; consumption trends, 412

  Production: softwood, 295; hardwood, 297; plywood, 296, 297, 298; particleboard, 299; fiberboard, 298, 299; pulp, 300; newsprint, 300; printing and writing papers, 301; paper and paperboard, 301; among top ten producers, 302

  Trading partners: Europe, 27, 96, 122, 123, 350, 358, 404; Pacific Rim, 95-96, 98-99, 122; Middle East, 96, 97, 108-9, 122-23; Oceania, 96, 97; China, 6, 22, 99, 122, 419, 420 (table); United Kingdom, 104, 122; Belgium-Luxembourg, 105; Scandinavia, 27, 105-6, 108; Netherlands, 106-7; West Germany,

107, 122; France and Italy, 107, 122; Africa, 108-9; Caribbean and Latin America, 109, 123-24, 127, 435; Southeast Asia, 262-68, 434, 435; Japan, 27, 65, 66, 67, 69, 70, 71, 73-74 (tables), 84, 98, 108, 122, 137-42, 149-56, 180, 202, 206, 207, 235, 284, 316, 320, 472, 508; anti-Japanese protectionism, 6, 138, 149, 153, 175, 202; Japanese tariffs, 71, 108, 123, 126, 139-40, 149, 153, 176-82; antitrust action against Japan, 150, 155-56; meetings with Japanese, 22, 70, 284; competition for Japanese market, 68, 209, 448, 508; understanding Japanese preferences, 70, 98, 99, 236; Western Wood Products Association and Japan, 98; West Coast trade with Japan, 137, 474-75; Japanese investment in, 481-82 See also North America; Oregon; Pacific Northwest; Washington State
U.S. Hardwood Plywood Institute, 309
U.S.-Japan Forest Products Committee, 500, 501
U.S.-Japan Lumber Trade Promotion Committee, 70
U.S. Products Standards, 101, 108
University of Washington, 1, 518
University of Wisconsin, 518

Value added, 312-13, 455-56
Vander Jagt, Guy, 170, 171
Veneer: tariffs, 177, 179-80; size and quality standards, 178; softwood used for plywood, 180; peeling system, 206; Japanese standards, 221-30; European standards, 236; European exports and imports, 343, 346-47; tropical hardwood, 434-35. See also Waferboard
Venezuela: indebtedness of, 37; potential for exports, 59; housing needs, 109; duty-free entry of some materials, 127; U.S. shipping agreement, 173
Vietnam, 433, 439
Vogel, Japan As Number One, 153

Waferboard, 51, 58, 101, 106, 181, 414
Washington Forest Protection Association, 274
Washington Mill Company, 279
Washington State: dependence on trade, 1, 6, 20; value of exports, 1; favorable conditions for trade, 2, 19; port facilities, 7, 133-34; forest products in the economy of, 19-20; trade programs, 21; study of restriction of trade, 89; Department of Commerce and Economic Development, 21, 109; federal government-owned land, 232-33; joint ventures, 258; state-owned land, 269-75; export of timber, 269, 270, 285, 311; federal trust land, 270; benefit to schools, 270, 273; competition, 271; trade barriers, 271-72, 312; reforestation, 272-73; resource management, 272-74; size of forest land, 274; assistance for small mills, 274; history, 276-86, 370; export of railroad ties, 287; volume of lumber exported, 283-84; export trading companies in, 430. See also Pacific Northwest
Washington State Department of Natural Resources, 89, 269, 272; Forest Land Management program, 272-73
Water quality standards, 132-33
Webb-Pomerene Act, 128, 150, 157, 386
Weldwood of Canada Limited, 257, 478
West Africa, 356, 433, 435, 436; prices, 443, 444
West Coast Lumberman, 280
Western Europe. See Europe, Western
Western hemlock, 269
Western Lumber Company, 281
Western Wood Products Association (WWPA), 79, 98, 126
Weyerhaeuser Company: exporting experiences, 7-8; assessments of U.S. supply and demand, 90-91; and environmental issues, 133-34; joint ventures, 258, 475, 482; and general trading companies, 472; and IIASA, 518
Willapa Lumber Company, 281
Wood consumption: world figures, 23-26, 95-96, 121, 333-37, 442, 455; Japanese imports as share of total consumption, 176; ten largest world consumers, 288-95, 303; per capita, 333-36; in Europe, 349-53, 410, 411; softwood consumers compared, 409; trends, 410-12, 441-42, 455
Wood products: importance to standard of living, 2; history of trade in, 3; volume of trade in, 3; exportable surpluses of softwood, 3; importance of soil, 5; projections of shortages, 5; production by the year 2000, 15; plywood symposia in Latin America, 21; plywood standards, 101-2, 123, 126; panel products, 101-10; pulp and paper products, 111-20, 332-33; solid-wood products, 121-28, 330-32; hardwood prospects, 122, 436-46; trade restrictions, 123-24, 237-39; diseases, 238; British Columbia production, 242-43; world consumption, 288-95, 440-42; world production, 295-302, 432; top ten countries, 303; largest importers

and exporters, 303-7; consumer preferences, 309; price advantages, 309-10, 329; financial and technological advantages, 310-329; trade restrictions, 310-15, 508-9; value added, 312-13; production problems for foreign markets, 316-27; compared to nonwood products, 328-39, 407, 408, 441; energy considerations, 330-33; consumption trends, 333-38, 406-7, 440-42; Europe's place in world trade, 340-49; European consumption, 349-53; Europe's sources of supply, 353-59, 402-15; softwood consumers compared, 409; trends in European markets, 410-15; hardwood vs. softwood, 413-14, 440-41, 471; small- and medium-size firms, 425-31; tropical hardwoods, 432-46; international joint ventures, 451-82; recycling, 455; costs of nonwood inputs, 506; supply and demand projections, 509-20. See also Forest resources; Wood consumption, individual products

Wood Technological Association of Japan, 216

Working Party supply and demand study for FAO, 114, 359

World Bank, 524

Yomiuri Shimbun, 258

Youde, Jim, 89

Yugoslavia, 37, 304

Zaire, 433, 439, 443

Zenith case, 155, 156

Zhao Ziyang, 422

Zivnuska, J. A., 510

THE GEO. S. LONG PUBLICATION SERIES

American Forest Policy in Development by Stephen H. Spurr

Man, Land, and the Forest Environment by Marion Clawson

Renewable Resource Management for Forestry and Agriculture
   edited by James S. Bethel and Martin A. Massengale

Tree Growth and Environmental Stresses by Theodore T.
   Kozlowski

World Trade in Forest Products edited by James S. Bethel

DATE DUE

1 5 1986